Monitoring Ecological Change

The state of ecosystems, biological communities and species are continuously changing as a result of both natural processes and the activities of humans. To be able to detect and understand these changes, and the factors that influence them, it is essential to have effective ecological monitoring programmes. This book offers a thought-provoking introduction to the topic of ecological monitoring and provides both a rationale for monitoring and a practical guide to the techniques available. Written in a non-technical style, the book begins by considering the relevance and growth of ecological monitoring and the organizations and programmes involved. Coverage then moves to the science of ecological modelling, including such aspects as spatial scales, temporal scales, indicators and indices. The latter part of the book focuses on the assessment of methods in practice, including many examples from monitoring programmes around the world. Building on the success of the first edition, this new edition has been fully revised and updated with two additional chapters covering the relevance of monitoring to reporting the state of the environment, and the growth of community-based ecological monitoring.

IAN SPELLERBERG is Professor of Nature Conservation at Lincoln University, New Zealand and Director of the University's Isaac Centre for Nature Conservation. He teaches in the areas of ecological monitoring, environmental sustainability and science and policy, and he has acted as an advisor for many tertiary education environmental programmes. He has published widely in the the areas of biogeography, ecology and nature conservation and is currently assisting in Core Project Three of the Diversitas Science Plan.

D1202648

Monitoring
Ecological Change

Second edition
IAN SPELLERBERG

CAMBRIDGE
UNIVERSITY PRESS

CAMBRIDGE UNIVERSITY PRESS

Cambridge, New York, Melbourne, Madrid, Cape Town, Singapore, São Paulo

CAMBRIDGE UNIVERSITY PRESS

The Edinburgh Building, Cambridge CB2 2RU, UK

Published in the United States of America by Cambridge University Press, New York

www.cambridge.org

Information on this title: www.cambridge.org/9780521820286

First published 2005

Printed in the United Kingdom at the University Press, Cambridge

A catalogue record for this book is available from the British Library

ISBN-13 978-0-521-82028-8 hardback
ISBN-10 0-521-82028-6 hardback

ISBN-13 978-0-521-52728-6 paperback
ISBN-10 0-521-52728-7 paperback

Contents

Foreword

Ten years ago, when I wrote the Foreword to the first edition of this book, monitoring of the environment was in a paradoxical state. On the one hand, it was an obvious necessity. On the other hand, monitoring all the species in an ecosystem is overwhelmingly demanding – especially since we have grounds for believing that only about a tenth of the species on Earth have been described by science. As I wrote then, in everyday life nobody would think of buying a car without speedometer, fuel gauge, temperature indicator and warning lights. Yet vastly more complicated environmental systems were being perturbed by human impact, and managed in what people hoped was the right way, with very little effective monitoring of how they were actually responding to our interference. Indeed, there was no agreement on what systems of monitoring were likely to be most effective – what warning lights should be fitted to the global juggernaut so that we could at least step on the brakes if it looked like running away. And – here lay the central paradox – while everyone saw that monitoring was essential, at the same time it was not highly regarded as a scientific activity. It appeared to lie a long way from the cutting edges of theoretical ecology and the unravelling of genomes.

The first edition of this book was very timely. It appeared a year before the unprecedented gathering of heads of state and government in Rio de Janeiro for the so-called 'Earth Summit' – the United Nations Conference on Environment and Development. That Conference adopted a massive action plan for global sustainable development – Agenda 21 – and also saw the signature of two global Conventions, on climate change and biodiversity. The impact of excessive combustion of fossil fuels on world climate was ringing alarm bells, and monitoring data, especially from Hawaii, which showed the steady increase in the concentration of carbon dioxide in the atmosphere, provided the evidence on which those concerns rested. The loss of rain forests across the tropics, again revealed by monitoring, was a major stimulus to the negotiation of the Convention on Biological diversity.

Since Rio, many countries have adopted biodiversity action plans. A number have also taken steps to evaluate what climate change may mean for their environment, agriculture, forests, wildlife and human lifestyles. Action against the environmental problems of previous decades – the pollution of air, waters, soil and sea and the wasteful degradation of natural resources – has continued. As Ian Spellerberg notes in the Preface to this second edition of his book, the result has been to swell the number of state of the environment reports, stimulate community-based environmental and ecological monitoring and tie monitoring closely to the need to meet the reporting requirements nations incur under international agreements and conventions. Several international organizations are publishing regular reports that purport to evaluate the overall state of the planet and the intensity of the human footprint. Many governments depend on national monitoring schemes to assess the efficacy of their laws, action programmes and enforcement measures. Systematic, dependable and coordinated monitoring has been recognized as an essential foundation for all these activities.

However, as I said in the Foreword to the first edition, it is easy to assert the need for systematic and well-designed monitoring, but much harder to do it. What should be monitored out of all the bewildering complexity of nature? It is not possible to measure everything: choice is imperative. Very often, that choice falls on physical attributes that are relatively easy to measure, such as temperature or the concentration of essential nutrients or pollutants. The result can all too easily be an impressive table of data whose biological meaning is obscure.

Biological monitoring starts at the other end. Its logic rests on the fact that living organisms integrate the impact of many variables and that their abundance, productivity and reproductive success can provide an indication of the overall health of the ecosystems of which they are part. Understanding of ecosystems can allow the choice of indicator species whose performance reflects that of the larger whole. Since we commonly manage the environment in order to sustain particular biological features – whether it be the production of crops, forests or fisheries or the diversity of wildlife – the direct surveillance of those biological parameters is likely to be the best way to establish whether management plans are working. The more we comprehend the system, the more effective we are likely to be in choosing the best things to record.

This book offers a practical approach to the evaluation of the state of ecosystems at local, national and global levels, and through them to an assessment of the viability of trends in human societies, economies and technologies. It provides both a rationale for biological monitoring and a guide to the techniques that are available and the parameters it makes best sense to monitor in particular situations.

There are no short-cuts in the sphere of environment, and biological monitoring is not a miraculous cure-all. Its results can be hard to interpret, because it can be a big step from observing change to understanding cause. However, biological changes are important indicators of the fundamental processes of our planet and can provide important warnings society needs to heed. The arguments and approaches set out in this book need to be taken very seriously, and used effectively, by those seeking to manage the environment. Good ecological monitoring can greatly increase the efficiency (and hence the cost effectiveness) of environmental management.

The monitoring of ecological change is directly relevant to fundamental decisions about the human future. With all our technological skills, we remain components of wider ecosystems and we depend totally on the biological productivity of the planet. As has been said many times, 'the world's economy is a wholly owned subsidiary of the world's ecology'. Today, 13 years after Rio's Earth Summit, and in the wake of its successor conference in Johannesburg, the emphasis at world level has shifted from protecting the environment to the alleviation of poverty and human suffering. Yet these social goals can only be achieved through sustainable development, which, in turn, can only be achieved if people live within the limits set by the environment and conserve the diversity and productivity of nature. The monitoring of ecological change remains fundamental, as a source of understanding of how well – or badly – we are doing in pursuing these fundamental and interlinked goals.

Martin Holdgate
Formerly Director General of the International
Union for the Conservation of Nature and
Natural Resources: the World Conservation Union

Preface

It is just over a decade since *Monitoring Ecological Change* was first published. During that time there have been a number of developments that have implications for ecological monitoring in theory and in practice. Three developments in particular stand out as being relevant to ecological monitoring. They are not new but there appears to have been a growth in activity of both. One is in the area of state of the environment reports; the second is in the area of community-based environmental and ecological monitoring; and, third, there is the extent of ecological and environmental monitoring that takes place because of international agreements and conventions.

The quality, health or state of the environment seems to be of increasing concern. Despite many effort over the last few decades to curb impacts on the environment, there seems little doubt that environmental degradation continues to get worse. Not surprisingly, many countries have published state of the environment reports and there has been much discussions about the nature and purpose of such reports. As well as national state of the environment reports, there has been a steady growth in the number of region reports that deal with environmental monitoring and environmental quality. Ecological monitoring has a role to play in these state of the environment reports.

Community-based environment monitoring has become widespread practice in many countries. In part, this has been the result of environmental organizations encouraging local communities to take a role in collection and analysis of environmental and ecological data. There also seems to have been considerable initiative on behalf of local communities to take ownership of monitoring programmes. Another contributing factor is the possibility that community-based environmental monitoring may be less expensive.

The 1992 Earth Summit gave rise to several products that in turn, have brought about demands on countries to monitor many environmental variables.

These range from carbon stocks to forest cover and the number of threatened species to levels of specific air pollutants. There has been a growth in international environmental and ecological reporting.

There have been other notable developments over the last 10 years. Long-term ecological monitoring sites have become firmly established along with an appropriate infrastructure to support such sites and networks of monitoring sites. The establishment of these sites has required the development and implantation of protocols and standards for monitoring.

Ecological monitoring and environmental monitoring appears more often in the curriculum and there are now many courses being offered in the area of assessment and monitoring. Interestingly, there has been a steep rise in the use of Environmental Change Network datasets in the UK by higher education. Furthermore, there seems to have been resurgence in the number of scientific meetings about ecological monitoring and long-term ecological research (although long-term ecological research is not necessarily ecological monitoring).

There have been many examples of the establishment of protocols for ecological monitoring. These include the adoption of standardized methods. In addition, objectives for ecological monitoring have often referred to the need to defend data scientifically and measure the effectiveness of the programmes.

Finally, there is no doubt that we live in a time of environmental change. That change is a combination of natural events and human intervention. It occurs at global, regional and local levels. The range of activities addressing the effects of that change is intriguing – from attempts to achieve international conventions about addressing the drivers of environmental change to investigations as to how climate change will affect private gardens. There is now no doubt that ecological monitoring has an ever increasing important role to play as we attempt to use nature in a sustainable and equitable manner in an ever-changing environment.

Acknowledgements

The kind help of many friends and colleagues has made this second edition possible. In particular I would like to thank the following: N.J. Aebischer, Rob Allen, Alan Andersen, Lee Belbin, Major Boddicker, Eleanor Ely, David Given, Richard Gregory, Paul Harding, Rochelle Hardy, Richard Harrington, Christine Heremaia, Brian Hopkins, Akemi Itaya, Ian Johnston, Elizabeth Kilvert, Keith Kirby, D. Klimetzek, James Lambie, Robert Leonard, John Ludwig, Tim Mallett, John Marchant, Patrick Meire, Glen Nolan, Christy Pattengill-Semmens, David Pearson, George Peterken, Andrew Plumptre, Gordon Ringius, John Sawyer, F. Schmidt-Bleek, John Stark, Benjamin Stout, Jane Swift, Rowan Taylor, Jo Treweek, Teja Tscharntke, Steve Urlich, Don Waller, Matthias Waltert, Jonet Ward, Emma Waterhouse, Tony Whitten and Arthur Willis.

Glossary of acronyms and abbreviations

ABC	abundance–biomass comparison
AFRC	UK Agriculture and Food Research Council
AMAP	Arctic Monitoring and Assessment Programme
ASPT	average score per taxon
BEWS	biological early warning system
BMWP	Biological Monitoring Working Party
BOD	biochemical oxygen demand
BRC	Biological Records Centre
BTO	British Trust for Ornithology
CALM	circumpolar active layer monitoring
CCAMLR	Convention for the Conservation of Antarctic Marine Living Resources
CCAS	Convention for Conservation of Antarctic Seals
CEH	Centre for Ecology and Hydrology
CEP	Committee for Environmental Protection (Antarctica)
CIS	Countryside Information System (software)
CITES	Convention on International Trade in Endangered Species
COMNAP	Council of Managers of National Antarctic Programmes
CORINE	Co-ordination of Information on the Environment
CRAMRA	Convention on Regulation of Antarctic Minerals Resource Activities
DDT	dichlorodiphenyltrichloroethane
DEFRA	Department for Environment, Food and Rural Affairs
EBCC	European Bird Census Council
ECE	Economic Commission for Europe
EcIA	ecological impact assessment

ECN	Environmental Change Network
EDU	Ecological Data Unit
EEC	European Economic Community
EIA	environmental impact assessment
EMAN	Environmental Monitoring and Assessment Network
EMAP	Environmental Monitoring and Assessment Programme
ENRICH	European Network for Research in Global Change
EPA	US Environment Protection Agency
EQI	environmental quality index
ESCAP	Economic and Social Commission for Asia and the Pacific
ESI	environmental sustainability index
EU	European Union
FAO	Food and Agriculture Organization of the UN
FATE	Feedback and Arctic Terrestrial Ecosystems
FOE	Friends of the Earth
FSC	Field Studies Council
GEMS	Global Environmental Monitoring System
GIS	geograptical information system
GMO	genetically modified organism
GPS	global positioning system
GRID	Global Resources Information Database
GTOS	global terrestrial observing system
IAIA	International Association for Impact Assessment
IASC	International Arctic Science Committee
IBAs	important bird areas
ICP	International Cooperative Programme
ICPB	International Council for Bird Preservation
ICSU	International Council of Scientific Unions
IEEP	Institute for European Environmental Policy
IGBP	International Geosphere–Biosphere Programme
ILO	International Labour Organization
ILTER	International Long-term Ecological Research
INFOTERRA	International Environmental Information Network
IOC	Intergovernmental Oceanographic Commission
ITEX	International Tundra Experiment
IUBS	International Union of Biological Sciences
IUCN	International Union for Conservation of Nature and Natural Resources
IUMS	International Union of Microbiological Sciences

IWC	International Whaling Commission
JNCC	Joint Nature Conservation Committee
LPI	living planet index
LTER	long-term ecological research
MAB	Man and the Biosphere (UNESCO)
MARC	Monitoring and Assessment Research Centre
MARMAP	marine monitoring and prediction
MCI	macro-invertebrate community index
MSC	Marine Stewardship Council
NBN	National Biodiversity Network
NC	Nature Conservancy
NCC	Nature Conservancy Council
NERC	UK Natural Environment Research Council
NGO	non-governmental organization
NOAA	National Oceanic and Atmospheric Administration
NSF	National Science Foundation
NSN	National Science Network
OECD	Organisation for Economic Co-operation and Development
RBP	rapid bioassessment protocol
RIVPACS	River Invertebrate Prediction and Classification System
RSPB	Royal Society for the Protection of Birds
SCAR	Scientific Committee on Antarctic Research
SCEP	Study of Critical Environmental Problems
SCOPE	Special Committee on Problems of the Environment
SEA	strategic environmental assessment
SIGNAL	stream invertebrate grade number average
SMRU	Sea Mammal Research Unit
SoE	state of the environment
SPOT	Systeme Probatoire d'Observation de la Terre
TEMS	terrestrial ecological monitoring sites
TERI	Terrestrial Ecosystem Research Initiative
UN	United Nations
UNDP	United Nations Development Programme
UNEP	United Nations Environment Programme
UNESCO	United Nations Economic, Social and Cultural Organization
UNSO	United Nations Statistical Office
VPA	Virtual Population Analysis
WCED	World Commission on Environment and Development
WCMC	World Conservation Monitoring Centre

WGMS	World Glacier Monitoring Service
WHO	World Health Organization
WHYCOS	World Hydrological Cycle Observing Systems
WMO	World Meteorological Organization
WQI	water quality index
WRC	Water Research Council
WWF	Worldwide Fund for Nature

1

Ecological monitoring

1.1 Introduction

The purpose of this chapter is to introduce the reader to the concept of monitoring ecological change and to some ecological monitoring programmes. Monitoring ecological change has considerable relevance at a time when humans are having an increasingly widespread and long-term impact on nature. I have drawn on personal experiences in defence of the value of ecological monitoring (to conservation and sustainable development) and also the value of long-term ecological research.

1.2 Terms and concepts

The aim here is not to undertake academic discussions about definitions. It is, however, necessary to distinguish between the various terms as used in this book. Recording, mapping, surveys and sampling are all methods of data collection that provide a basis for monitoring, that is the systematic measurement of variables and processes over time.

Census

The term census generally refers to population counts, which, in turn, can be used in monitoring programmes.

Surveillance

Surveillance is the systematic measurement of variables and processes over time, the aim being to establish a series of data in time.

Monitoring

Monitoring is also the systematic measurement of variables and pro-cesses over time but assumes that there is a specific reason for that collection of data, such as ensuring that standards are being met.

In a report of the Study of Critical Environmental Problems (SCEP, 1970) entitled *Man's Impact on the Global Environment*, there is a similar definition of monitoring: 'systematic observations of parameters related to a specific problem, designed to provide information on the characteristics of the problem and their changes with time'.

Ecological monitoring is, therefore, about the systematic collection of ecological data in a standardized manner at regular intervals over time. Some organizations and people recognize or have established different types or categories of monitor-ing. For example, the Department of Conservation in New Zealand recognizes three types; result monitoring, outcome monitoring and surveillance monitoring (Box 7.1, p. 224).

In another example, Vaughan *et al.* (2001) have described four categories of environmental monitoring:

- simple monitoring: recording the values of a single variable at one point over time
- survey monitoring: the absence of an historical record for an environmental problem in a particular area can be replaced by a survey of the current environmental conditions in both the affected area and the area not affected
- surrogate or proxy monitoring: compensating for the lack of previous monitoring by using surrogate information to infer changes
- integrated monitoring: using detailed sets of ecological information.

Three examples of ecological monitoring are shown in Box. 1.1. The first example shows temporal changes (10.5 years) in desert rodents; the second example comes from an estuary monitoring programme where levels of effluent have decreased and the final example is based on experimental planted grassland communities.

All examples serve to introduce the concept of monitoring ecological change. At the same time, these examples prompt some interesting questions and issues. The first example uses data from captured individuals and data are expressed simply as population size. Is the size of the captured population an indication of the total population size? In the second example there has been management of the pollutants entering the marshland community. While the data show decreased levels of pollutants and increased abundance of plant communities, there remains the challenge of demonstrating cause and effect.

Box 1.1   Examples of ecological monitoring

Example A

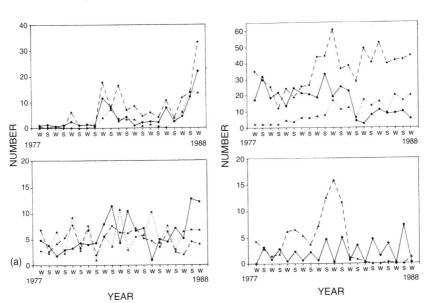

(a)

YEAR

Population dynamics of 11 common rodent species at a Chihuahuan study site over 10.5 years. Numbers are given as six-month averages for summer (S) and winter (W). Left, murid rodents; right heteromyid rodents. (With permission from Brown & Heske, 1990.)

Example B

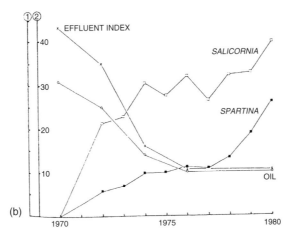

(b)

Comparison of effluent quality index, the area of damaged marsh recolonized by *Spartina* and *Salicornia* spp. (as a percentage) (vertical axis 1) and oil content (ppm) (vertical axis 2 between 1970 and 1980. (With permission from Dicks & Iball, 1981.)

Example C

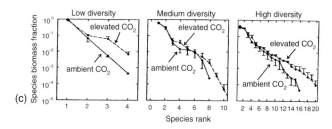

(c)

Plant species dominance at ambient and elevated CO_2 in plots of differing plant species diversity. (With permission from Niklaus *et al.*, 2001.)

The third example describes some experiments with planted grassland communities and elevated levels of CO_2. The authors make the observation that short-term effects may be misleading when attempting to predict long-term effects (Niklaus *et al.*, 2001).

Sampling, recording, mapping, surveying, inventories and long-term ecological research can contribute to ecological monitoring. Regular counts or census of a population of birds can form the basis of ecological monitoring – or monitoring ecological change in the bird populations. For example, Dunwiddie & Kuntz (2001) described long-term trends of the Bald Eagle in winter on the Skagit River, Washington, based on data from weekly counts between 1978 and 2000. Peak one day counts varied from 77 in 1983–1984 to 506 in 1991–1992.

Surveillance is also the systematic measurement of variables and processes over time, the aim being to establish a series of data in time. Similarly, monitoring is the systematic measurement of variables and processes over time but assumes that there is a specific reason or objective for that collection of data such as ensuring environmental standards are being met. The 1970 SCEP report carried a similar definition of monitoring (see p. 2).

Phenology has some relevance to ecological monitoring. It is the study of the times of recurring natural events: a calendar of nature's events. For examples the date on which the first salmon reappear, when frogs first arrive at a pond or when the first spring wild flowers bloom.

Recurring events in nature have long been recorded and some records go back many centuries. In the UK, phenology was well recognized by 1875 when the Royal

Meterological Society established a national recorder network. Later, in the 1990s, a pilot scheme for a phenology network was established at the Centre for Ecology and Hydrology (CEH) in Cambridge. In 2000, the UK Woodland Trust together with the CEH launched a project to promote phenology. Now, many thousands of people from around the UK are registered on-line with the phenology network.

How exciting it would be to see similar activities in many other countries. At the same time there needs to be care with the analysis of any changes that may be detected in reoccurring natural events. For example, although climate warming may appear to be affecting the seasonal behaviour of some amphibian species, such conclusions may be premature (Beebee, 2002).

1.3 Why ecological monitoring?

Ecological monitoring has to be resourced and financed. There may be long-term resourcing implications and some ecological monitoring can be relatively expensive. What, if any, is the justification for ecological monitoring? Four reasons come to mind.

1. The processes of many ecosystems have not been well researched and monitoring programmes could provide basic ecological knowledge about those processes.
2. Management of ecosystems, if it is to be effective, requires a baseline, which can only come from ecosystem monitoring.
3. Anthropogenic perturbations on the world's ecosystems have long-term effects, some synergistic and some cumulative: therefore, it follows that long-term studies are required.
4. The data from long-term studies can be a basis for early detection of potentially harmful effects on components of ecosystems.
5. With the ever-increasing loss of species, loss of habitats and damage to biological communities, ecological monitoring is needed to identify the implications of these losses and damage.

Interestingly enough, the reports to the US National Science Foundation (NSF) regarding long-term observations of ecosystems made similar recommendations for the data that could be obtained and the nature of the measurements: that such data could concern:

- cyclic changes
- time lags in ecosystem responses to outside influences
- test of ecological theories concerning stability, community structure and system development
- sensitive indicators of ecological change.

There are international pressures and requirements for environmental and ecological monitoring. For example, many countries have signed international conventions and agreements (e.g. the 1992 Convention on Biological Diversity (see the Appendix) and the Montreal Process) and consequently there is a requirement for monitoring to take place and for respective countries to report on that monitoring. That monitoring could range from forest cover and condition through to population levels of marine organisms as a monitor of carbon stocks (Coomes *et al.*, 2002).

The Convention on Biological Diversity (Appendix), draws attention to the need to identify and monitor ecosystems, habitats, species, communities, genomes and genes. Article 7 of the Convention is about identification and monitoring. However, it is not possible nor practical to monitor all species, communities and ecosystems. Therefore, there has to be some kind of prioritzation. There are criteria for prioritizing ecosystems, habitats, species and communities for ecological monitoring (see p. 229). Priorities for monitoring biological diversity have also been drawn up by the United Nations Environment Programme (UNEP, 1993).

Within some countries, there may also be national requirements for monitoring, particularly with regard to environmental management and reporting the state of the environment. Many countries have established legislation requiring certain standards of environmental quality and in most cases the legislation is enforced mainly to prevent unacceptable levels of pollution and to ensure appropriate quality of the environment. In New Zealand, the 1992 Resource Management Act has the purpose of promoting sustainable management of natural and physical resources with the emphasis on the effects of activities on the environment. Environmental monitoring or surveillance is required to determine those effects.

Whether or not there is a legal requirement, management of nature resources and services cannot take place successfully or in a sustainable manner unless we know what is happening. Ecological and environmental monitoring is, therefore, a prerequisite for environmental management and sustainable development.

Ecological monitoring and research are intertwined. For example, the relevance of environmental monitoring to research and environmental management was highlighted in a 1992 discussion document produced by the Scientific Committee on Antarctic Research (SCAR) and the Council of Managers of National Antarctic Programmes (COMNAP).

> Environmental monitoring is a fundamental element of basic research, environmental management and conservation. The organized and systematic measurement of selected variables provides for the

establishment of baseline data and the identification of both natural and human-induced change in the environment. Monitoring data are important in the development of models of environmental process, which in turn facilitate progress towards a predictive capability to detect environmental impact or change .....

1.4   Personal reflections and experiences of ecological monitoring

I have included the following because it provides a context for the material that I have selected for this book. It reveals some details about me and hints as to the bias that has been adopted in this book.

During the 1980s, I often advocated that marine, aquatic and terrestrial 'sites' should be set aside as long-term ecological monitoring sites to provide baseline information. The questions most commonly asked of me were: 'What would be the purpose of such sites?' 'Why monitor the ecology of a site?' I believed that such sites would provide baseline data and the important background information for measuring change in biophysical conditions and that such sites could be used to assess any improvement or degradation in the state of the environment. How else would it be possible to distinguish between natural ecological changes and those changes brought about by human impacts? I was not advocating monitoring for the sake of monitoring. There was a specific reason and application for the proposed ecological monitoring.

In 1986, I established the basis for a long-term ecological monitoring site on a private nature reserve in the south of England. Unfortunately, that project was not supported and did not continue.

On a Monday evening, 21 April 1986, I chaired the usual weekly environmental sciences seminar for students at the University of Southampton. That evening, our guest speaker was an expert on energy and he spoke very strongly about the importance of nuclear power and talked much about its safety record. By the following Monday, the news had broken about the Chernobyl nuclear accident, which took place in the Ukraine on 26 April. The contamination spread northwest over Europe and affected many areas including the grazing land in Wales and the west of England. For many years after, the affects of the contamination were observed in the grassland ecosystems.

The purpose of this story is not about the benefits or dangers of nuclear power. Before Chernobyl, the idea of establishing long-term ecological monitoring sites to assess the state of the environment for the purpose of background or baseline information was not widely supported. Then, all of a sudden, throughout western Europe there was a realization that there was little or no background or baseline information on which to base comparisons or identify human-induced changes. Concern was expressed about the effects of radiation on agricultural ecosystems including grazing lands in parts of the UK such as the Welsh uplands.

Does it always take an environmental disaster to put theory into practice? Perhaps not; but I cannot help but suspect that ongoing changes in the biophysical environment have had a significant role to play not only in heightening concern about the state of plant and animal populations and communities but also in helping to win more support for the need to have long-term ecological monitoring sites. But where should those sites be?

If there was to be only one biogeographical region that was to be used for monitoring the state of the Earth and changes in the biophysical environment, it could be the polar regions. Since the early 1990s, the media have reported the shrinking of ice caps, the appearance of open water at the North Pole, and huge icebergs breaking off from the Antarctic ice shelves. Whatever the cause, these events clearly demonstrate that changes are taking place, even in those regions such as Antarctica that are relatively remote from human impact. Or should that be remote from *direct* human impact?

1.5   Priority areas for ecological monitoring

Thinking globally, considering all the Earth's biogeographical regions, are there areas that should be a high priority for ecological monitoring? The answer of course depends on what the objectives are. It could be argued that ecological monitoring needs to occur in:

- regions where there are greatest impacts caused by humans so that the effects of land use can be managed in a sustainable manner
- regions not greatly affected by humans so that baseline information can be obtained; this would include biological communities for which there were comparable communities that had been affected by human activities
- regions where there has previously been little ecological monitoring but where we need to know if environmental degradation does occur; for example, there, ranging from sites around effluent discharges to deep-sea locations, that are many marine regions could justifiable be subject to ecological monitoring programmes.

There is one region that would at first sight seem remote from human impacts yet has a central role in global environmental processes: this is the Antarctic region. This region includes a great ice-covered land mass and biologically rich oceans would appear to be a region and that is relatively safe from

Table 1.1 *Some environmental impacts (deliberate, incidental or accidental) in the Antarctic*[a]

Area	Impacts
Terrestrial (including inland waters)	Habitat destruction/modification
	Destruction/removal/modification of biota, fossils, ventifacts, etc.
	Modification of vital rates of biota (disturbance to production and/or growth)
	Modification of distribution of biota
	Introduction of alien biota
	Pollution by biocides and noxious substances, nutrients (eutrophication), radionuclides, electromagnetic radiation, noise
	Modification of thermal balance of environment
	Aesthetic intrusion
Marine (including shoreline, enclosed waters, benthos)	Habitat destruction/modification
	Destruction/removal/modification of biota
	Modification of vital rates of biota
	Pollution by biocides and noxious substances, nutrients, radionuclides, inert materials (dumping), noise, heat
Atmospheric	Pollution by sulphur oxides, nitrogen oxides, carbon monoxide, carbon dioxide, hydrocarbons, radionuclides, dusts, microbiota, electromagnetic radiation
	Ozone: local excess at ground level, depletion in stratosphere

[a] Some very unlikely impacts and impacts of negligible severity have been ignored.

exploitation and sufficiently remote not to be harmed by pollution. The Antarctic region might also be considered to be one of the last locations on earth where there was a need to undertake any kind of monitoring or surveillance of the wildlife. However, as long ago as 1985, there were reports on the impacts of human activities in the Antarctica (Table 1.1).

For over 200 years, the southern oceans (the broad band of water that circles the southern hemisphere between latitude 40° and Antarctica) have been exploited for whales, fish and plankton. Pollutants from the industrialized world have reached and penetrated the Antarctic ecosystems and the operational activities of Antarctic research and exploration have had their deleterious impact on the coastal populations of birds and mammals (Bonner, 1984; Wilson *et al.*, 1990). Supply ships have spilled petroleum fuel in Antarctica and sadly we know very little about the way in which oil pollution affects ecosystems in polar regions.

During the 1950s, a number of scientific programmes were established in relation to research on global environmental change. The International Geophysical Year of 1957 was a landmark in this respect and marked the establishment of environmental monitoring in Antarctica. During the 1960s, there was a considerable increase in research alongside a greatly increased support for environmental, especially geophysical, investigations. Antarctic biological research has long included a wide programme of activities, some of which have been directed at population ecology of mammals and birds. For example, surveillance and census of seals, penguins and other birds has been undertaken for many years but few results from that surveillance and census work will make any significant contribution to any Antarctic conservation strategy. This is because the data have not been collected in a systematic and standardized manner and because of the lack of an infrastructure for long-term ecological monitoring. This is a missed opportunity, of which I have had some small personal experience.

It was during the 1960s that I was a member of a research team in Antarctica and part of my research included an analysis of the population ecology of the Adélie Penguin (*Pygoscelis adeliae*) and the McCormick Skua (*Catharacta maccormicki*). The populations of these two birds were occasionally recorded in the area of Cape Royds, Ross Island, as part of an ongoing surveillance programme. My field work over three years was a very small contribution to the population records that had been kept before the 1960s and that continue to be maintained.

At that time, there seemed to be little concern about the potential value of data from the census of those birds, especially where it was to be undertaken on a systematic basis. Although we saw the value of those records as providing a 'watchful eye' on the status of the populations, no one seemed to be sure of any long-term objectives of the surveillance programme. The recording was not administered so as to ensure continuity: records were not kept in a central depository and recording methods were not uniform.

During that time (1960s), the logistical and research activities had various impacts on the Antarctic birds in and around the area of Cape Royds. Helicopters bringing tourists to the area regularly flew close to penguin colonies, causing havoc. Perhaps not surprisingly, the people involved in the research were concerned even if the helicopter flight was on a mercy mission (Box 1.2).

There was one occasion when I recorded what appeared to be an outbreak of disease amongst the skuas. Birds were observed dying. Throughout each summer, there would always be a few dead skuas found in the study area. On this occasion, the incidence of death had greatly increased. It is always difficult to demonstrate cause and effect. One possible contributing factor may have been the fact that rubbish from the large bases nearby attracted scavaging skuas. Could the birds have been poisoned or contracted some kind of disease?

Box 1.2 Surveillance and recording of penguins and skuas in relation to human activities.

This brief extract (from Hayter, 1968) illustrates the tensions that can sometimes exist when a small Antarctic research team in the field has a medical emergency requiring the evacuation of one of its members.

Bill M. agreed that Gregor should be evacuated and kindly agreed to take him to the Sick Bay for a check-up. Jack asked no unnecessary questions and met my request for a chopper immediately, so after telling Ian to have Gregor ready for evacuation I took off for [Cape] Royds.

Ian met us at the pad at Royds, about 200 yds from the hut, and we could see the others emerging carrying Gregor in his sleeping bag, not the best form of transport for a man with perhaps a bursting gut. The pilot yelled down to us through the noise of the blades that he could lift and settle on snow right next to the hut.

'Oh no' shouted Ian, 'it will scare the hell of the penguins'. Something of their sanctity had been imbued in me also, and for one second I hesitated. How mad can a man get?

'To hell with the penguins' I shouted back; and signed to the pilot to take her away. In the resulting confusion maybe fifty chicks and eggs out of several thousand fell prey to the marauding skuas, so where one lost the other gained and perhaps Gregor most of all.

Photographs.

(a) The Cape Royds' Adélie Penguin rookery.

(b) Helicopter landing at Cape Royds.

Literature reviewing this area includes Kerry & Hempel (1990), Young (1994), Norman (2000) and Stonehouse (2002).

Research parties also had their impacts and although efforts were made to keep impacts to a minimum, there was never any formal procedure for identifying and avoiding or minimizing impacts. There were no protocols for surveillance or monitoring.

Today, the value and importance of a long-term ecological monitoring programme on those two species in the McMurdo Sound area is obvious. Physical disturbances, pollution and other perturbations seem to have affected the populations of those and other birds in McMurdo Sound as well as populations of other bird species around the coastal regions of Antarctica. Without long-term data from good monitoring programmes, the ecology and population dynamics of many species remain unstudied and, therefore, the extent and nature of the effects of pollution and physical disturbance cannot easily be assessed. It is easy to comment with hindsight, but I mention these experiences because they 'set the scene' and introduce some fundamental aspects of ecological monitoring that are discussed throughout this book.

Over the last 45 years, an Antarctic Treaty (1959) with its 27 consultative parties and over 20 signatories has afforded only moderate protection to the biota of this region of the world. This is despite the 1991 Protocol on

Environmental Protection to the Antarctic Treaty, which calls for regular and effective monitoring.

Some agreed nature conservation measures for Antarctica were agreed in the 1970s and 1980s (Convention for the Conservation of Antarctic Seals (CCAS), 1972; Convention for the Conservation of Antarctic Marine Living Resources (CCAMLR), 1980). These measures were based on the results from several working groups of the International Union for the Conservation of Nature and Natural Resources (IUCN) and the work of SCAR which is one of the major environmental, units of the International Council of Scientific Unions (ICSU).

The attraction of possible extractable minerals and oil once posed a considerable threat to the wildlife and physical environment of Antarctica. The Antarctic Treaty included no specific provisions on mining. The 1988 Convention on Regulation of Antarctic Minerals Resource Activities (CRAMRA), would have allowed exploration for minerals and controlled mining, had it not been unsuccessful because of a lack of signatories.

More recently, there was the establishment of the Protocol on Environmental Protection to the Antarctic Treaty, 1991. This prohibited all mineral resource activities but also called for regular and effective monitoring to allow assessment of the impacts of ongoing activities on the Antarctic environment and associated ecosystems.

Environmental monitoring is now fairly commonplace in many areas of Antarctica. Environmental databases combine with the potential of Geographic Information Systems (GIS) to help to implement the Protocol (Cordonnery, 1999) and have brought about a new age of environmental monitoring. However, there are relatively few examples of ecological monitoring, that is monitoring biological communities, animal populations or plant species. The 1980 CCAMLR resulted in some ecological monitoring programmes.

In September 1988, there was a landmark in the reporting on change in Antarctic ecosystems. The fifth of a series of symposia held under the auspices of SCAR addressed ecological change and conservation in Antarctica Ecosystems. That symposium included papers on long-term changes in Antarctic mammal and bird populations (Kerry & Hempel, 1990).

From the early 1990s, long-term ecological monitoring became, for some countries, an integral part of environmental management in the Antarctic (see, for example, the Summary of COMNAP, 1998).

Ecological monitoring has become a routine component in assessments of large-scale impacts resulting from activities such as drilling programmes and the construction of crushed rock airstrips (SCAR, 1996). Interestingly enough, environmental impact assessments or EIAs (see Ch. 10) were carried out in Antarctica as early as 1973–1974 (Parker & Howard, 1977) and more recently there have been formal and detailed evaluations of the procedures for EIAs in Antarctica. One such assessment by Benninghoff & Bonner in 1985 drew attention to the importance of monitoring key indicators of environmental change.

However, it was not for another decade that serious thought was given to formal reporting on the state of the Antarctic environment as a whole. In 2000, work began on an Antarctic component of the Australian State of the Environment report: a reporting system specifically designed to monitor the state of the Antarctic (Belbin *et al.*, 2003). At about the same time, the New Zealand Antarctic Institute published the first comprehensive state of the environment report written for any region of Antarctica. The region was the Ross Sea (Waterhouse, 2001) and, prompted by the Protocol, which requires the Committee for Environmental Protection (CEP) to report on the 'state of the Antarctic environment', the text provided an extensive and detailed account. It included an analysis of the pressures on the environment; the state of the soil, air, water and living organisms; and an account of the responses that have taken place. Unresolved aspects have been identified, such as 'few, monitoring programmes have been established to assess impacts away from stations' and 'cumulative impact at frequently visited sites are unknown'.

Hopefully, very soon there will be a follow-up to this report in the form of a process to identify targets for ecological monitoring. The data from many of the wildlife monitoring projects together with the known causes could be used as a basis for establishing targets and implementing a policy to achieve those targets. If this does not happen, then there will continue to be surveillance rather than ecological monitoring.

1.6   Long-term ecological studies and ecological monitoring

Long-term ecological research or studies and ecological monitoring are not the same thing. However, although not all long-term ecological studies could be said to be examples of monitoring, there are nevertheless characteristics that are common to both monitoring and long-term studies.

The phrase long term is subjective and it might be asked exactly what is a long-term study. Perhaps the term is best put into the context of the life history and lifespan of the species being studies. How many generations or how many seasons of activity need to be recorded to make the study long term? There can be no generic answer to the question of what is meant by long term.

The applications of long-term monitoring studies have often been discussed since the early 1970s, notably in 1970 at the Marine Pollution Conference in Rome and again in 1977 in the USA, where a recommendation was made to the NSF that long term monitoring of ecosystems should be funded and supported. Obviously,

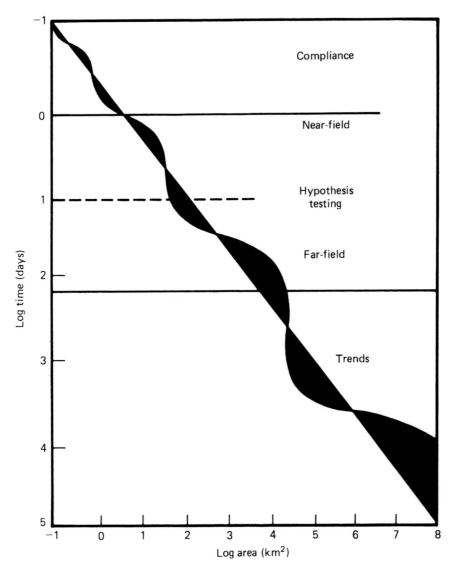

Fig. 1.1 Conceptual relationships of scale in space and time for different categories of monitoring. The temporal scale represents the duration of sampling (with respect to each perturbation) and the spatial scale represents the potential sampling area. (Redrawn from Wolfe & O'Connor, 1986.)

the decision about the most appropriate timescale for a monitoring programme needs to be linked to the objective of the monitoring activity. The importance of linking objectives to time scales can usefully be shown in a conceptual manner (Fig. 1.1), a manner that also emphasizes an important aspect of monitoring, that is the difficulty of identifying natural changes in the absence of long-term data.

The conceptual approach shown in Fig. 1.1 is based on that aspect of monitoring which deals largely with environmental quality (Wolfe & O'Connor, 1986; Wolfe et al., 1987) and three main aspects of that monitoring have been identified: compliance monitoring, hypothesis testing and trend monitoring. Compliance monitoring attempts to ensure that activities are carried out to meet statutory requirements. Hypothesis testing or model verification checks the validity of assumptions and predictions. Trend monitoring identifies large-scale changes that have been anticipated as a possible consequence of multiple activities.

At about the same time (late 1970s) that recommendations were being made to the NSF about the applications of long-term environmental monitoring, there was a recommendation that the Ecological Society of America should establish a standing committee on long-term ecological measurements. The idea was that this committee would serve as a liaison between the NSF and the ecological community. In the mid 1980s, members of the Ecological Society of America also agreed to promote a network for rare plant monitoring.

The importance of long-term ecological studies was again the focus of attention in the USA in 1984 and in 1987. In 1984, Likens set up a committee to gather information about existing long-term studies in ecology and, subsequently, 100 ecologists contributed to an evaluation of long-term ecological studies (Strayer et al., 1986). This report confirmed that long-term studies had made essential contributions to ecology but it also identified some interesting and essential components of successful long-term ecological studies. Apart from financial support, leadership (dedicated guidance by a few project leaders), simple design and clearly defined objectives were considered to be essential.

In 1987, the Cary Conference (Likens, 1989) discussed the great value and applications of long-term ecological studies and that conference seems to have been influential in subsequent monitoring initiatives in the USA. Those initiatives have been continued, with some major discussions about long-term ecological research that appeared in the July/August 1990 issue of *BioScience*. By that time, so it was reported, the programme of long-term studies sponsored by the NSF was well established (Franklin et al., 1990).

Much attention continued to be devoted to ecological monitoring and long-term research in the USA, as shown by the contents of the *Proceedings of the Fourth Symposium on Environmental Monitoring and Assessment Programme* (EMAP) held in 1999 (Wiersma, 2000). Today there is a network of long-term ecological research (LTER sites) sites around the world (see more details in Ch. 2).

The British Ecological Society (established in 1913) has, from time to time, supported the case for long-term studies of ecosystems. At a workshop in 1989 held for young ecologists by the British Ecological Society, long-term ecological monitoring was considered to be one of three main research priorities. However, it was not until the late 1980s that there was serious discussion about the establishment of a network of long-term ecological 'reference' sites. In 2003, the Environmental Change Network (ECN) marked its tenth anniversary with a session at the British Ecological Society Annual Meeting entitled *Long-term Ecological Research and Monitoring: Current Approaches and Future Directions*. Currently, the UK monitors 260 variables at 12 terrestrial sites and 42 freshwater sites.

Elsewhere in the world, the applications of long-term ecological studies and monitoring have begun to be recognized. In Canada for example there is the Environmental Monitoring and Assessment Network (EMAN) with sites across the country that include NatureWatch (frogs, ice, plants etc.) and input from scientists, community groups and schools.

In New Zealand, a symposium was held in 1988 on environmental monitoring with emphasis on protected natural areas. Time and time again, speakers at that symposium argued for more support of environmental monitoring programmes while at the same time noting that very little biological monitoring, especially marine biological monitoring, had previously taken place. Some years later, the Centre for Resource Management hosted a meeting (at Canterbury University) about long-term ecological research sites in New Zealand (Ward, 1990).

More recent initiatives in New Zealand include a nationwide carbon and biological diversity monitoring system sponsored by the Ministry for the Environment. In terms of indigenous woody vegetation, this means remote sensing mixed with a network of permanent plots throughout the forests and shrublands (Coomes *et al.*, 2002). The New Zealand Department of Conservation has been considering common kinds of ecological variable that could be used in monitoring (Table 1.2). It can be seen from this table that monitoring programmes have made use of a wide range of variables at different levels: at the ecosystem level through to populations and habitats and then to the physiological and cellular attributes of specific organisms.

1.7   Monitoring ecology in the future and in the past

Monitoring ecology in the future has included many applications concerning models of population change and models of predator–prey numbers. Awareness of climate change has led to many large research projects on its likely effects on the distribution of organisms. Monitoring ecological change in the future has implications for more than simply ecologists. For example, those of us who have gardens may become involved in programmes that look at the

Table 1.2 *Biological variables and processes used in monitoring and surveillance*

A. Variables

Biomass
  Area of cover or percentage cover
  Production
  Amount of dead material, litter
  Vegetation structure
  Lichenometric studies
Species lists (species composition)
  Species richness
  Species diversity
  Species frequency
  Proportion of all samples in which a species occurs
  Occurrence of 'indicator species'
  Occurrence of rare species
Phenology of selected species
  Spatial patterns of distribution
  Population density
  Relative abundance of predators and prey
  Trophic position
Population age–class distribution
  Diameter (of trees) at breast height
  Birth, recruitment and death rates
  Size
  Growth rates
  Reproductive state
  Number in reproductive condition
  Plants in flower
  Size of breeding colony
  Chemical content of living and dead material
  Soil structure and composition

B. Processes
Productivity
Litter accumulation
Decomposition
Consumption rates
Carbon, nitrogen fixation
Respiration
Colonization
Succession
Bioaccumulation

effect of climate change on gardens. In the UK, the Royal Horticultural Society together with the National Trust have commenced work on the likely effects of climate change on gardens. This is being undertaken because of the possibility that, by approximately 2080, temperatures may have increased by as much as 3.5 °C.

Ecological monitoring can also consider the past: that is, some biological monitoring can be done retrospectively using historical data. Historical monitoring or retrospective studies are important for the establishment of baselevels that occurred during preindustrial times and can be used to determine changes in pollution concentrations as a result of human activities. The material used for historical monitoring is especially interesting and includes aqueous sediments, ice and snow, peat, plant tissues, museum collections, herbarium specimens and human remains. One of the richest sources used for historical monitoring is the sedimentary record, providing data on pollution from both effluents and atmospheric deposition.

The sediments of the Great Lakes of North America, for example, have possibly been the source of more data for historical monitoring than anywhere else in the world. The combination of suitable sedimentary features for accumulating pollutants and the industry and agriculture surrounding the Great Lakes has resulted in marked concentrations of metals in the more recent sediments. For instance, the inputs of zinc and copper, which have increased markedly this century, can be attributed to the burning of coal and, in some instances, to the electroplating industry.

Materials of a biological origin that have been used in historical monitoring include herbarium specimens of mosses and lichens and zoological specimens of fish, bird feathers and eggs, animal hair, horns and teeth. The use of mosses and lichens as indicator species for the detection of airborne pollutants is particularly important, and past records could be established using herbarium collections. The use of bird feathers has been shown to be very effective in the historical monitoring of mercury in marine and terrestrial environments.

Witt *et al.* (2000) in Australia have described retrospective monitoring of rangeland vegetation change. In the absence of long-term records, they based their research on analysis of sheep faeces that had accumulated (1930s to 1990s) near a shearing shed in the semi-arid rangelands of southwest Queensland. The results indicated significant changes in the sheep's diet. This novel method may indeed prompt more research into historical vegetational changes.

Pollen analysis (Davis, 1989) has often been used in retrospective monitoring. An analysis of the relative proportion and incidence of pollen from different plant species found in deposits of peat or other substrata, together with other sources of information, can be used to construct a picture of ecological processes that have occurred over the last few thousand years (Fig. 1.2).

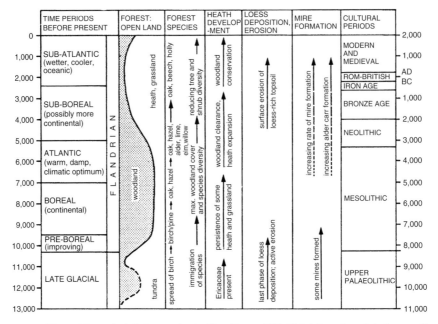

Fig. 1.2 Post-glacial changes in the New Forest (UK). This is based on data from pollen analysis, radiocarbon dating and other sources. (Reproduced from Tubbs (1986) with kind permission of the late Colin Tubbs.)

Pollen analysis was one of several techniques used by Engstrom *et al.* (1985) in a fascinating study of the limnological history of Harvey's Lake, Vermont in North America. Today, phytoplankton of that lake is dominated by the blue-green alga *Oscillatoria rubescens* and it is that species which has left behind pigments in the sediments. These sedimentary pigments were used as indicators of changes in primary production and a detailed, accurate chronology was made possible with the combined use of lead-210, caesium-137 and carbon-14 dating as well as stratigraphy of pollen and sawmill waste deposits. It was found that two periods of increased sedimentary anoxia (1820–1920 and 1945 to present) could be attributed to sawmill wastes and later also to increased levels of nutrients coming from dairy wastes.

Monitoring the effects of agricultural chemicals on organisms has also been done retrospectively. One classic example, which commenced in 1966, was Ratcliffe's (1980) study of thinning of Peregrine Falcon (*Falco peregrinus*) eggshells. Rather like a detective, Ratcliffe examined museum collections of eggs taken from various regions in Britain and calculated an eggshell index based on the weight of the egg divided by the length multiplied by the breadth. The change observed in the shell index (Fig. 1.3) from 1947 onwards has since been

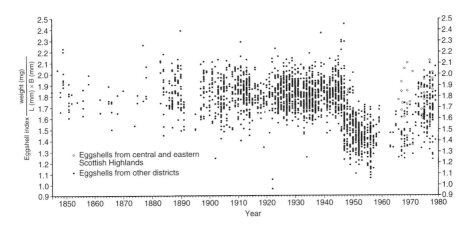

Fig. 1.3 Decrease in eggshell thickness for Peregrine Falcons. (From Ratcliffe (1980) with kind permission of Derek Ratcliffe and the publishers (T. & A. D. Poyser).)

shown beyond doubt to have been caused by accumulation of dichlorodiphenyl-trichloroethane (DDT) in peregrine tissues. This study used material that was not collected specifically for the purpose of monitoring variations in eggshell thickness.

On hindsight and bearing in mind the previous rapid growth in the use of agricultural chemicals (for example, the US total production of DDT from 1944 to 1968 was 1225 million metric tons (SCEP, 1970)), a monitoring programme could have been initiated to look at the effects of those chemicals on non-target organisms. That is, the data could have been sampled in a representative manner and assembled in a systematic manner, thus providing a sound basis for analysis. Much the same could be said for the many agricultural and horticultural chemicals currently in use today.

1.8   The relevance of ecological monitoring

Interest in the changes in populations of plants and animals (especially birds) is both widespread and topical. Not only is there an interest in such changes but there is also concern about declines in populations. Pollution, loss of habitat, disturbance and changes in land use may all be implicated in the decline of populations, but cause and effect cannot easily be demonstrated without data from monitoring and surveillance studies. For without long-term studies, it is difficult to distinguish between natural changes and those changes caused directly or indirectly by pollution and other impacts.

Surprisingly, we know very little about natural, temporal changes in plant and animal populations. We know even less about the long-term processes in ecosystems, especially undisturbed ecosystems. There are virtually no undisturbed biological communities left on Earth but there have been some attempts to establish long-term monitoring programmes in relatively undisturbed areas. The information from these programmes becomes more and more valuable with time, not only in terms of understanding natural processes but also in terms of providing a baseline for comparisons when disturbance or perturbation occurs.

The four major classes of ecological phenomenon identified by Strayer *et al.* (1986) as being appropriate for long-term ecological studies clearly illustrate the value of monitoring by way of long-term ecological studies.

1.   Slow processes (ecological phenomena such as forest succession and some vertebrate population cycles that occur on scales longer than the traditional three year funded research or indeed longer than a human life time).

2.   Rare events (perturbations such as fires and outbreaks of pests or disease).

3.   Subtle processes (ecological processes where the year-to-year variance is larger than the long-term trends).

4.   Complex phenomena (intricate ecological relationships in biotic communities).

The process in some ecosystems function very slowly indeed. For example, polar plant communities are very slow growing and some plants in the Arctic may take as long as three years for a flower to bloom and set seed. When thinking about monitoring effects of pollutants in ecosystems, such as radionuclides in grassland communities, we may all too easily forget that there is considerable variation in the rate at which different ecosystems function. Concern has been expressed recently at the high levels of pollutants in Arctic ecosystems and we may have to think very carefully how to monitor pollutants in ecosystems that function very slowly indeed.

We perhaps start to see that the answer to the question (what's the relevance of ecological monitoring?) can best be answered with reference to different timescales, different spatial scales and different biological scales. The relevance of ecological monitoring is best seen at many different scales: from small to large and from single species to biological communities and ecosystems.

Some ecological monitoring programmes have immediate value with regard to management of natural resources. For example, one early example of monitoring and one that was established with much foresight was Hardy's (1956) continuous plankton recorder study. In 1929, Hardy proposed to survey and monitor the distribution and movements of plankton in the North Sea with the object of relating the results to movements of fish populations. The work of surveying plankton, conceived many years ago by Hardy, is now continued in a less than systematic manner despite important applications in relation to population dynamics of commercially valuable fish.

Ecological monitoring has played and continues to play an important role in conservation of species, natural communities and landscapes. World organizations have, for example, been established to monitor population levels and the distribution of endangered plant and animal species. Monitoring changes in land use and the effects of loss of habitats and isolation of populations are other important applications of monitoring in conservation.

There have been some most interesting developments in the methods used for insect pest monitoring programmes but equally interesting is the behavioural and ecological information that has come from the long-term studies of insect pests. However, as we shall see below, funding even for this kind of monitoring has been reduced. By way of contrast, there seems to be continued support for pollution monitoring.

Chemical monitoring of pollution has become very highly developed, with the use of very powerful computers and sophisticated technology. It is now possible to detect very small quantities of some pollutants in a relative short space of time (Table 1.3). This would imply that biological monitoring has become redundant, but nothing could be further from the truth. Chemical monitoring tells us what is there but it does not tell us what the effects are, especially the long-term effects on ecosystems. For this reason, biological monitoring has a very valuable and interesting role to play in monitoring pollution. The use of living organisms to monitor pollution has become very sophisticated and there have been some recent exciting advances in the methodology for automated biological monitoring of pollution as well as monitoring the quality of drinking water (e.g. Gruber & Diamond, 1988).

The use of biological indicators or bioindicators in monitoring programmes has been of particular interest to the International Union of Biological Sciences (IUBS) who, at their 21st General Assembly in 1982, initiated the Bioindicators Programme in Ottawa. As well as recommending that the IUBS organize biological monitoring programmes, that Assembly advocated biological indicators in monitoring by way of the following:

- encourage scientific and national bodies to develop and improve methods for monitoring environmental change
- promote interdisciplinary and international cooperation in standardizing methods

Table 1.3 *Detection limits and analysis time for some pollutants*

Type of pollutant	Analytical technique	Detection limits	Analysis time[a]	Some applications
Volatile organic compounds	GC/MS with MSD	10 µg/l (50–200 µg/kg)	50 min	Waters, sediments
Phenols and nitrocompounds	GC/MS with MSD	25 µg/l	1 h	Waters and wastewaters
Amines	LC/MS	100 ng	30 min	Soil and water samples
Pesticides	GC/MS or GC/ECD	1–10 ng/l (2–10 µg/kg)	30 min	Liquids, plants, animals
Polychlorophenols	HPLC	25 µg/l	35 min	Wastewaters
Polyaromatic hydrocarbons	HPLC	10–25 µg/l (0.2–0.5 mg/kg)	30 min	Waters, soil samples
Halocarbons	GC with headspace sampler	25 µg/l	20 min	Waters and wastewaters
Polychorinated biphenyls	GC/ECD or GC/MS	10 ng/l (2 µg/kg)	30 min	Waters, soil samples
Heavy metals	DPP	0.1–5 µg/l	45 min	Waters

GC, gas chromatography; MS, mass spectrometry; MSD, mass selective detector; LC, liquid chromatography; HPLC, high pressure liquid chromatography; ECD, electron capture detector; DPP, differential pulse polarography; AAS, atomic absorption spectrophotometry; PCBs: polychlorinated biphenyls.
Data for the instrumental step of the analysis from Orio (1989) with permission of Academic Press.

- encourage the exchange of research results amongst laboratories of the world
- support conferences dealing with bioindicators at cellular, individual, population and ecosystem level.

Pollution knows no political boundaries and, therefore, we might expect cooperation and collaboration between countries when it comes to monitoring the effects of pollution. Countries within the European Economic Community (EEC) and Scandinavian countries have often discussed collaborative monitoring projects but it was only as recently as 1979 that the Convention on Long-range Transboundary Air Pollution was signed within the framework of the European Commission. This cooperative monitoring programme, which came into force in 1983, provides regular information on selected air pollutants from many monitoring stations.

Ecological monitoring within the EEC has only slowly become recognized by member governments following the establishment of community-wide mechanism for protection of important wildlife areas. In 1988, for example, there was a proposal for a Council Directive on the protection of natural and semi-natural habitats of wild fauna and flora. That proposal mentioned monitoring in Article 24.

> Member states shall take all necessary measures to ensure the monitoring of the biological communities and the populations of the species specified in accordance with Annex 1 and in the areas classified under Article 5. Member States shall send the Commission the information resulting from monitoring, so that it may take appropriate initiatives with a view to the coordination necessary to ensure fulfilment of the objectives of the Directive.
>
> The Community and Member States shall cooperate to ensure consistency of monitoring and measurement methods.

If for no other reason, there was an overwhelming case for ecological monitoring in Europe because of the EEC's ambitious plan to encourage development in the poorest regions of Europe. The Worldwide Fund for Nature (WWF) and the Institute for European Environmental Policy (IEEP) warned that governments were hastily assembling major but possibly unviable schemes, the impacts of which had not been considered sufficiently. Certainly, the long-term ecological impacts of most of these major projects were not considered; often short-term economic advantages and political gain overrode objections to environmentally damaging schemes.

Towards the end of the millennium, similar issues arose in eastern Europe. Growth in major road developments became commonplace and while such developments may meet the needs of developing economies, there are important implications for the ecology of the area. Roads, for example, pose an incremental threat to the biogeography and ecology of many terrestrial forms of wildlife (Spellerberg, 2002).

The range of applications of ecological monitoring is broad. Examples could include the following:

1. *Adaptive management.* Monitoring data via adaptive management provides a basis for managing nature.
2. *Environmental planning.* Monitoring land use and landscapes as a basis for better use of the land: that is, combining conservation with other uses.
3. *Monitoring the state of the environment.* Using organisms to monitor pollution and to indicate the quality of the environment.

4. *Ecological science.* Monitoring as a way of advancing knowledge about the dynamics of ecosystems.
5. *Pests and diseases.* Monitoring of insect pests of agriculture and forestry in order to establish effective means of control of those pests.
6. *Climate change.* Monitoring the effects of climate change.

These are all important applications but they can not be achieved without the appropriate funding, staffing and logistical support. Nor can they be achieved without enforcement of statutory requirements, at least where such statutory requirements apply.

A considerable challenge for future monitoring has arisen from research on genetically modified organisms (GMOs). There seem to be many unknowns about the implications of releasing GMOs and the possible long-term effects on ecosystems seem not to have been considered in depth. There is, therefore, a clear and urgent need for development of rigorous biological monitoring techniques as well as some careful thought being given to the logistics of such monitoring programmes. Who would do the monitoring and how would the data be stored and communicated? The potential applications of GMOs and the competitive research now taking place will not make it any easier to ensure that effective and rigorous monitoring does take place. There is, however, a potentially very important contribution yet to be made by ecologists if release of GMOs becomes even more common.

The environment is ever changing, and most of those changes result from the pressure of human activities and human population density. Recording and monitoring the state of the environment is becoming commonplace. Ecological monitoring is an essential component of studies of environmental change and ecological monitoring can play a role in policy formulation. Many scientists seem to support the view that defining and understanding changes will be necessary for sustainable development (Brydges, 2001).

Although environmental and ecological monitoring programmes have increased since the late 1980s, there remain some very important contemporary issues (Table 1.4). These issues have to be addressed. Resourcing is a major issue. There are few agencies prepared to commit resources for long periods of time. Without that commitment, the sustainability of the monitoring is in doubt. Perhaps linked to effective use of resources is the growing acknowledgement of the need for collaboration. One example is the environmental and ecological monitoring that is taking place in the San Francisco estuary (Hoenicke *et al.*, 2003). This partnership of many agencies has resulted in a collaborative monitoring programme that appears to have been very effective in the systematic adjustment of research questions.

Table 1.4 *Contemporary issues in long-term research and ecological monitoring*

Issues	Factors
Resources and policies	Funding: few agencies are prepared to commit funds for long-term research or long-term ecological monitoring
	Infrastructure problems arising out of changes in priorities, personnel and information technology
	Who pays: the developer or the regulator?
	Mitigation failure: who pays if monitoring shows that mitigation has not worked or that the environmental impact assessment was not accurate? Who takes responsibility?
	Assessments of monitoring programmes: how effective are the programmes?
	Collaboration moves towards greater collaboration
Methods	Establishment of protocols
	How many sites are required?
	Frequency and timing of data recording: how often and for how long?
Scientific credibility	Demand for improved standards of practice
Community and political pressures	Increasing participation by local communities
	Demands for environmental standards
	Increased reliance on 'action plans' and 'indicators'
	Rejection of technology (e.g. reluctance to accept genetically modified foods)
Changing environments	Effects of climate change

Ecological monitoring is an integral component of EIAs yet there continues to be debate about who should pay. Little has been done to address the issue of who pays if the ecological monitoring shows that the mitigation failed or the EIA was inadequate.

The protocols for monitoring are essential as are the standardized methods. A common question asked is how often should the data be collected and for how long? From a scientific point of view, this is probably one of the most difficult questions to answer.

Ecological monitoring must be based on rigorous and credible science and it must be undertaken by people who are competent and who meet professional standards.

Community and political pressures are affecting what is monitored and how it is monitored. Community pressures may result in demands for certain kinds of monitoring to be undertaken. There are also examples of community pressure blocking new initiatives, such as the sale of genetically modified food.

Finally, ecological monitoring is about change over time. The environment is forever changing, but there are overarching changes or trends taking place that have implications for ecological monitoring: the most important being climate change and effects of climate change on ecology and biogeography of organisms throughout the world.

2

Environmental monitoring programmes and organizations

2.1 Introduction

This chapter has two aims: to describe the pressures that have led to environmental monitoring and ecological monitoring and to provide a brief introduction to some environmental and ecological monitoring organizations and programmes. Some of these are international; others are national or regional. The examples given here are but a few of the many monitoring organizations and programmes that exist today. The selection of organizations and programmes has been undertaken on the basis of providing as great a variety as possible of the scope and scale of the monitoring activities. In pursuit of these two aims, this chapter discusses the successes and difficulties that monitoring programmes have had throughout their history.

2.2 An historical perspective

Looking back through episodes of human impact on the environment, it is difficult to know where to commence with an introduction to national and international environmental monitoring activities and organizations. Many historians could possibly identify examples of environmental monitoring and ecological monitoring that took place many centuries ago, but which cannot be considered here for reasons of space.

Very significant to the development of the science of ecology were the early 1900 studies of bird populations in Europe (Table 2.1). The late David Lack

Table 2.1 *Some events and publications of significance to the development of environmental and biological monitoring*

Year	Event
1912	Counting of the number of breeding Great Tits commences in the Netherlands
1928	Counts of heronries commence in England and White Storks in Germany
1929	Alister Hardy proposes to survey and monitor the distribution of plankton in the North Sea
1946	International Convention for the Regulation of Whaling establishes the International Whaling Commission
1948	UN Charter
	International Union for the Protection of Nature (IUPN) established
1955	The *Wenner Gren Conference on Man's Role in Changing the Face of the Earth*, Princeton, NJ
1956	*Man's Role in Changing the Face of the Earth* (Thomas 1956) published.
1957	The IUPN becomes the International Union for the Conservation of Nature and Natural Resources (IUCN)
	International Geophysical Year
1958	Law of the Sea: the first UN Conference on the Law of the Sea approves draft conventions
1959	Antarctic Treaty: Economic and Social Council of the UN adopts resolution to publish a register of national parks and equivalent reserves of the world
1961	Establishment of World Wildlife Fund (Worldwide Fund for Nature)
1962	*Silent Spring* (Carson, 1962) published.
1964	International Council of Scientific Unions (ICSU) established the International Biological Programme (IBP)
1966	IUCN *Red Data Books* first published
1968	UNESCO *'Biosphere' Conference*
1969	Friends of the Earth (FOE) founded
1970	The US National Environmental Policy Act (NEPA) requires preparation of Environmental Impact Assessments
1971	Man and the Biosphere (MAB) Programme of UNESCO launched
	Greenpeace International founded
1972	UN Stockholm *Conference on the Human Environment*
	Concept of a global monitoring system (GEMS) endorsed by the Stockholm Conference
	UN Environment Programme (UNEP) established
	'Blueprint for Survival' sponsored by the journal *Ecologist*
	Limits to Growth (Meadows *et al.*, 1972) published
1973	UNEP introduces Earthwatch
1974	UNEP Regional Seas Programme established
1975	Convention on International Trade in Endangered Species of Wild Fauna and Flora (CITES); the Kenya Rangeland Ecological Monitoring Unit (KREMU) established as a result of collaboration between Kenya and the Canadian International Development Agency

Table 2.1 (*cont.*)

Year	Event
1976	The Special Committee on Problems of the Environment (SCOPE) reports to ICSU on global trends in the biosphere most urgently requiring international and interdisciplinary scientific effort
1977	UN *Conference on Desertification*
1979	*World Climate Conference* organized by the World Meteorological Organization recognizes the 'greenhouse effect'
1980	World Conservation Strategy (IUCN) launched
	IUCN Conservation Monitoring Centre (now UNEP-WCMC) established
	The *Global 2000 Report* to US President Carter
1982	UN Nairobi Conference. World Charter for Nature Conservation adopted.
	The World Environment 1972–1982 (Holdgate *et al.*, 1982 published).
	International Union of Biological Sciences (IUBS) initiated the Bioindicators Programme
1983	UN General Assembly calls for the establishment of an independent commission: the World Commission on Environment and Development
1984	*The Resourceful Earth. A Response to Global 2000* (Simon & Kahn, 1984) is published
	First of a series of symposia on monitoring programmes organized by the IUBS
1985	European Community Environmental Impact Assessment Directive adopted
	First Cary Conference: status and future of ecosystem science was discussed
	Conference held in Venice on *Man's Role in Changing the Global Environment* (Orio & Botkin, 1986).
1987	*Our Common Future* (WCED, 1987) is published
	Second Cary Conference (New York): endorsement of the need for long-term sustained ecological research
1989	The 'Green Summit' in Paris of seven industrial countries; final communique says 'urgent need to safeguard the environment'
	From Strategy to Action (IUCN, 1989a) published: a response to *Our Common Future*
1990	*BioScience* (July/August 1990, 40, issue 7) publishes three major articles on long-term ecological research (Franklin *et al.*, 1990)
1992	UN *Conference on Environment and Development* at Rio de Janeiro
	Convention on Biological Diversity
	The Framework Convention on Climate Change
	Agenda 21
	Global Biodiversity: Status of the Earth's Living Resources (Cambridge, 1992) is published using data compiled by the World Conservation Monitoring Centre
1995	*Global Biodiversity Assessment* (Heywood, 1995) published
1999	*GEO 2000* (UNEP, 2000), a millennium report on the environment, published
2000	*State of the World 2000* (Brown *et al.*, 2000) is published; this is 17th annual review published by the Worldwatch Institute
2002	The UN Earth Summit at Johannesburg
	Diversitas science plan published
	The first *World Atlas of Biodiversity: Earth's Living Resources for the 21st Century* is launched by UNEP-WCMC 2002

played a central role in the analysis of that data and two quotes have a place today as they did in the 1960s (Lack, 1966):

> It is important to continue the study of a natural population for many years
>   ... it may be stressed that long-term ecological studies are not so difficult as their paucity might suggest, and provided that the initiator feels that the work is worthwhile, ways can normally be found to continue it.

The monitoring studies of bird populations in the early 1900s were preceded by similar studies on mammals in North America in the nineteenth century. We owe much to the naturalists of that time for raising the awareness of change in nature.

Significant events in the history of the exploitation of marine populations also seem a useful place to start; it was in 1946 that an International Fisheries Convention was signed and the International Whaling Commission (IWC) was established (Table 2.1). The need for marine biological monitoring at that time was evident but any practical use of monitoring data in the exploitation of marine populations was not to take place until many years later.

The year 1948 was a very important year for wildlife conservation and monitoring of endangered plants and animals. In that year the International Union for the Protection of Nature was established at a conference at Fontainebleau, convened through the initiative of Sir Julian Huxley during the time when he was Director-General of the United Nations (UN) Economic, Social and Cultural Organization (UNESCO). Later, in 1957, that organization was renamed the International Union for the Conservation of Nature and Natural Resources (IUCN).

Commencing in the early 1960s, a number of very important publications highlighted the need for ecological monitoring programmes. First, *Silent Spring* was published (Carson, 1962; second edition in 2002) and, despite criticisms about the emotive and less than scientific approach used, the book prompted much concern about the effects of pesticides, particularly the effects of pesticides on non-target organisms as a result of very widespread use of agriculture chemicals in the USA.

The Man and the Biosphere (MAB) programme was established in 1970 by the general conference of UNESCO. Since then, many hundreds of biosphere reserves have been established by MAB. They serve as multipurpose areas, dedicated to conserving ecosystems and species. In 1986, the Smithsonian Institution joined with UNESCO-MAB to create the SI/MAB Biodiversity Programme under the US National Museum of Natural History. The SI/MAB programme combines long-term biodiversity measurment and monitoring with professional training courses covering the principles and procedures of monitoring and data collection.

As well as playing a key role in education, SI/MAB has a commitment to the establishment of a global network of permanent, long-term biodiversity monitoring plots in biosphere reserves and other protected areas. The SI/MAB process is outline in Figure 2.1

Over the years, MAB has developed a framework for assessment and monitoring of biological diversity as well as protocols for baseline assessments. For example the Gamba Project in Gabon and the Camisea Project in Peru are directed at assessment and monitoring of biological diversity. The Camisea Project has a network of biological diversity monitoring sites alongside a proposed pipeline route. This may well provide a model to show how different groups can work together and a model for baseline studies, but hopefully it will not be ecological monitoring just for the sake of monitoring.

A 'milestone' report in 1980 was the *World Conservation Strategy* (IUCN, 1980). Prepared by the IUCN, UNEP and WWF, this report focused on three objectives:

- maintenance of ecological processes and life-support systems
- preservation of genetic diversity and the conservation of wild species
- ensuring the sustainable utilization of species and ecosystems and the use of all our natural resources carefully giving due consideration to the needs of future generations.

A successor volume to the *World Conservation Strategy* was published in 1991 (IUCN, 1991). Seven years after the publication of the *World Conservation Strategy*, the report of the World Commission on Environment and Development, (WCED, 1987) was heralded as the most important document of the decade on the future of the world. This report reexamined the critical environment issues and development problems with the aim of formulating realistic proposals to solve the issues and problems but with sustainable development. The report also highlighted the need for environmental monitoring in its many forms as noted by Maxwell Cohen of the University of Ottowa (WCED, 1987, p. 275).

> We need a kind of new earth/space monitoring system. I think that it goes further than simply an earth environmental system. It's a combined earth/space monitoring system, a new agency that would have the resources to be able to monitor, report, and recommend in a very systematic way on the earth/space interaction that is so fundamental to a total ecological view of the biosphere.

Mention of human population growth and distribution can hardly be avoided, and quite rightly so, when discussing environmental issues and diminishing

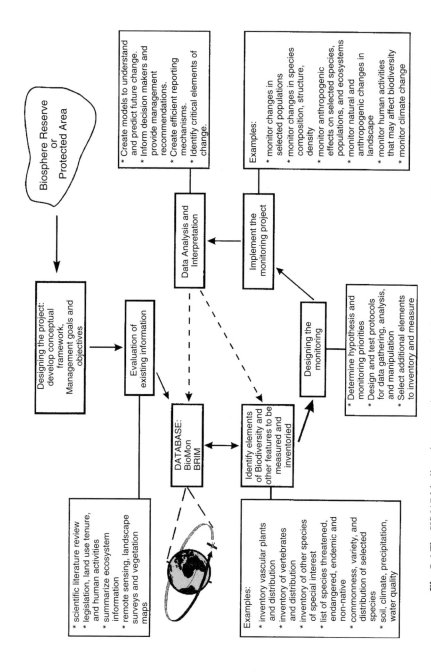

Fig. 2.1 The SI/MAB biodiversity measuring and monitoring process. (Reproduced with kind permission of MAB.)

biological resources. Statistics on and monitoring of world population trends, health, agriculture, energy, fisheries and forests is undertaken by the UN Statistical Office (UNSO), the Food and Agriculture Organization of the UN (FAO) and the World Health Organization (WHO). A most important international event, which was to act as a stimulus for global environmental monitoring, was the 1972 UN *Conference on the Human Environment* (popularly called the 'Stockholm Conference') held at Stockholm in Sweden. This was the first 'Earth Summit'.

Following the 1972 Earth Summit, the UNEP was established. Worldwide research on the environment as well as monitoring and information exchange were considered to be essential and an action plan agreed at Stockholm had three major components: environmental assessment, environmental management and supporting measures.

In 1973, UNEP established Earthwatch to coordinate all environmental monitoring and assessment activities throughout the UN organization. Later in 1994, the first interagency Earthwatch Working party agreed on the following mission.

> ... to coordinate, harmonize and integrate observing, assessment and other reporting activities across the UN system in order to provide environmental and appropriate socio-economic information for national and international decision-making on sustainable development and for early warning of emerging problems requiring international action. This should include timely information on the pressures on, status of and trends in key global resources, variables and processes in both natural and human systems and on the responses to problems in those areas.

Ten years on from the Stockholm Conference, the UN adopted UN Resolution 37/7, the World Charter for Nature (Box 3.2, p. 85) on 28 October 1982 and paragraph 19 in that Charter was confirmation of the worldwide acceptance of a need for biological and ecological monitoring. 'The status of natural processes, ecosystems and species should be closely monitored to enable early detection of degradation or threat, ensure timely intervention and facilitate the evaluation of conservation policies and methods.'

Unfortunately the Earth Charter was no more than a charter and was not binding. The charter had been adopted by a vote of 111 with one against (USA) and 18 abstentions.

The 1992 Earth Summit in Rio resulted in five important documents:

- The Rio Declaration
- Forest Principles
- Agenda 21

- the Framework Convention on Climate Change
- The Convention on Biological Diversity.

The first two documents were statements of principles. Agenda 21 was a statement about worldwide sustainable development and the other two documents were important conventions. The Convention on Biological Diversity is particularly relevent to ecological monitoring. The website for the Secretariat of the Convention is http://www.biodiv/org (see Appendix for the full text). The Convention refers to monitoring and in Article 7 it says under the heading 'Identification and monitoring':

> Each contracting party shall, as far is possible and as appropriate, in particular for the purposes of Articles 8–10:
>
> a.   Identify components of biological diversity important for its conservation and sustainable use having regard to the indicative list of categories set down in Annex 1;
>
> b.   Monitor, through sampling and other techniques, the components of biological diversity identified pursuant to subparagraph (a) above, paying particular attention to those requiring urgent conservation measures and those which offer the greatest potential for sustainable use;
>
> c.   Identify processes and categories of activities which have or are likely to have significant adverse impacts on the conservation and sustainable use of biological diversity, and to monitor their effects through sampling and other techniques; and
>
> d.   Maintain and organize, by any mechanism data, derived from identification and monitoring activities pursuant to subparagraphs (a), (b) and (c) above.

Following the 1992 Earth Summit, Agenda 21 groups were set up all around the world and they continue to operate today. Many countries prepared national biological diversity strategies and some of these are now being implemented. The Convention on Climate Change had a chequered history and few advances have been made.

In the years following the 1992 summit, there were further initiatives, which led to publications about the state of the world's ecosystems and the state of biological diversity. These publications drew attention to the pressures and the continuing decline in the state of ecosystems.

In addition, there has been much effort directed at compiling information on ecosystems and biological diversity; not only on the condition but also what is there. For example, the *World Resources 2000–2001* report (UNDP, UNEP, World

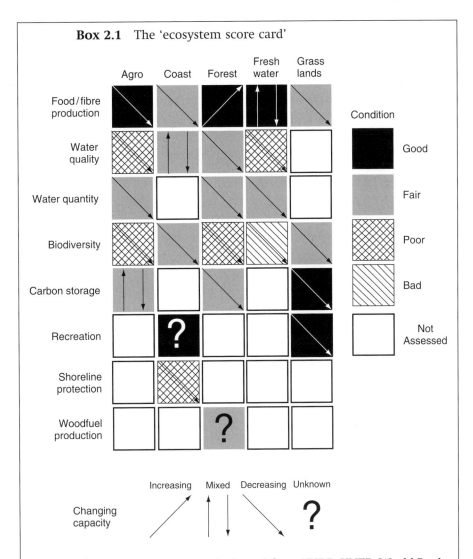

Box 2.1 The 'ecosystem score card'

The ecosystem score card adapted from UNDP, UNEP, World Bank and World Resources Institute (2000). The score card assessed the current (2000–2001) output and quality of the ecosystem, goods or service with that of 20–30 years previous. Note that none rated excellent. Changing capacity assessed the underlying biblogical ability of the ecosystem to continue to provide the goods or service.

Bank and the World Resources Institute, 2000) is a moving account about the degradation of the world's ecosystems. The 'ecosystem score cards' (Box 2.1) provide an interesting way of conveying the information in a simple way to a very wide audience.

Other initiatives include the Millennium Ecosystem Assessment, launched in 2001, which is a project over four years to survey the health of the world's ecosystems. There is the All Species Foundation, which is dedicated to the complete inventory of all species on Earth. The hope is to achieve this by 2027. This is a formidable task and certainly much larger than any genome project. Finally there is BioNet-International, which is the global network for capacity building in taxonomy for ecologically sustainable development.

The 2002 Earth Summit in Johannesburg had far less success than the 1992 Earth Summit. However, the main outcomes were directed at the following five topics:

- water sanitation
- sustainable energy
- health care in the developing world
- improving agricultural methods
- biological diversity.

2.3   International environmental monitoring organizations

In the space available, it is not possible to do justice to even a sample of environmental and biological monitoring organizations. The UN *Conference on the Human Environment* was a landmark in the development of world environmental monitoring initiatives. Credit is due, however, to the many previous efforts that set about establishing global monitoring, such as efforts to establish a network of major monitoring stations in natural areas throughout the world (Massachusetts Institute of Technology, 1970). Many of these efforts are usefully outlined in a directory published in 1970 by the Smithsonian Institution. The Special Committee on Problems of the Environment (SCOPE) and other components of the ICSU were also to play an important part in bringing these initiatives to the attention of the UN Conference. The 2003 UNEP list of monitoring and research programmes is now very impressive (Table 2.2).

The United Nations Environment Programme and the Global Environmental Monitoring System

The concept of a global system of environmental monitoring was endorsed by the UN *Conference on the Human Environment* in 1972; later in 1975, UNEP embarked on the most ambitious programme to monitor the quality of air, water and food on a global scale. The Global Environmental Monitoring System (GEMS) was established (with a programme activity centre at the UNEP headquarters at Nairobi, Kenya) to acquire data for the better management of

Table 2.2 *The 2003 United Nations Environment Programme list of monitoring and research programmes*

Arab Centre for the Studies of Arid Zone and Dry Lands
Arctic Monitoring and Assessment Programme
African Ministerial Conference on the Environment
Centre for Global Environmental Research (NIES)
Coordinated Programme on Marine Pollution Monitoring and Control in the South-East Pacific
Coral Reef Monitoring Network
Diversitas (IUBS/SCOPE/UNESCO)
Earth Observation Programmes (ESA)
Earthwatch
Environment Research Programme
European Environment Agency
Environmental Monitoring and Assessment Programme (US EPA)
Cooperative Programme for the Monitoring and Evaluation of Long-Range Air Pollutants in Europe (UN and ECE)
Environmental Programme for the Danube River Basin
Earth Observation Programmes (EC, ESA)
EUREKA Environmental Projects (ECE)
European Experiment on Transport and Transformation of Environmentally Relevant Trace Constituents in the Troposphere over Europe
Global Atmosphere Watch (WMO)
Global Atmosphere Watch Background Air Pollution Monitoring Network
Global Atmosphere Watch Global Ozone Observing System
Global Climate Observing System (WMO, IOC, UNEP, ICSU)
Global Environmental Epidemiology Network (WHO)
Global Environment Monitoring System (UNEP)
Global Environment Monitoring System Urban Air Quality Monitoring Project
Global Environment Monitoring System Food Contamination Monitoring Project
Global Environment Monitoring System Human Exposure Assessment Location Project
Global Environment Monitoring System Assessment of Freshwater Quality
Global Environmental Radiation Monitoring Network (UNEP/GEMS, WHO)
Global Investigation of Pollution in the Marine Environment
Global Sea Level Observing System (UNESCO/IOC)
Global Network Isotopes in Precipitation (IAEA)
Global Ocean Observing System (UNESCO/IOC)
Global Terrestrial Observing System
Human Dimensions of Global Change (IGBP)
International Centre for Integrated Mountain Development
International Cooperative Programmes (UN and ECE)
International Geosphere–Biosphere Programme (ICSU)
Integrated Global Ocean Services System
International Hydrological Programme

Table 2.2 (*cont.*)

Environmental Programme of the International Institute for Applied Analysis
International Joint Commission
Integrated Monitoring
Intergovernmental Panel on Climate Change (WMO, UNEP)
International Programme on Chemical Safety (UNEP, ILO, WHO)
International Plant Genetic Resources Institute
International Satellite Land Surface Climatology Project
International Space Year
Tsunami Warning System in the Pacific
International Waterfowl and Wetlands Research Bureau
Man and the Biosphere Programme (UNESCO)
Interim Committee for Coordination of Investigations of the Lower Mekong Basin
Mussel Watch (NOAA)
OECD Environment Committee
Regional Seas Programme Committee (Oceans and Coastal Areas/Programme Activity Centre)
Scientific Committee on Problems of the Environment
South Pacific Regional Environment Programme Marine Pollution Programme
Sahara and Sahel Observatory
System for Analysis, Research and Training (ICSU)
United Nations Industrial Development Organisation Environment Programme
World Climate Programme
World Climate Programme Water
World Climate Research Programme
Working Group on Climate Change Detection (WMO)
World Glacier Monitoring Service
World Weather Watch Programme

See p. xiv–xvii for all acronyms and abbreviations.

the environment. The aim (then and now) is to monitor and assess the global environment. As a first step, the expertise of the WHO was enlisted because of that organization's experience in air monitoring.

The first, and demanding, task to be tackled by GEMS was that of coordination, collection and dissemination of information from environmental monitoring programmes, particularly at the international level and through the services of the International Environmental Information Network (INFOTERRA). This information network was formerly known as the International Referral System for Sources of the Environment. The task of assembling and disseminating of data on a global scale seemed then, as is now, an extremely daunting task despite the great improvements in methods of data collection and communication that have taken place over many years. The GEMS programme has responded

Table 2.3 *Topics treated in Annual State of the Environment Reports submitted to the Governing Council of the United Nations Environment Programme from 1972 to 1982*

Subject area	Topics
The atmosphere	Climate changes and causes
	Effects of ozone depletion
Marine environment	Oceans
Freshwater environment	Quality of water resources
	Groundwater
Terrestrial environment	Land resources
	Raw materials
	Firewood
Food and agriculture	Food shortages, hunger, and degradation and losses of agricultural land
	Use of agricultural residues
	Pesticide resistance
Environment and health	Toxic substances and effects
	Heavy metals and health
	Cancer, malaria, schistosomiasis
	Chemicals in food chains
Energy	Energy conservation
Environmental pollution	Toxic substances
	Noise pollution
Man and the environment	Human stress and social tension
	Demography and populations
	Tourism
	Transport
	Environmental effects of military activity
	The child and the environment
Environmental management achievements	The approach to management
	Protection and improvement of the environment
	Environmental economics

well to these demanding tasks and there seems no doubt that it is now a significant world environmental monitoring agency.

The Governing Council of UNEP required annual reports and comprehensive five-yearly reports on the state of the environment, a requirement laid down by the UN General Assembly, namely to 'keep under review the world environmental situation in order to ensure that emerging environmental problems of wide international significance receive appropriate and adequate consideration'. The broad scope of topics in the Annual State of the Environment reports published during the first few years following the Stockholm Conference was impressive (Table 2.3) and showed that an international monitoring strategy had been achieved. For example, global air monitoring uses data from more than 60 countries and monitoring water quality took place at 344 monitoring sites in 42 countries.

By 1982, monitoring activities supported by UNEP were firmly established under the auspices of GEMS and included the following:

- climate-related monitoring
- monitoring of long-range transport of pollutants
- health-related monitoring
- ocean monitoring
- terrestrial renewable-resources monitoring.

The principal activity of GEMS is environmental monitoring, an activity assisted by the expertise of specialized agencies; for instance, there are several agencies that contribute data from terrestrial renewable-resource monitoring (Table 2.4). The objectives now include:

- strengthening monitoring and assessment capabilities of participating countries
- increasing the validity and comparability of environmental data
- producing global and regional reports
- increasing cooperation within UN specialized agencies
- promoting the collection of data
- providing local and national authorities with tools and methods
- increasing the use of indicators
- providing early warnings on emerging issues of potential international importance.

Note the use of 'indicators', which has now become an integral part of many environmental monitoring programmes. Note also the point about 'validity', which has become so important in the light of recent environmental backlashes.

An early example of 'terrestrial renewable-resources monitoring' was provided by the work carried out from 1980 to 1985 on the Sahelian pastoral ecosystems in Senegal (GEMS, 1988), under the auspices of both FAO's 'Ecological Management of Arid and Semi-arid Rangelands' and UNEP's GEMS. The Sahel region of Africa is the semi-arid region to the south of the Sahara where there have been pronounced droughts since the late 1980s. Several countries made up the Sahel region (Burkina Faso, Chad, Djibouti, Ethiopia, Mali, Mauritania, Niger, Senegal, Somalia and the Sudan).

The GEMS programme had previously specified that 'monitoring activities will be undertaken and expanded following the recommendations of the group

Table 2.4 *Examples of monitoring projects implemented by the United Nations Environment Programme*

Project title	Location of headquarters	Date founded	Participants
Global Environment Monitoring System (GEMS)	UNEP Nairobi	1975	Global
Biological monitoring pilot project	WHO Geneva	1978	10 countries
Tropical forests resources assessment	FAO Rome		76 countries
Tropical forest cover monitoring	FAO Rome		3 countries
Pastoral ecosystem monitoring in West Africa	LNERV/ISRA Daker		
Desertification monitoring in Latin America	ONERN Lima		
Soil degradation in North Africa and Middle East	FAO Rome		
Monitoring status of mammals	WCMC Cambridge	1980	Global
Monitoring status of birds	ICBP Cambridge	1980	Global
Trade in endangered species	WCMC Cambridge	1980	Global
Coral reefs	WCMC Cambridge	1980	Global
Parks and protected areas	WCMC Cambridge	1981	Global

ONERN, Peruvian National Office for National Resources Evaluation; ISRA-LNERV, Senegelese Institute for Agricultural Research-National Laboratory of Veterinary Research; all other acronyms given on p. xiv–xvii.

of government experts which will examine monitoring as a means of evaluating problems resulting from agricultural and land-use practices'. The main objectives, therefore, of the Sahelian programme included the provision of baseline data, the establishment of standard methods on an international level for the monitoring of the rangelands and the suggestion of actions to combat desertification (GEMS, 1988).

A test area of $30\,000\,km^2$ was selected for the Sahelian Pastoral Ecosystem monitoring project and monitoring was undertaken with satellite imagery, aerial photography, low-altitude flights and ground control validation. Three main lines of activity developed from the project; satellite evaluation of green biomass on the range at the end of the rainy season, livestock census and evaluation of erosion, and control of remotely sensed data via ground sampling. This made it possible to forecast the management needs for the nine months of the dry season and, therefore, to improve management of livestock in an area subject to drought and desertification.

As well as the GEMS programmes in the Sahel, the IUCN (with generous support from Nordic countries) became involved in the challenging environmental problems and the desperate plight of millions of people in the Sahel. Following the establishment in 1987 of a Sahel Coordination Unit at IUCN headquarters in Gland, one of the largest IUCN field programmes was established. The IUCN Sahel programme's objectives included:

- to develop ways to manage living natural resources that better correspond to prevailing climatic conditions and which permit sustainable development
- to help to preserve the biological diversity of the Sahel
- to monitor the changes taking place in the Sahel region.

A report prepared by IUCN and the Norwegian Agency for International Development (IUCN, 1989b) provided an assessment of the major issues affecting the Sahel region, a report that usefully combined studies on both the people and the environment as a basis for commencing plans leading to sustainable development.

Remote sensing by satellite imagery, a method which is now undertaken by GEMS, enlarged the scope of monitoring. In 1985, UNEP established a new element of GEMS in the form of the Global Resource Information Database (GRID) and the key to this was the successful use of satellites. Satellite data for environmental monitoring is now commonplace: another example is the Advanced Earth Observation Satellite 11, which is used in Asia and the Pacific. The Economic and Social Commission for Asia and the Pacific (ESCAP) has hosted many meetings on the use of satellites in environmental monitoring.

By 2003, GRID had become a network of environmental data centres. The control facility is in Nairobi with particular emphasis on Africa but there are 12 other centres. Advanced technology and global datasets from numerous organizations (including UNEP, UNESCO, FAO, WHO and the IUCN) enable GRID to assess and examine interactions between different environmental datasets.

The applications of the GRID system are many. One application was in connection with monitoring and conservation of African Elephants, leading to the conclusion that the off-take of ivory was more than the elephant populations could sustain and that many populations were in danger of extinction (UNEP/

GEMS, 1989). Following on from this, the recent action by the Convention for International Trade in Endangered Species (CITES) highlighted the serious damage inflicted on elephant populations by ivory poaching. The reports from CITES have, in turn, forced many nations to give urgent consideration to the serious implications of sustaining the ivory industry (see Caughley *et al.* (1990) for an interesting account on the decline of the African Elephant). Monitoring elephants and the ivory episode was just one small example of the potential of GRID, a potential which seems underappreciated and inadequately understood by the international community. Indeed, much can be learnt from the African experiences of monitoring; a wealth of monitoring has been undertaken by international agencies as well as by national centres of research such as the National Programme for Ecosystem Research based in Pretoria.

Many reports have now been published by UNEPS and GEMS but one landmark publication came in 1987 when the UNEP Environmental Data Report was published. Prepared by the Monitoring and Assessment Research Centre (MARC) in cooperation with other agencies, this report summarized the best environmental data currently available at that time. A computer database on environmental information based on the UNEP Environmental Data Report is maintained by MARC (see below) and this database is updated as new information becomes available. A section on natural resources is included in the report along with eight other sections: environmental pollution, climate, populations/settlements, human health, transport/tourism, wastes and natural disasters. Concentrating mainly on land-use statistics, the section on natural resources provides basic data with broad categories for monitoring changes in usage.

International Union for Conservation of Nature and Natural Resources

The IUCN is the world's largest conservation organization and was founded in 1948 at Fontainebleau, France. It is a unique assemblage of over 890 state government agencies and non-governmental organizations (NGOs) from over 135 countries around the world.

The headquarters is in Switzerland and the main components of the IUCN (Fig. 2.2) include the Secretariat, the Council, the World Conservation Congress (which meets every three years), national and regional committees and six commissions. The commissions link the expertise of many specialists from around the world. The Commissions are:

Ecosystem Management
Education and Communication
Environmental law
Environmental Strategy and Planning

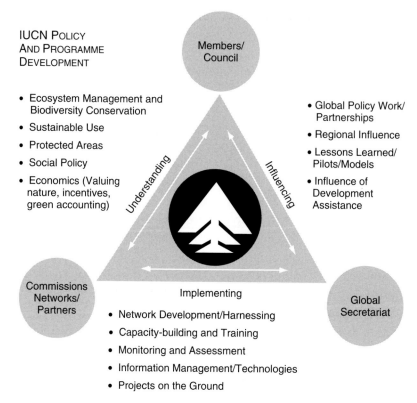

Fig. 2.2 IUCN policy and programme development. (From the *Pocket Guide to IUCN 1996–1997* with permission of IUCN.)

National Parks and Protected Areas
Species Survival.

The IUCN thematic programmes and networks include biodiversity policy, tropical forests, wetlands, marine ecosystems, plants, populations and sustainable development, and indigenous peoples and conservation. This range of themes enables IUCN and its member organizations to develop sound policies and programmes for conservation of biological diversity and sustainable development of natural resources throughout the world.

The United Nations Environment Programme World Conservation Monitoring Centre

The World Conservation Monitoring Centre (WCMC) was established in 2000 as the world biodiversity information and assessment centre of the UNEP. There are three divisions: information services, assessment and early warning, and conventions and policy support.

The origins of the centre go back to 1979–1980 when the WCMC (previously the Conservation Monitoring Centre) was established to provide an improved environmental information service for use by conservation bodies, development agencies, governments, industry, scientists and the media. The WCMC went on to develop integrated databases covering several themes including:

- data on plant and animal species of conservation value
- data on habitats of conservation concern and sites of high biological diversity
- data on the world's protected areas
- data on international trade in wild species and their derivatives
- a conservation bibliography.

WCMC worked in close collaboration with other organizations such as GEMS and GRID and the WCMC has also provided data for the effective operation of major international conventions including CITES, Ramsar (wetlands and wetland birds), and the Bern Convention (European endangered habitats and species).

The inventory of threatened species held by WCMC is based on information on many thousands of animal and plant taxa. Detailed accounts of some taxa have been published in the IUCN *Red Data Books*. The list of protected species held by the WCMC Wildlife Trade Monitoring Unit was an important basis for the implementation of the 1975 CITES. Over 1000 animal species, many of them mammals, have been recognized by CITES as being threatened by extinction. Current developments arising from the implementation of CITES are published in the *Traffic Bulletin*, funded by the Peoples' Trust for Endangered Species (TRAFFIC, USA) and a programme of WWF (USA) and published by the Wildlife Trade Monitoring Unit. The CITES convention has been ratified by about 100 countries and has played an important role in species conservation at an international level.

Protection of habitats is of greatest importance for wildlife conservation. Information on protected areas was collected for more than 20 years by IUCN'S Commission on National Parks and Protected Areas. The information on thousands of protected areas was held and managed by the Protected Areas Data Unit, which became part of the WCMC.

The development of WCMC in 2000, with the transition to UNEP, was undertaken with full support of IUCN and WWF. Today the UNEP WCMC provides information for policy and action to conserve the living world. The programmes include:

- species
- forests

- protected areas
- marine areas
- mountains.

Diversitas

Diversitas was established in 1991 by UNESCO, IUBS and SCOPE. A partnership with the Convention on Biological Diversity was established in 1996. In 2001, a new period of activity was initiated for Diversitas. This new phase followed a recommendation from five organizations that are now the sponsors of Diversitas (ICSU, IUBS, International Union of Microbiological Sciences (IUMS), SCOPE and UNESCO).

The overall goals of Diversitas are to:

- promote an integrative biological diversity science, linking biological, ecological and social disciplines in an effort to produce socially relevant new knowledge
- provide the scientific basis for the conservation and sustainable use of biological diversity.

In September 2002, Diversitas published its Science Plan. This plan has three core projects:

project 1: discovering biological diversity and predicting its changes
project 2: assessing impacts of biological diversity changes
project 3: developing the science of conservation and sustainable use of biological diversity.

Core project 1, amongst other things, will address the development of scientific tools for monitoring biological diversity.

International Whaling Commission

The IWC was established in 1946 under the International Convention for Regulation of Whaling. It was set up primarily to support commercial whaling and that role, responsible for overexploitation, continued unchanged until the 1960s. Initially the IWC could not realistically be called a biological monitoring organization but it has been a forum for debates about whale populations and has commissioned investigations into whale stocks. In 1972, following the Stockholm Conference, the IWC rejected a moratorium on commercial killing of whales, which had been supported by 53 nations. Since that time, attempts to implement a moratorium have had a chequered history: an indefinite moratorium on killing Sperm Whales commenced in 1981, a general moratorium in 1986 and an end to commercial whaling in 1988. Years later, however, Norway and Japan continue to argue for commercial and/or research programmes in which Minke Whale would be taken.

If ever there was a strong case for heeding advice then whaling could not be a better example. The diminishing whale catches since the 1940s have demonstrated very clearly that there was overexploitation of these mammals. Yet, in the absence of 'better monitoring' of whale populations, the folly of whaling has continued until very recently, sometimes under the guise of 'scientific whaling'. It is ironic that in 1989, the most accurate estimate of Blue Whales ever undertaken has shown that populations of that species are lower than previously feared by conservation organizations.

The IWC Scientific Committee has tried to prepare 'comprehensive assessments' of whale stocks. For example, the first of the comprehensive assessments of North Pacific Gray Whales and Southern Hemisphere and North Atlantic Minke Whales were undertaken in the 1990s. Today the IWC encourages and coordinates whale research and publishes results of scientific research. There is also ongoing work to assess the effects on whales of environmental change such as global warming and pollution and of activities such as whale watching.

The Sea Mammal Research Unit

The Sea Mammal Research Unit (SMRU) is one of the foremost research institutions carrying out research on marine mammals in the world (Johnston, personal communication). The mission of the SMRU is to carry out fundamental research into the biology of upper trophic level predators in the oceans and, through this, to provide support to the UK Natural Environment Research Council (NERC) so that it, in turn, can carry out its statutory duty to advise government about management of seal populations.

In 1966, SMRU was relocated from Cambridge to St Andrews, where it is part of the Gatty Marine Laboratory at the University of St Andrews. This brought the unit closer to its main sites, which contain bottlenose dolphins, porpoises and harbour and grey seals.

Central to the research strategy of SMRU is underpinning technological and modelling developments. A unique quality of the SMRU research is that it can, within the constraints of resources, offer a full range of skills that allow the unit to operate effectively from the development of ideas and hypotheses through to the delivery of fully developed science programmes. Its unique combination of satellite tags and web-based visualization software can provide scientists and the public with insights into the lives of these animals that would not otherwise be possible. Sophisticated biophysical modelling, that is informed by the data coming from instruments on marine mammals, is being used to develop predictions of the effects of environmental change on marine mammals.

Worldwide Fund for Nature, Friends of the Earth and Greenpeace

As well as the IUCN and the Centre for Marine Conservation, there are many other organizations actively involved in the surveillance and census of wildlife and natural resources. These include the WWF (previously the World Wildlife Fund, founded in 1961), the Friends of the Earth (FOE) founded in 1969 and Greenpeace, which was founded in 1971. These three organizations have sponsored research for biological monitoring, commissioned monitoring reports and have taken an active part in the international surveillance of wildlife and natural resources.

The WWF is an international organization with headquarters in Switzerland and national groups in about 40 countries. Mainly a fund-raising organization, WWF has sponsored international research on species and wildlife throughout the world. For example, there is the ecological monitoring of the Masai Mara ecosystems with the objective of monitoring vegetation and rainfall effects on wildlife especially the cheetah and its prey.

Environmental monitoring and ecological monitoring in North America

Environmental monitoring has been undertaken in the USA for many years: for example, hydrological monitoring by the US Geological Survey and climatic studies by the National Oceanographic and Atmospheric Administration (NOAA). The monitoring undertaken by the latter is directed largely at contaminant levels in sediments and tissues, and on measures of associated biological effects in organisms (as opposed to the effects at the ecosystem level). One such programme is in marine monitoring and prediction (MARMAP). A useful source of information on these monitoring activities may be found in annual reports (*Reports to Congress on Ocean Pollution, Monitoring and Research*, US Department of Commerce, NOAA) and in a bibliography of ocean pollution monitoring (NOAA, 1990).

The US Environmental Protection Agency (EPA) established EMAP in 1989. The aim was to provide data for monitoring environmental conditions at both regional and national scales. For this purpose, North America was divided into grids of $40\,km^2$ squares. A first stage of EMAP was the Prairie Pothole Study (Austin *et al.*, 2001). This study evaluated selected landscapes and fields as indicators of environmental quality. The numbers of breeding ducks were also used as an indicator of landscape condition.

In the early days of EMAP, it was recognized that there was a need for a process to develop standard methods of sampling. Without consistent methods, results would not be comparable from state to state. One example of the process was the EMAP for lakes and streams (Hughes *et al.*, 1992). More information about this activity is given on p. 168.

Long-term ecological programmes and biological monitoring have also been undertaken or promoted in the USA for many years by several agencies (including the EPA, the NSF, the US Fish and Wildlife Service, the Smithsonian Institution and the Nature Conservancy (an NGO)).

The NSF established the LTER network in 1980. The purpose was to support research on long-term ecological phenomena. Consequently, there has been some long-term ecological monitoring programmes established within this network (examples in Table 2.5).

The initiative for LTER, which embraces aspects of ecosystem monitoring and surveillance, came largely from a series of workshops and meetings about fundamental issues in ecology. This long-term approach, although a pilot programme, was supported because it was recognized that many ecological phenomena occur on scales much longer than that normally supported by traditional research, and because it was agreed that long-term trends in ecosystems were simply not being monitored. The data collected from the recording sites address the following:

- pattern and control of primary productivity,
- dynamics of populations of organisms selected to represent trophic structures
- pattern and control of organic matter accumulation in surface layers and sediments
- patterns of inorganic inputs and movements of nutrients through soils, groundwater and surface waters
- patterns and frequency of disturbances.

The sites that have been established for the US LTER network are fairly representative of ecosystems in North America. They include examples of forest, tundra, agricultural and prairie ecosystems.

In 1987, over 60 scientists from 11 nations met at second Cary Conference and endorsed the need for sustained ecological research and agreed that such research is a necessary prerequisite for monitoring and interpreting ecological data.

From about 1990 onwards there was a growing interest in long-term ecological research throughout the world. In 1993, the US LTER network facilitated a meeting on international networking in long-term ecological research. Consequently, the International Long-term Ecological Research (ILTER) network

Table 2.5 *Examples of long-term ecological research sites in North America*

Site name	Institutional affiliation	LTER research topics	Principal biome
H. J. Andrews Experimental Forest	Oregon State University, US Forest Service Pacific Northwest Research Station	Successional changes of composition, structure and processes Nature of forest–stream interactions Population dynamics of forest stands Effects of nitrogen fixers on soils Patterns and rates of log decomposition Disturbance regimes in forest landscapes	Coniferous forest
Arctic LTER site	Marine Biological Laboratory University of Alaska Universities of Massachusetts, Clarkson, Minnesota, Cincinnati and Kansas	Movement of nutrients from land to stream to lake Changes caused by anthropogenic influences Controls of ecological processes by nutrients and by predation	Arctic tundra, lakes and streams
Central Plains experimental range (CPER) – Shortgrass Steppe	US Department of Agriculture, Agricultural Research Service (ARS) and Colorado State University	Hydrologic cycle and primary production Microbial responses Plant succession Plant and animal population dynamics Plant community structure Organic matter aggregation or degradation Influence of erosion cycle on redistribution of matter, nutrients and	Grassland

Table 2.5 (*cont.*)

Site name	Institutional affiliation	LTER research topics	Principal biome
Harvard Forest	Harvard University	pedogenic process Influence of atmospheric gases, aerosols and particulates on primary production and nutrient cycles Long-term climate change, disturbance history and vegetation dynamics Comparative ecosystem study of anthropogenic and natural disturbance Community, population and plant architectural response to disturbance Forest–atmosphere trace gas fluxes Ecophysiology and micrometeorology Organic matter accumulation, decomposition, and mineralization Element cycling, the root dynamics and forest microbiology	Temperate-deciduous coniferous forest
Konza Prairie Research Natural Area	Kansas State University	Role of fire, grazing, and climate in a tallgrass prairie ecosystem	Tallgrass prairie
North Inlet (Hobcaw Baronyl)	University of South Carolina Belle W. Baruch Institute for Marine Biology and Coastal Research	Patterns and control of primary production Dynamics of selected populations Organic accumulation Patterns of inorganic contributions Patterns of site disturbances	Coastal marine

was launched. At that time, the mission statement for the ILTER network included:

- to promote and enhance the understanding of long-term ecological phenomena across national and regional boundaries
- to promote comparability of observations and experiments, integration of research and monitoring, and encourage data exchange
- to enhance training and education in comparative long-term ecological research
- to contribute to the scientific basis of ecosystem management.

By 2003, as many as 25 countries had established formal national LTER programmes and joined the international network. A recent account of the scientific accomplishments of the LTER programme can be found in Hobbie (2003) and Hobbie *et al.* (2003).

In addition to the ecological science supported by LTER, and because education and training is included in the mission, 'schoolyard' long-term ecological research has been established. Several schools have sites for long-term ecological research. Some will include ecological monitoring. What better way is there to introduce the concept of ecological monitoring into education at all levels? Such an initiative could well be taken up by schools in many other countries.

Observatory Earth Inc.

Developed at the Quetico Centre in central Canada, Observatory Earth Inc. is a non-profit organization combining international research and organizations monitoring global change. It was founded on the principles of sustainability science, that is 'meeting fundamental human needs while preserving the life-support systems of planet earth' (Kates, 2001). Observatory Earth Inc. has ten founding statements (Box 2.2), one of which points to the need for long-term monitoring.

At an international conference on ecosystem health held at the Quetico Centre in 2002, there was much discussion about ecological monitoring and the role of communities in monitoring and long-term research (see Ch. 8). An outcome of the Conference was a 120 page report entitled *A Blueprint for Design and Implementation of a Global Climate Change Monitoring Node*.

The Nature Conservancy

The Nature Conservancy (NC) was founded in 1951 in the USA and is a very successful private conservation organization. It operates a network of over 1000 nature reserves and administers many important monitoring programmes, especially with the aim of providing good data for management and conservation of plant and animal species. The NC has sponsored detailed assessments of monitoring on nature reserves with the objective of identifying the most important monitoring needs and appropriate cost-effective methods for monitoring.

Box 2.2   The Observatory Earth Inc. ten founding statements.

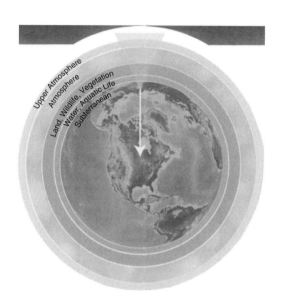

1.   Global change scientists, policy makers and business leaders around the world are concerned about the **rapidly escalating changes** in the Earth's ecology.

2.   To date, tracking and monitoring of global change in the Earth's natural environment have been intermittent, discrete and short-term. Scientists agree that our **monitoring must become long-term and ongoing**, and coordinated and integrated to ensure that we have the necessary information to sustainably manage human activity on the planet.

3.   To address this need and further Canada's leadership role in global change monitoring, a group of concerned citizens living and working at Quetico Centre in central Canada has developed an innovative notion. They've **founded Observatory Earth Inc., which is to be a node in a global network of organization** to monitor, assess, discuss and share data on global change.

4.   Observatory Earth will facilitate global change research by **bringing researchers and other interested parties together** to present,

discuss and make plans for future research, to coordinate research efforts in the region, to communicate the results of this research to those who have need of it, and to educate people on issues related to global change.

5. A primary reason for the creation of Observatory Earth is **to provide decision-makers with timely, useful, and accessible information** which will enable them to develop informed policy and management plans.

6. Observatory Earth Inc. is founded on the principles of sustainability science. Sustainability science is a newly emerging problem driven science that intends to put a scientific foundation under the concept of sustainable development. This initiative offers the potential **to place Canada at the forefront of international sustainability science**.

7. The **Quetico region is ideal for setting up a global change monitoring node**; large tracts of protected wilderness, and natural inter-sections of major watersheds, biomes, and climate zones occur here in Northwestern Ontario.

8. Fundamental to the Observatory Earth concept is **Quetico Centre's conference and education complex**. Shaping and communicating change is not new to Quetico Centre. From its inception in 1958, Quetico Centre has helped people to plan futures constructively.

9. **Observatory Earth is a partnership** based on the facilitating platform offered by Quetico Centre and the efforts and contributions of partner institutions interested in the effects of global change.

10. Observatory Earth Inc. and similar groups around the world are in position to promote the collection of critical global change data, and to make them accessible to those who require them. Furthermore, **Observatory Earth will promote greater scientific, corporate, political and public dialogue** concerning the issues and impacts of global change.

Ecological Monitoring Assessment Network

In Canada, EMAN was established in 1994 and is made up of a network of organizations and individuals involved in ecological monitoring in Canada (Brydges & Lumb, 1998; Vaughan *et al.*, 2001). The aim is to improve understanding, detection, description and reporting on ecological change. Interestingly, EMAN was established because of a need for coordination of the many and varied ecological monitoring programmes.

The main objectives of EMAN include the following (note the similarity between these objectives and those of GEMS):

- to provide a national perspective on how Canadian ecosystems are being affected by pressures on the environment
- to provide scientifically defensible rationales for pollution control and resource management policies
- to evaluate the effectiveness of policies
- to identify new environmental issues at an early stage.

There is a partnership between EMAN and the Canadian Nature Federation, which supports community-based environmental monitoring including NatureWatch (see p. 263). Partnerships are key to the activities of EMAN and this has prompted a need for monitoring protocols, that is standardized protocols for ecological monitoring. The agreement of protocols including agreements about what to monitor is a prerequisite for the success and indeed survival of any ecological monitoring programme.

Environmental monitoring in Europe

There are many environmental monitoring programmes based in Europe. For example there is the Swedish Environmental Monitoring Programme (National Swedish Environmental Protection Board, 1985). This programme monitors various environmental variables, including ecological variables, at a range of sites.

Awareness of marine pollution problems are not new in Europe but cooperative, regional monitoring has taken many years to become established. Under the Helsinki Convention on the Pollution of the Marine Environment of the Baltic, marine biological monitoring was based on data gathered on species abundance, fish population densities, biomass and diversity. Changes in the phytoplankton and other communities have also been monitored at several sites but the value and reliability of these variables for monitoring have been questioned (Morris, et al., 1988). Some of these variables are difficult to monitor in the field and are difficult to equate with specific pollutants, but of more importance has been the difficulty in obtaining complete comparability of results from cooperative programmes.

The European Network for Research in Global Change (ENRICH) has facilitated the establishment of an inventory of monitoring programmes and the Global Terrestrial Observing System (GTOS) has prepared a revised list of terrestrial ecological monitoring sites (TEMS).

Table 2.6 *Synthesis of major monitoring and research initiatives with their target ecosystems, issues, geographical coverage and measurements*

Organization and initiative	Target ecosystem/ compartment	Target issue	Target geographical coverage	Measurements
WMO-WHYCOS	Freshwater	Hydrological cycle	Global	Water chemistry, meteorology
WHO/UNESCO/ WMO/UNEP GEMS water	Freshwater/ groundwater	Water quality	Global	Water chemistry
WGMS	Glaciers	Climate, glacier variation	Global	Physical properties, energy flux
UN ECE ICP-waters	Freshwater	Acidification	Europe, North America	Chemical, physical, biological indicators
IASC FATE	Arctic, terrestrial	Global change issues	Europe, North America	Active layer thickness, soil properties, etc.
MAB-NSN ITEX	Arctic, terrestrial (tundra)	Climate	Europe, North America	Meteorology, plant performance
US NSF CALM	Arctic, terrestrial (permafrost)	Climate	Global	Active layer, physical properties
AMAP	Arctic, terrestrial and freshwaters	Pollution	Arctic environment	Pollutants in media
ICSU IGBP	Freshwaters/ oceans/ atmosphere/ terrestrial	Hydrology, land surface processes, global change	Global	Chemical, physical, biological indicators
FAO/UNEP/ UNESCO/ICSU GTOS	Terrestrial	Changes from anthropogenic impacts	Global	Forest cover, land degradation, soil moisture, etc.
ILTER	Terrestrial, freshwaters	Ecosystem dynamic, global change, pollution	Global	Chemical, physical, biological indicators
EU and UN ECE ICP	Forests	Pollution	Europe (North America)	Chemical, physical, biological indicators

Table 2.6 (*cont.*)

Organization and initiative	Target ecosystem/ compartment	Target issue	Target geographical coverage	Measurements
Forests Level I and II UN-ECE ICP integrated monitoring	Terrestrial, freshwaters	Pollution, acidification	Europe, North America	Chemical, physical, biological indicators
EU TERI	Terrestrial, freshwaters	Nitrogen deposition, biodiversity, global change issues	Europe	Chemical, physical, biological indicators
MAB-NSN EuroMAB	Terrestrial, coastal, marine	Sustainability	Europe, North America	Various

See p. xiv–xvii for all acronyms.

Parr *et al.* (2002) have listed some of the major monitoring programmes dealing with freshwater, glaciers, arctic and alpine environments, forests and other natural and semi-natural systems (Table 2.6). They suggest that the need for a consistent pan-European programme of long-term integrated monitoring of terrestrial systems is recognized by the scientific community. They argue that the design of such a system can be made difficult not least because of the constraints imposed by the need to make maximum use of the many existing sites and networks.

The Coordination of Information on the Environment Biotopes Programme

The acronym CORINE (Coordination of Information on the Environment) stands for an experimental programme of the Directorate-General for the Environment, Nuclear Safety and Civil Protection of the Commission of the European Communities (Schneider, 1989). In 1982, a European mapping case-study entitled *Biotopes of Significance for Nature Conservation* was undertaken and this led in 1985 to the Commission adopting the CORINE Biotopes Programme. The objectives for the first four years were as follows.

- to make an inventory of biotopes of major importance for nature conservation in the community
- to collate and make consistent data on acid deposition, and in particular establish a survey on emissions into the air

- to evaluate natural resources in the southern part of the community, in particular in those regions which are eligible for support
- to work on the availability and comparability of data.

The intention was to develop and make available information as a basis for nature conservation policies. Several additional aims were established and these are all common to the basic needs of ecological monitoring programmes. These aims included:

- establish a team of experts who would provide the data
- determine criteria to ensure consistency in selecting sites important for nature conservation
- set up standards
- design or use existing computer systems for data compilation, storage, retrieval and analysis and display
- make inventories available to users of the information

By 1991, the CORINE biotopes manual was published with extensive information on habitats found throughout the European Community. This information has proved to be important baseline data for nature conservation, monitoring environmental change and for establishing estimates of carbon stored. For example, Cruickshank *et al.* (2000) took the CORINE land-cover database for Ireland and used it to estimate the amount of carbon stored be each land-cover (vegetation) type. This is a first step towards monitoring changes in carbon sequestration from, and emissions to, the atmosphere by terrestrial vegetation. Such research has been prompted by climate change (Parks, 2002) and global warming and by an acknowledged requirement for inventories of carbon stores and fluxes.

The NERC in the UK was involved very early on with the planning of CORINE. The Environmental Information Centre (based at Monks Wood) undertook a pilot study to catalogue important nature conservation areas. This was part of the programme of work to establish biotopes databases, using sophisticated computer programs. Despite the use of sophisticated computing facilities, the initial problems encountered by CORINE were basic, such as the need for compatibility between data-gathering systems and good communication between those involved. A conceptual framework for monitoring within the CORINE programme addressed most if not all these basic problems.

Monitoring surveillance and census of biological systems in the UK is undertaken or sponsored by both government and NGO. The information is not centralized and there has been a long evolution of changes in organizations and responsibility for long-term monitoring. Nevertheless, biological monitoring has for many years been of increasing interest. The same could be said of ecological monitoring, especially with regard to nature conservation. For example, the UK Countryside Commission (as it was) for many years carried out habitat surveys and sponsored monitoring projects directed at management of wildlife and habitats on demonstration farms. The Countryside Commission has now merged with the Rural Development Commission to form the Countryside Agency.

In 1986, the UK Department of the Environment, as it was called then, updated a document originally published in 1974 entitled *The Monitoring of the Environment in the United Kingdom* (UK Department of the Environment, 1986). This was mainly concerned with monitoring pollution, but environmental monitoring in the UK has also included landscape monitoring and monitoring of land-use change. The decentralized aspect of environmental activities in the UK has always made it difficult to maintain an overview of the various programmes. For that reason, the Department of the Environment established in 1986 a detailed, computer register of environmental monitoring schemes sponsored or carried out by central government, including some 400 schemes covering mainly air-pollution-monitoring schemes, some water-monitoring schemes and programmes of radioactivity monitoring. At the time, it was planned to include schemes on fauna and flora, especially habitat quality and endangered species and extinctions.

Countryside Survey 2000

The Countryside Survey has been funded both by the NERC and the UK Department for Environment, Food and Rural Affairs (DEFRA) as well as several other sponsors. This survey is conducted every several years (1978, 1984, 1990, 2000) to follow ecological changes. The recording is undertaken by staff from the NERC CEH. Detailed observations have been made in a random sample of 1 km grid squares across Britain. Data have been collected on habitat types and extent, field boundaries, plant species and distribution, and freshwater invertebrates.

The most recent results can be found on the Countryside survey 2000 (CS2000) website and in a 2000 report *Accounting for Nature: Assessing Habitats in the UK Countryside* (Haines-Young *et al.*, 2000). The results have been used to monitor changes in the UK at many levels of biological organization from changes in landscape features to plant diversity and also in terms of 'quality of life indicators' (Box 2.3).

Natural Environment Research Council

The main aim of NERC is to understand the behaviour of complex biological, physical and chemical interactions for the purpose of detecting change and for the purpose of analysis of change. The organization includes research centres (such as the British Antarctic Survey and CEH) and

Box 2.3 The 'headline results' from the UK Countryside Survey 2000 and the Northern Ireland Countryside Survey 2000

1.   Plant diversity increased in arable fields, especially in the field boundaries. Plant diversity increased by 38% in some arable field boundaries in England and Wales.

2.   Plant diversity continued to decline in the least agriculturally improved grasslands in Great Britain. Plant diversity in some meadows fell by 8%, including losses of meadow species important for butterflies. The area of neutral grassland in Northern Ireland decreased by 32%.

3.   Following marked losses in the 1980s, there was no significant difference in the 1990 and 1998 estimates of hedgerow length in England and Wales. There is some evidence that losses in the early 1990s have been reversed.

4.   Road verges showed evidence of increasing nutrient levels and losses in plant diversity. Plant diversity fell by 9% in some road verges in England and Wales.

5.   Broadleaved woodland expanded by 4% in England and Wales and 9% in both Scotland and Northern Ireland between 1990 and 1998. The total area of coniferous woodland in the UK was unchanged.

6.   The area of semi-natural acid and calcareous grasslands feel by 10% and 18% in the UK. The area of bog fell by 8% in Northern Ireland. There was evidence of increasing nutrient levels or eutrophication in dwarf shrub heath and bog.

7.   The number of lowland ponds increased by about 6% between 1990 and 1998 in Great Britain.

8.   The biological quality of streams and small rivers improved in Great Britain. Over 25% of sites improved in condition and only 2% were downgraded.

9.   Streamside vegetation became more overgrown, and plant diversity decreased by 11% in England and Wales. Fen, marsh and swamp expanded by 27% in England and Wales and 19% in Scotland but declined by 19% in Northern Ireland.

10.   More broadleaved woodland was created on formerly developed land than was lost to new development in Great Britain in 1990s.

From Haines-Young et al. *(2000).*

collaborative centres (such as Plymouth Marine Laboratory, the Sea Mammal Research Unit, and the Tyndall Centre for Climate Change Research). The NERC was instrumental in the establishment of the Environmental Information Centre (including the Biological Records Centre (BRC)).

A relatively new initiative, which arose out of discussions between the NERC and the Agricultural and Food Research Council (AFRC) is the ECN. This commenced with a proposal to designate a limited number of sites for which there would be long runs of environmental data and either constant or well-recorded regimes. Early in development it was believed that such a network would have considerable value, for example in detecting the effects of global warming and the effect of land-use change in the UK. Potential monitoring sites were identified and these included both nature reserves and some permanent vegetation plots.

The ECN was formally launched in 1992. The year 2003 marked 10 years of ECN operations. The ECN is now a UK multi-agency research programme coordinated by the NERC.

A very significant product of the ECN is the publication of protocols for standard measurements at terrestrial sites (Sykes & Lane, 1996). These protocols cover a wide range of standardized core measurements, including physical, chemical and biological variables, instituted at each of the network's sites. The systems used in ECN for data recording, database management and quality assurance are also included. There is a similar publication for freshwater sites.

Centre for Ecology and Hydrology

The merger of the Institute of Terrestrial Ecology (which was established in 1973) and freshwater biological laboratories gave rise to CEH, one of the centres of NERC. Research in CEH is managed in nine interdisciplinary programmes, all of which address contemporary environmental themes and issues.

Previously, terrestrial monitoring programmes had included the monitoring of heathland fragmentation and monitoring pesticide residues in predatory birds. The Predatory Birds Monitoring Scheme is one of the longest running schemes of its kind in the world and results have been published in various scientific papers.

The Ecological Data Unit (EDU) was established in 1984 within the Institute of Terrestrial Ecology with the aim of collating data relating to ecological change and promoting ecological monitoring. One of the first tasks of the EDU was to prepare a register of permanent vegetation plots.

Biological Records Centre and mapping and monitoring programmes

Monitoring trends in populations, monitoring the distribution of a species and monitoring the extent of habitats and communities can be based only on good (reliable and accurate) recording and survey methods. One example of such a biological recording scheme, which has, for many years, researched recording and survey methods for monitoring the distribution and abundance of various species, is the BRC.

The BRC originated as the Distribution Maps Scheme of the Botanical Society of the British Isles, which commenced in 1954 following a conference in 1950 on the distribution of British plants. For many years, botanists around the country contributed many records and it is these records that provided the basis for the distribution maps. After much effort (particularly on the part of Franklyn Perring), the first edition of *Atlas of British Flora* (Perring & Walters, 1962) was produced. That atlas was not only the culmination of many years of data collection and research but was also a 'landmark' in the history of recording and survey methods. It is also a tribute to the skills of those who coordinated the many recorders throughout the UK.

Initially based at the University Botanic Gardens in Cambridge, the BRC staff and records were transferred in 1964 to the Monks Wood Experimental Station (at that time part of the Institute of Terrestrial Ecology) in the formal establishment of the BRC. The funding status of the BRC has been chequered and in 1989 it was incorporated within the Environmental Information Centre at Monks Wood. This brought together considerable expertise in remote sensing, GIS and ecological databases, including biological recording.

One of the main aims of the originators of the BRC was the production of distribution maps for the study of the history and biogeographical relationships of the British flora (Fig. 2.3). That main aim continues today, and distribution maps are available for several taxa including plants. The Countryside Survey 2000 using the Countryside Information System software has recently been directed at plant distributions and a new *Atlas of the British and Irish Flora* was published in 2002.

In the UK, the BRC has based much of its recording on units of 10 km squares of the national grid, partly because the squares are the same size irrespective of latitude and because they are marked on all ordnance survey maps. The status of a species expressed in terms of its occurrence in the number of 10 km squares can provide a basis for monitoring the change in status of a species. For example, the change in number of 10 km squares over a certain time period was one criterion on which the threat numbers (an index of conservation need) was based in Britain's first *Red Data Book* (Perring & Farrell, 1983). The species *Campanula rapunculus* in Fig. 2.3 declined by as much as 10% up to about 1980 and was, at that time, allocated a threat number of 11 (highest recorded for any species in Britain in 1983 was 13 and the maximum potential threat number was 15).

The BRC pioneered the mapping of national species distributions from a computer database and now the BRC has the largest computerized database

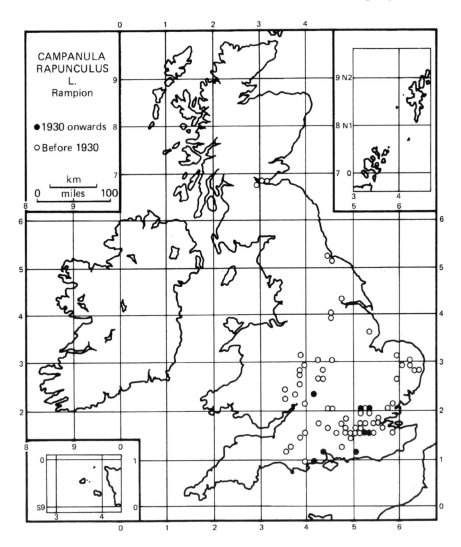

Fig. 2.3 An early example of a species distribution map from the *Atlas of British Flora*.

for animals and plants in the UK. The database, in the long term, is the most valuable asset of the BRC, particularly in relation to biological monitoring, environmental assessment and modelling. The accuracy of species distributions is being recorded with more and more precision as advances in computerized mapping continue. National recording schemes were established for many taxa including algae, lichens, fungi, beetles (14 groups), flies (22 groups), fishes, amphibians, reptiles, birds and mammals.

Biological mapping and monitoring programmes have been established in several countries including the USA (e.g. Stein *et al.*, 2000). The advances in information technology and use of web pages has made it much easier for more people to contribute to these plant recording and monitoring programmes. Recorders are now able to log their records on-line and modern electronic databases are able to maintain up-to-date distribution maps. Plant conservation networks such as the New Zealand Plant Conservation Network (Box 2.4) have provided a cost-effective method of bringing together much expertise. These networks provide access to experts in plant conservation and allow ease of access to the latest information on all aspects of plant conservation including the changing status of plant species. These programmes also contribute to the work of the Global Strategy for Plant Conservation.

Box 2.4   The New Zealand Plant Conservation Network

One of the global targets (to be achieved by 2010) for the Global Strategy for Plant Conservation refers (p. xvi) to establishment of networks for plant conservation: 'Networks for plant conservation activities established or strengthened at national, regional and international levels'. The terms and technical rational were given as:

> Networks can enhance communication and provide a mechanism to exchange information, knowhow and technology. Networks will provide an important component in the coordination of effort among stakeholders for the achievement of all the targets of the strategy. They will also help to avoid duplication of effort and to optimize the efficient allocation of resources. Effective networks provide a means to develop common approaches to plant conservation problems, to share policies and priorities and to help disseminate the implementation of all such policies at different levels. They can also help to strengthen links between different sectors relevant to conservation, e.g. the botanical, environmental, agricultural, forest and educational sectors. Networks provide an essential link between on-the-ground conservation action and coordination, monitoring and policy development at all levels. This target is understood to include the broadening of participation in existing networks, as well as the establishment, where necessary, of networks.

The New Zealand Plant Conservation Network (www.nzpcn.org.nz/) was established in 2003 following discussions between many individuals and groups involved in plant conservation. The network is intended to:

- improve plant conservation practice
- increase efficiencies and minimize overlaps in achieving plant conservation outcomes
- increase resources to focus on priorities
- make more information freely available to those involved in plant conservation.

There are more than 119 indigenous vascular plant species at risk of extinction in New Zealand. In addition, there are 102 species listed as declining. Many native plant communities are also threatened. These include wetlands, coastal dunes and lowland forests.

The main focus of the New Zealand Plant Conservation Network is the protection and recovery of those threatened species and their associated plant and animal communities.

There are many ways for the conservation Network to achieve its vision and these come under the following main headings:

- education and advocacy, e.g. focus attention and raise awareness of the status of New Zealand indigenous plants
- strategy, prioritization and resources, e.g. provide a national body to coordinate plant conservation projects
- technical expertise, e.g. maintain a directory of people involved in plant conservation.
- information, e.g. Collate and disseminate information about plant conservation, activities and events, and provide electronic communications.

The information below is the details about the organization given on its website.

The Vision

The vision of the New Zealand Plant Conservation Network is that no indigenous species of plant will become extinct nor be placed at risk of extinction as a result of human action or indifference, and that the rich and unique plant life of New Zealand will be recognized, cherished and restored.

The Mission

To promote indigenous plant conservation in New Zealand and throughout Australasia. To collaborate to protect and restore New Zealand's indigenous plant life and their natural habitats and associated species. To disseminate information about the taxonomy, biology, ecology, and status of indigenous plant species and communities in New Zealand and to promote activities to protect them throughout their natural range.

The Membership

Any organisation or individual for which the above mission statement is an important objective is eligible. Membership of the network is voluntary and free. It comprises botanists, landowners, botanic gardens, museums, schools, government agencies, horticulturalists and tertiary institutions. Details of membership are held in a directory published annually by the network.

Background

Collaboration between people is proving to be a powerful tool in the global move to implement plant conservation initiatives. In Canada, Europe and Australia plant networks have already been established to bring together people with a range of skills to work together on plant conservation programmes. In New Zealand there are regional networks in Wellington and Auckland. These initiatives have brought about important changes in the way people help each other to conserve native plant life. This has led naturally to the development of a national network in New Zealand.

The Importance of Plant Conservation

Plant biodiversity is a key component of nature's life-support systems. Most of New Zealand's indigenous plant species are unique – they are not found growing in the wild anywhere else in the world. The plant communities that they are part of and the animal communities that they support are also endemic to New Zealand. Our responsibility is to protect these natural resources not only because of their many uses (such as medicines, fuel, clothes and building materials) but also because of their ecosystem services (such as soil conservation and oxygen production).

The Network's Resources

The network comprises individuals and agencies with a wide range of expertise and resources that it can utilise to achieve plant conservation in New Zealand. Members of the network include some of the countries leading botanists and ecologists.

The Focus for the Network

In 1999 the Threatened Plant Committee of the New Zealand Botanical Society listed over 100 indigenous vascular plant species at risk of extinction in the wild. In addition many other non-vascular plant species are believe to be endangered. It is those species and their associated plant and animal communities that are focus for the New Zealand Plant Conservation Network.

What the Network will do

The network will maintain a bibliography of plant conservation literature. It will assist with the development of a list of sites that are of national importance because they support populations of threatened species, are centers of plant diversity or are centers of plant endemism.

The network will assist with the maintenance of a list of nationally threatened plant species and lists of species that are regionally uncommon. It will provide a directory of people involved in plant conservation with their contact addresses.

The network will disseminate information about plant conservation activities or events and provide electronic communications to its members. In the first two years of its life the network will co-ordinate the preparation of a New Zealand Plant Conservation Strategy to provide a framework for plant conservation activities nation-wide and a focus for its members.

The many years of using grid squares to record species has come under scrutiny and there has been much discussion as to whether these established methods of recording need to change. Alongside such discussion is the fact that there is a growing demand for the information and the inevitable question has to be asked 'does the user of the information pay?' These are just two of the many questions that are being addressed by those taking part in the National Biodiversity Network (NBN).

In 1999, the BRC entered a new phase when it redeveloped its own role and its work with national societies and recording schemes. A Management Advisory Group was established and the main organizations that funded BRC (NERC and the Joint Nature Conservation Commission (JNCC)) agreed a six-year partnership with additional funding. Highly relevant was the fact that the main objective and increased funding was to make the BRC more effective in its long established role as the national focal point for species recording.

In 2004, the BRC is the national focus for all species recording (apart from birds) in the UK. It has the capability of addressing global problems, such as monitoring deforestation, climate change and desertification but it also has the opportunity and experience to address national problems such as habitat loss. The BRC databases contain nearly 12 million records of more than 12 000 species.

Despite the extensive biological recording undertaken by the BRC and numerous other regional organizations and despite the extensive use of data from these recording schemes, for many years, the UK had no effective system for overall coordination of recording and monitoring wildlife let alone for accessing information. The BRC and JNCC established a NBN gateway for accessing information on the internet about the biological diversity of the UK. This provides access to the 10 km distribution maps.

Nature Conservation Organizations in the UK

The year 1949 saw the establishment of the Nature Conservancy in Britain. In 1973 that Council was split into two bodies, the Nature Conservancy Council (NCC) and the Institute of Terrestrial Ecology. The statutory functions of the NCC, under the Nature Conservancy Council Act 1973, come under three main headings: establishment and management of nature reserves, provision of advice about nature conservation, and the support of research for nature conservation.

The NCC was directly involved in, or provided support for, a range of biological and ecological monitoring activities. For example, one particularly important national monitoring programme, which is still running, is the National Countryside Monitoring Scheme (jointly funded by the NCC and the Countryside Commission for Scotland). Other NCC monitoring programmes were directed at particular taxonomic groups. For example, monitoring the abundance of butterflies (a joint project by the NCC and the Institute of Terrestrial Ecology) was started in 1976. The two aims of this scheme were very soon achieved and both regional and national trends (Fig. 7.2, p. 244) could be seen.

In 1990, the NCC was divided into separate organizations for England, Scotland and Wales (linking the last two with the Countryside Commissions in those countries). The 1990 split (under the 1990 Environmental Protection Act) gave rise to:

- English Nature (in addition to the Countryside Commission)
- Scottish Natural Heritage
- Countryside Council for Wales
- the JNCC (coordinating role and responsible for European and international matters).

One of the special functions of the JNCC is to establish common standards throughout the UK for monitoring of nature conservation and for research into nature conservation and the analysis of the results. In this context, JNCC in 2001–2002 developed the 'marine recorder' software application to make national marine databases as widely available as possible. At this time, JNCC also completed guidance for common standards for monitoring for three habitat types of European interest (under the Habitats Directive). There was also a contribution to English Nature's recently established reef-monitoring project.

Together with the British Trust for Ornithology (BTO), JNCC has continued the ornithological time series monitoring. This includes responses of birds to changes in the lowlands and to changes in the uplands.

Rothamsted Experimental Station

Rothamsted Experimental Station, founded in 1843, is the oldest agricultural research station in the world. It is divided into several divisions including agronomy and crop physiology, biomathematics, crop protection, molecular sciences and soils. A number of now classic long-term research programmes have evolved at Rothamsted, two examples of which are Lawes and Gilbert's Broadbalk experiment (involving wheat) and Lawes and Gilbert's park grass experiment (involving grass grown for hay). Both of these research programmes, which were initially ecological investigations into plant nutrition and elements limiting plant growth, have since led to a wide range of applied ecological and agricultural research.

Rothamsted Insect Survey

The distribution and abundance of insects has been monitored by agencies throughout the world for many years and there are many classic insect monitoring and census programmes. For example, commencing in 1950 (and lasting for 12 years), there was a winter moth census on five oak trees in Wytham Wood, Oxford (Varley *et al.*, 1973).

Monitoring the status of insects has been undertaken on a wide scale for both pest and non-pest species. Such data provide information that can contribute towards an understanding of insect dynamics and may lead to more effective control or conservation strategies.

The Rothamsted Insect Survey had its beginnings in 1959, first with moths and then later with aphids. The survey has been using 12.2 m suction traps (Macaulay *et al.*, 1988) to monitor aphids since 1965. In 2002, 16 traps were operated throughout the UK. The traps are emptied daily from April until November and weekly at other times and the aphids identified to species level where possible, although with the rapid throughput required, some morphologically similar species can only be recorded to species group or genus. All the aphids and other insects are stored in case there is a need to examine them in the future. At some traps, a glycerol-based collecting medium allows aphids to be assayed for insecticide resistance status. The presence of persistent viruses in individual aphids can also be determined. DNA analyses can be carried out in

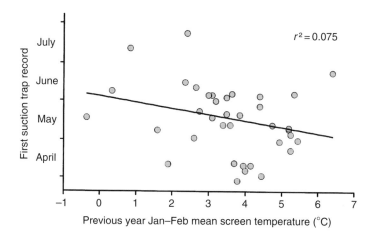

Fig. 2.4 Relationship between temperature and trap records for *Myzus persicae* at Rothampsted (1965–2002). (Information kindly provided by Richard Harrington.)

order to study the population genetics of particular species. A weekly bulletin is published on the worldwide web (www.iacr.bbsrc.ac.uk/insect-survey) detailing the numbers of 23 species of economic or other special interest at each site. This information is accompanied by an interpretation relevant to the needs of particular contractors and is used to decide on the necessity for and timing of control strategies.

An interactive system allows internet users to tailor the bulletin so that they only receive information about traps and species of interest. The data can be used in interactive decision-support systems such as that for control of barley yellow dwarf virus, which is transmitted by aphids (Northing *et al.*, 2002).

There are strong relationships between winter temperature and the timing and size of the spring flight of a range of aphids that pass the winter in the active stages rather than as an egg (Harrington *et al.*, 1990). These relationships (Fig. 2.4) are used to provide forecasts that help to determine the need for early season aphid control and the need to begin inspecting crops. They can also be used to assess the likely impact of climate change on aphid dynamics (Harrington *et al.*, 1995, 2002).

In 2002, 70 suction traps of the Rothamsted design were operating in 19 European countries (Fig. 2.5). Many of these traps have a long history. Data from all the traps have been brought together in a single database, which forms what is probably the most comprehensive, standardized information for any terrestrial invertebrate group in terms of its spatial and temporal coverage and number of species (Northing *et al.*, 2002). Combined with long-term data on weather, land use and gaseous pollutants, the system offers an unparalleled

Fig. 2.5 Distribution of suction traps of the Rothamsted design operating in 19 European countries in 2002. (Information kindly provided by Richard Harrington.)

opportunity to examine the dynamics of aphids in relation to ecological variables.

Organizations monitoring birds

Many organizations monitor bird populations and the conservation status of single bird species. Most are regional or local organizations. Birdlife

International (formerly the International Council for Bird Preservation) is a worldwide organization with partner organizations or representatives in 88 countries. Birdlife International has identified areas that support high numbers of unique species and these areas are called endemic bird areas.

Some sites are particularly important because of the large concentrations of birds that occur there or for the unique species or for the distinct habitats. About 20 000 of these sites have been identified and they are called important bird areas (IBAs). The criteria for IBAs have been agreed internationally and are based on species that are threatened, endemic or restricted to key biomes. This work has been completed in Europe, the Middle East and Africa and is now under way in other regions.

More recently, Birdlife International has proposed a protocol to the Convention on Biological Diversity. A key objective is to help to safeguard global biological diversity and to ensure that adequate and long-term funding is made available to developing countries.

In Europe, the European Bird Census Council (EBCC) is a focal point for all European countries that have national bodies responsible for monitoring bird populations and their distribution. The EBCC encourages and promotes rigorous bird monitoring work aimed at better conservation and management of bird populations and at providing indicators of the changing ability of European landscapes to support bird populations. The EBCC produced the first ever *European Atlas of Birds* and it promotes national or regional atlases at finer scales.

The BTO is the sole organization responsible for bird ringing in Britain and is the major contributor to bird census work and monitoring of populations via a number of schemes including the Common Bird Census (which commenced in 1961) and more recent programmes (see p. 288).

Other organizations in Britain that monitor birds include the Royal Society for the Protection of Birds (RSPB) and JNCC. One example of a collaborative projectis the Wetland Bird Survey, involving the BTO, the Wildfowl and Wetlands Trust, the RSPB and the JNCC. A census of breeding seabirds in Britain and Ireland is coordinated by the JNCC; Seabird 2000 was the third census undertaken.

The Game Conservancy, which administers the National Game Census, was established in 1961. The aim of this census is to monitor game bird population trends for species such as Red Grouse (*Lagopus lagopus*), Grey Partridge (*Perdix perdix*) and Wood Pigeon (*Columba palumbus*) (p. 295).

Other bird-monitoring programmes have been undertaken around the UK on Skomer, Fair Isle and other islands. The bird-monitoring programmes on these islands owes much to the work of the Fair Isle Bird Observatory Trust.

Wildlife and natural history organizations

In Britain, the RSPB, the Mammal Society, the British Herpetological Society, the Botanical Society of the British Isles, the British Butterfly Conservation Society and the British Lichen Society are just some of the NGOS who have sponsored or supported monitoring, surveillance or census of wildlife and habitats.

The British Lichen Society's lichen-mapping scheme started in 1965 and is based on the distribution of species within 10 km grids. The lichen-mapping scheme is proving to be a particularly valuable scheme, partly because lichens are good indicator species and partly because of the well-organized manner in which the scheme is operated. An atlas of the bryophytes is also soon to be produced for the British Isles.

Coordination of views and interests amongst conservation NGOs is undertaken by the Wildlife and Countryside Link. This is a liaison or umbrella organization for all the major voluntary conservation groups in the UK who are concerned with the protection of wildlife. The Wildlife and Countryside Link brings together environmental voluntary organizations in the UK united by their common interests. The work can be split broadly into information management and coalition building. Of the many reports sponsored by the Wildlife and Countryside Link, some have been directed at a record of losses in protected areas such as Sites of Special Scientific Interest. Similar umbrella organizations are found in other countries.

3

State of the environment reporting and ecological monitoring

3.1 Introduction

The main aim of this chapter is to provide an introduction to what has become known as 'state of the environment' (SoE) reporting. Ecological monitoring is a requirement for such reporting. The two are inextricably linked. One objective of this chapter is, therefore, to assess and comment on the information being used to monitor for SoE reporting. Indices and indicators are, in particular, being incorporated into SoE reports. Examples of indicators and indices are included, some of which may yet to have a role in SoE reporting.

3.2 Terms and terminology: state, condition or what?

Several terms have been used when referring to the environment and environmental change. These terms include:

- state
- condition
- quality
- integrity
- health.

Environmental quality, soil quality, etc. are common terms in SoE reports. High quality implies something that is better than low quality. Smith (1987) has referred to 'quality' as 'suitable for use'. Later, Johnson *et al.* (1997) suggested a definition of environmental quality as 'a measure of the condition of an environment relative to the requirements of one or more species and/or human need or purpose'.

Integrity in the sense of ecological integrity is a measure of the extent to which an ecosystem has been affected by humans and human activity. Similarly, biological integrity is a condition reflecting the extent to which humans have altered the composition, structure and function of biological communities.

What about environmental health and ecosystem health? We sometimes hear about 'economic health'. Such expressions are commonly used and are seemingly easy to understand. Indeed, there are publications that are devoted to environmental health. However, can ecosystem health be measured and does the term describe the property of the environment? Some people have questioned the use of the term environmental health. Suter (1993), for example, talked about economic health as a figure of speech and was critical of the attempts to turn figures of speech into operational goals.

Carignan and Villard (2002) noted that some authors have argued that ecological integrity and ecosystem health are fundamentally different even though they may seem interchangeable and may relate to the same goal. Furthermore, some authors have argued that these terms are subjective and, therefore, vulnerable to variations in interpretation. This may be true; nevertheless, surely agreement can be reached about how to measure SoE in an objective and quantifiable manner.

3.3　The changing state of the environment

The 'state' of the environment has been observed and has been of interest and concern to many people and many organizations for hundreds of years. John Keats (1795–1821) wrote the poem *La belle Dame sans Merci* in 1819. In that poem, there are some lines that conjure up despair and seem to reflect the state of the environment:

> Oh, what can ail thee, Knight at arms
> Alone and palely loitering;
> The sedge is wither'd from the lake,
> And no birds sing.

From poetry in the nineteenth century to research in the twenty-first century, there have been observations that all is not well with nature and the environment. Most of these changes are the direct effects of human activities. Pollution, deforestation, intensive agriculture, urbanization and exploitation of marine animal populations have resulted in changes to our environment. However, there seem now to be many other subtle changes taking place. Some may reflect long-term changes that have been taking place at low levels for may years. For example, understorey plants appear to be vanishing from some northern forest stands in North America. Donald Waller at the University of Wisconsin and his research students have found a reduction in understorey plants over a 50 year period (Fig. 3.1).

There has been, and continues to be, a general decline in amphibian populations around the world (Collins & Storfer, 2003). There have been dramatic declines in bumblebee populations. The number of House Sparrows in some urban areas

Fig. 3.1 Views of study sites in northern Wisconsin forests and Michigan's upper peninsula. Over 50 years, nearly one out of every four species has gone. Displacement by exotic species may account for some losses. Other factors are being investigated. (Photographs kindly provided by Donald Waller.)

Fig. 3.1 (Cont.)

of the UK has fallen 'catastrophically' according to preliminary findings of an 18-month government survey. Reports of weather extremes have consistently captured the headlines (Parks, 2002).

What is the reason for these changes? Are there any factors common to all these changes? Are they isolated trends or do they form part of a larger pattern or trend? Are they the result of natural ecosystem processes? Has poor environmental management contributed to any changes? Is there evidence of exploitation of natural resources and degradation of the environment? Before we can assess whether or not something has changed, we need to know a lot about the natural processes that take place over time. That is, we require sound, scientifically defensible information on the environment at different stages in time so that we can make temporal comparisons.

Over a long period of time, there has been a steady increase in the number and membership of organizations and an increase in the membership of those organizations with an interest in some aspect of the condition of the environment. More importantly, there has been a steady growth in the level of concern about the degradation of the environment. The increasing impacts of human populations, the declining levels of natural resources and the increasing levels of pollution are but a few areas of concern.

While some may dispute claims that there are reasons for concern and may argue that the reports of environmental degradation are questionable, there seems to be consensus that the condition, state or 'health' of the environment is declining and that decline is cause for concern.

3.4   Sceptics and the green backlash

Sceptical environmentalists are not new. Many years prior to the well-publicized book by Lomborg (2001), there were publications that questioned the often-made claims about extinction rates, rates of loss of tropical forests and levels and extent of pollution. All environmental events are subject to different interpretations. There are those who believe, for example, that there is a climate change taking place. There are others who dispute this. There are those who advocate that we are facing a species extinction crisis. However, there are those who are highly critical of the calculations used to assess rates of loss of species. There has been much written about the depletion of the world's resources yet there are those who advocate that the environment could support a population higher than it is today without poverty. As Jones *et al.* (2002) remind us, we need to be smarter environmentalists and be more critical of environmental activists. Good reviews of Lomborg's book can be found in the January 2002 issue of the *New Scientist* (Vol. 286, No. 1).

Ever since environmental concerns were expressed there have been alternative views. Those alternative views can lead to bitter arguments or even challenges to claims made about environmental degradation. This is a kind of a green backlash. This green backlash could be said to be good debate but it goes further than that. It is more about challenges to interpretation of data and how data are collected. The green backlash is more than just an alternative view: it is a critical appraisal of environmental claims. The green backlash is fuelled by sloppy and poor data collection and equally poor analysis of data. It is also fuelled by apparent disputes between experts.

There is also a more sinister form of green backlash, and this is well explored in a book of the same title (*Green Backlash*; Rowell, 1996). This book explores the global anti-environmentalism movement, which has emerged partly as a result of successful ventures by environmental groups.

3.5　Reports on the state of the environment

There have been many 'milestone' publications that provide information of the state of the world's environment (Table 3.1). The Organisation for Economic Co-operation and Development (OECD) was one of the first organizations to publish such reports (in 1979). The more recent publications of 2001 are particularly relevant to measuring and tracking environmental change.

Table 3.1 *Milestone publications on the state of the environment (excluding the many national state of the environment reports)*

Topic	Reference
Global Biodiversity. Status of the Earth's Living Resources	Groombridge (1992)
Global Biodiversity Assessment	Heywood (1995)
The World Environment 1972–1982	Holdgate *et al.* (1982)
The State of the World Atlas	Kidron & Segal (1995)
The State of the Environment in Member Countries	OECD (1979)
State of the environment reports	OECD
Environmental Outlook	OECD (2001b)
The State of the Environment Atlas	Seager (1995)
Global Environment Outlook 2000	UNEP (1999)
World conservation strategies	IUCN (1980, 1991)
World Resources 2000–2001	UNDP, UNEP, World Bank and World Resources Institute (2000)
State of the world reviews are published regularly, e.g. *State of the World 2000* is the 17th annual review	Worldwatch Institute

One (OECD, 2001a) deals with environmental indicators and the other (OECD, 2001b) puts forward the concept of traffic lights as signals of key finding (there are green, yellow and red issues).

Notable is the series (almost an institution) of reports published since 1984 by the Worldwatch Institute: *State of the World* reports. These are directed towards progress towards a sustainable society. The objective of *State of the World 2000* is to provide the most up-to-date diagnosis, and prognosis, of the state of the earth. The seventeenth review appeared in 2000.

There are World Conservation Strategies (IUCN, 1980, 1991) and very readable ('classic') books such as *Our Common Future* published in 1987 by the WCED. The end of the twentieth century seemed to be a prompt for some substantial reports on the world's environment. In 1999, UNEP published the *Global Environmental Outlook 2000* (a millennium SoE report).

More recently, there was *World Resources 2000–2001; People and Ecosystems: the Fraying Web of Life* published by the World Resources Institute with support from the UN and the World Bank (2000). There are some relatively new and unique kinds of report (of which there are many) that provide information about the state or the condition of different components of the biophysical environment. These are called SoE reports (Box 3.1). and are intended to provide an overview of current knowledge on the environment, the pressures on the environment and how these pressures can be addressed. This is based on the premise that 'people make better decisions if they are better informed about the possible consequences of their actions'.

Box 3.1   Some State of the Environment Reports

National reports State of the Environment have been prepared by the following countries: Albania, Armenia, Australia, Austria, Azerbaijan, Bosnia & Herzegovina, Bulgaria, Canada, China, Czech Republic, Denmark, Estonia, Finland, France, Georgia, Germany, Hungary, Italy, Japan, Kyrgyz Republic, Latvia, Lithuania, Macedonea, Malaysia, Moldova, Netherlands, New Zealand, Norway, Poland, Romania, Russia, Slovak, Slovenia, Sweden, Switzerland, Tajikistan, Turkmenistan, Ukraine, UK, USA and Uzbekistan (see www.grida.no/soe/links.htm). The World Resources Institute has published several volumes of the World Directory of Country Environmental Studies. These are essentially a catalogue of these reports (related information is available at http://www.wri.org/sdis/).

The Antarctic became the focus for the first State of the Environment Report in 2001 Waterhouse (2001).

Five examples of 'state of the environment' books that were published in the last decade of the twentieth century.

Quality of the Environment in Japan 1994

Contents: long-term trends in population and socioeconomic activities (population increase, size of the economy, energy consumption, production and consumption of agricultural, forest and marine products); the current state of issues related to the environmental load from the perspective of material and energy cycles (development of economic activities and the materials and energy cycle, the nitrogen cycle, the energy cycle, the forest resource cycle). The final chapter is on the current state of the environment (atmosphere, water quality, soil, waste, natural ecosystems, diversity of wildlife, contact with nature).

State of the Environment in Asia and the Pacific 1995

Contents: environmental conditions and trends (land, forests, biodiversity, inland waters, coastal and marine environment, atmosphere and climate); causes and consequences (population, consumption and poverty, human settlements, agriculture, industry, energy, transport, tourism, solid waste, trade, human health, natural hazards and disasters); responses (public authorities environmental action, responses of industry and private sector, contributions of NGOs and major groups, information, monitoring and research, communication for the environment, follow-up of Agenda 21);

future outlook (economic development and environment: trends and prospects, environmental challenges).

Europe's Environment 1995

Contents: the context (reporting, environmental change and human development, the European continent); the assessment (air, inland waters, the seas, soil, landscapes, nature and wildlife, the urban environment, human health), the pressures (population, exploitation, emissions, waste, noise and radiation, chemicals and genetically modified organisms, natural and technological hazards); human activities (energy, industry, transport, agriculture, forestry, fishing and aquaculture, tourism and recreation, households); problems (climate change, stratospheric ozone depletion, loss of biodiversity, major accidents, acidification, tropospheric ozone and other photochemical oxidants, management of freshwater, forest degradation, coastal zone threats and management, waste production and management, urban stress, chemical risk).

Australia State of the Environment 1996

Contents: portrait of Australia, human settlements, biodiversity, the atmosphere, land resources, inland waters, estuaries and the sea, natural and cultural heritage, towards ecological sustainability.

The State of New Zealand's Environment 1997

Contents: the place and the people, production and consumption patterns, environmental management, the state of or atmosphere, the state of our air, the state of our waters, the state of our land, the state of our biodiversity.

In the last decade of the twentieth century there appeared to be a sudden rush to prepare such SoE reports and reports were issued from more than 40 countries ranging from Albania to Uzbekistan (Box 3.1). Many countries or regions, including small parts of Antarctica, Australia, Europe, Japan, New Zealand, Asia and the Pacific are the focus of published books (most being several hundred pages in length) about the condition, quality or state of the environment.

3.6   Why state of the environment reports?

The reason for producing SoE reports can be found in the products of the 1992 UN Earth Summit. However, there is some interesting prompting for such reports that can be found in events taking place over the previous 20 years.

In Stockholm in 1972, the first 'earth summit meeting' took place. This was the UN Stockholm *Conference on the Human Environment*. Later in 1982 there was the UN Conference in Nairobi. In that year, the UN adopted the World Charter for Nature. How often do we hear about the World Charter for Nature and indeed where can the text be found? The main part of the text is shown in Box 3.2.

Box 3.2   World Charter for Nature: UN Resolution 37/7, 1982

(i) General Principles

1.   Nature shall be respected and its essential processes shall not be impaired.
2.   The genetic viability on the earth shall not be compromised, the population levels of all life forms, wild and domesticated, must be at least sufficient for their survival, and to this end necessary habitats shall be safeguarded.
3.   All areas of the earth, both land and sea, shall be subject to the principles of conservation, special protection shall be given to unique areas, to representative samples of all the different types of ecosystem[s] and to the habitats of rare or endangered species.
4.   Ecosystems and organisms, as well as the land, marine and atmospheric resources that are utilised by humans, shall be managed to achieve and maintain optimum sustainable productivity, but not in such away as to endanger the integrity of those other ecosystems or species with which they coexist.
5.   Nature shall be secured against degradation caused by warfare or other hostile activities.

(ii) Functions

6.   In the decision making process it shall be recognized that human's needs can met only be ensuring the proper functioning of natural systems and by respecting the principles set forth in the present Charter.
7.   In the planning and implementation of social and economic development activities, due account shall be taken of the fact that the conservation of nature is an integral part of those activities.
8.   In formulating long-term plans for economic development, population growth and the improvement of standards of living, due account shall be taken of the long-term capacity of natural

systems to ensure the subsistence and settlement of the populations concerned, recognising that this capacity may be enhanced through science and technology.

9.   The allocation of areas of the earth to various uses shall be planned, and due account shall be taken of the physical constraints, the biological productivity and diversity and the natural beauty of the areas concerned.

10. Natural resources shall not be wasted, but used with a restraint appropriate to the principles set forth in the present Charter, in accordance with the following rules:

(a) Living resources shall not be utilized in excess of their natural capacity for regeneration;

(b) The productivity of soils shall be maintained or enhanced through measures which safeguard their long-term fertility and the process of organic decomposition and prevent erosion and all other forms of degradation;

(c) Resources (including water) which are not consumed as they are used shall be reused or recycled;

(d) Non-renewable resources, which are consumed as they are used, shall be exploited with restraint taking into account their abundance, the rational possibilities of converting them for consumption and the compatibility of their exploitation with the functioning of natural systems.

11. Activities, which might have an impact on nature, shall be controlled, and the best available technologies that minimise significant risks to nature or other adverse effects shall be used, in particular:

(a) Activities which are likely to cause irreversible damage to nature shall be avoided;

(b) Activities which are likely to pose a significant risk to nature shall be preceded by an exhaustive examination; their proponents shall demonstrate that expected benefits outweigh potential damage to nature, and where potential adverse effects are not fully understood, the activities should not proceed.

(c) Activities which may disturb nature shall be preceded by assessment of their consequences, and environmental impact studies of development projects shall be conducted sufficiently in advance; and if they are to be undertaken, such activities shall be planned and carried out so as to minimise potential adverse effects;

(d) Agriculture, grazing, forestry and fisheries practices shall be adapted to the natural characteristics and constraints of given areas;

(e) Areas degraded by human activities shall be rehabilitated for purposes in accord with their natural potential and compatible with the well being of affected populations.

12. Discharge of pollutants into natural systems shall be avoided:

(a) Where this is not feasible, such pollutants shall be treated at the source using the best practicable means available;

(b) Special precautions shall be taken to prevent discharge of radioactive or toxic wastes.

13. Measures intended to prevent, control or limit natural disasters, infestations and disease shall be specifically directed to the causes of these scourges and shall avoid adverse side effects on nature.

(iii) Implementation

14. The principles set forth in the present charter shall be reflected in the law and practice of each State, as well as at the international level.

15. Knowledge of nature shall be broadly disseminated by all possible means, particularly by ecological education as an integral part of general education.

16. All planning shall include, among its essential elements, the formulation of strategies for the conservation of nature, the establishment of inventories of eco-systems and assessments of the effects on nature of proposed policies and activities; all of these elements shall be disclosed to the public by appropriate means in time to permit effective consultation and participation.

17. Funds, programmes and administrative structures necessary to achieve the objective of the conservation of nature shall be provided.

18. Constant efforts shall be made to increase knowledge of nature by scientific research and to disseminate such knowledge unimpeded by restrictions of any kind.

19. The status of natural processes, eco-systems and species shall be closely monitored to enable early detection of degradation or threat, ensure timely intervention and facilitate the evaluation of conservation policies and methods.

20. Military activities damaging to nature shall be avoided.

21. States and, to the extent they are able, other public authorities, international organisations, individuals, groups and corporations shall:

(a)   Co-operate in the task of conserving nature through common activities and other relevant actions, including information exchange and consultations;

(b)   Establish standards for products and manufacturing processes that may have adverse effects on nature, as well as agreed methodologies for assessing these effects;

(c)   Implement the applicable international legal provisions for the conservation of nature and the protection of the environment;

(d)   Ensure that activities within their jurisdictions or control do not cause damage to the natural systems located within other States or in the areas beyond the limits of national jurisdiction;

(e)   Safeguard and conserve nature in the areas beyond national jurisdiction.

22.   All persons, in accordance with their national legislation, shall have the opportunity to participate, individually or with others, in the formulation of decisions of direct concern to their environment, and shall have access to means of redress when their environment has suffered damage or degradation.

23.   Each person has a duty to act in accordance with the provisions of the present Charter, acting individually, in association with others or through participation in the political process, each person shall strive to ensure that the objectives and requirements of the present Charter are met.

In 1992 the UN Rio Earth Summit (see Ch. 2) gave rise to five main products (three non-legally binding and two legally binding). These were:

- the Rio Declaration
- the Forest Principles
- Agenda 21
- the Climate Change Convention
- the Convention on Biological Diversity.

The last two were established as legally binding for the signatories.

The 21 principles of the Rio Declaration define the rights and responsibilities of nations as they pursue human development and well-being. The Forest Principles were established to guide the management, conservation and sustainable development of all types of forest.

Agenda 21 was developed as a blueprint to encourage sustainable development socially, economically and environmentally into the twenty-first century and it contained significant prompt for SoE reporting. Chapter 40 of Agenda 21 specifically called for improved environmental decision making. To achieve this, we need to know what is happening, why it is happening, the implications of the changes and what can be done (by who and by when).

In the years following Agenda 21, there has been significant action throughout the world via local Agenda 21 groups. Local people and communities have made substantial contributions to Agenda 21 groups but it is disappointing to find that so many young people today know nothing about Agenda 21.

It is intriguing that Agenda 21 should be a significant prompt for SoE reporting because it is not legally binding and much of its content concerns communities and groups working together. Although Agenda 21 and the 1992 Earth Summit have prompted the publication of SoE reports, there have been other prompts. The UNEP encourages all nations to measure the status of their national environment. During the early 1990s, the UNDP was providing financial assistance to Pacific Island nations to help them produce SoE reports.

Other prompts come from agencies such as the OECD. In 1979, the OECD recommended that member countries develop SoE reports. All OECD members have agreed to produce regular SoE reports and indeed most have started to do this. The OECD programme of environmental performance reviews has the principal aim of helping member countries to improve their individual and collective performances in environmental management. The primary goals are:

- to help individual governments assess progress by establishing baseline conditions, trends, policy commitments, institutional arrangements and routine capabilities for carrying out national evaluations
- to promote environmental improvements and a continuous policy dialogue among member countries, through a peer review process and by transfer of information on policies, approaches and experiences of reviewed countries
- to stimulate greater accountability from member countries' governments towards public opinion within developed counties and beyond.

3.7   The content and objectives of state of the environment reports

Many SoE reports follow a standard format based on the OECD's pressure–state–response framework. That is, there is information on the state of the data, the nature of the environment, the pressures on it, its current state and society's responses to it. Such a framework is a common format, but there is some argument to expand this to a form such as driving force–pressure–state–exposure–effect–action.

The environmental issues and variables contained in SoE reports are fairly standard and include biological diversity, ecosystems, global warming, ozone depletion and the quality of air, soil and water. Examples of some of these variables are shown in Table 3.2 and are arranged within the pressure–state–response format.

The objectives of the SoE reports are essentially two-fold: first, to provide information on which to base actions and to establish appropriate policies and, second, to provide greater awareness about the condition of the environment. More recently, there has been a noticeable shift towards streamlining SoE reports and increasing the accessibility of information for the wider community. A great many reports are now on the Internet and resources are available via UNEP and other agencies to enable countries to produce and display SoE information for the Internet.

Common to all SoE reports is a recognition that in many areas there is a lack of information or at least insufficient information on which to establish any actions. Linked to this is the overall gloomy picture of increasing levels of

Table 3.2 *Examples of environmental issues and variables in state of the environment reports*

Issue	Pressure	State	Response
Biological diversity	Exploitation of populations	Number of endangered species, population decline	Legislation to control exploitation
Ecosystems	Invasive species, habitat degradation	Area remaining or area undisturbed	Establishment of protected areas, restoration
Global warming	Emissions of CO_2, CH_4, N_2O	Global CO_2 levels	Legislation to limit greenhouse gas emissions
Ozone depletion	Emissions of chlorofluorocarbons, hydrochlorofluorocarbons, etc.	Level of ozone concentrations	Legislation to limit ozone depleting emissions
Soil quality	Intensive agriculture	Extent of erosion; loss of nutrients	Changes in agricultural practices

environmental degradation. Some of the reports go to considerable lengths to try to draw attention away from the gloomy picture of the environment to that of the value of the information provided and the opportunities to improve the state of the environment.

So who contributes to SoE reports and perhaps more importantly who uses SoE reports? If we accept the premise that 'people make better decisions if they are better informed about the possible consequences of their actions', then perhaps there is a role for lay people to contribute data to SoE reports. The role of communities in environmental monitoring is discussed in Ch. 8.

3.8   Biological and ecological information in state of the environment reports

The sources of information in SoE reports can be extremely varied and indeed this raises a number of issues. In some countries, most of the information is collected and held by central government departments or regional government. Additional information can be held by schools and universities, wildlife and environmental societies, etc. One of the challenges has, therefore, been to identify the sources of the information. In addition, there may be issues of ownership and reliability. Integration of the widely scattered information into one SoE report has presented many challenges.

One example of a set of variables for SoE reporting is that proposed for the Australian Antarctic environmental reporting (Table 3.3). This set of variables is the result of a decision to develop a reporting system specifically for Antarctica (Belbin *et al.*, 2003). A simple template and data for each 'indicator' has been incorporated into a web-accessible database system called system for indicator management and reporting (SMIR). It is useful to compare the list of variables in Table 3.3 with the list of impacts in Table 1.1.

There is considerable variation in the content of biological information provided in SoE reports. The state of the country's nature or biological diversity is sometimes based on an ecosystem or biological community approach while in other examples the emphasis is more about the state of the knowledge rather than the wildlife. There is not information about change in all cases. Indeed, there appears to be a tendency to talk about the current status of nature rather than what has happened over a certain time period. This is partly because some data do not exist for trend analysis.

The Dobris Assessment of Europe's environment has an ecosystem and biological community approach. That is the state of the forests, agriculture, seas, soils, etc. is each considered in turn in relation to condition, function and change.

Table 3.3 *Australian Antarctic set of environmental indicators (as of September 2002)*

SIMR indicator number	Indicator title	Indicator type[a]
1	Monthly mean air temperatures at AAT stations	C
2	Highest monthly air temperatures at AAT stations	C
3	Lowest monthly air temperatures at AAT stations	C
4	Monthly mean lower stratospheric temperatures above AAT stations	C
5	Monthly mean mid-tropospheric temperatures above Australian Antarctic stations	C
8	Monthly mean atmospheric pressure at AAT stations	C
10	Daily broadband ultraviolet radiation observations using biologically effective ultraviolet/radiation detectors	C
11	Atmospheric concentrations of greenhouse gas species	C, P
12	Noctilucent cloud observations at Davis	C
13	Polar stratospheric cloud observations at Davis	C
14	Midwinter atmospheric temperature at altitude 87 km	C
15	Stratopause region parameters for Davis	C
27	The annual population of Adélie penguins at colonies in the vicinity of Casey, Davis and Mawson and on Shirley Island and Whitney Point	C
28	Standard demographic parameters for Adélie Penguins at Mawson	C
29	Breeding population of the Southern Giant Petrel at Heard Island, the McDonald Islands and within the AAT	C
30	Breeding population of King Penguins at Heard Island	C
31	Annual population estimates of Southern Elephant Seals at Macquarie Island	C
33	Annual catch in tonnes of marine species harvested in Australian Antarctic and sub-Antarctic waters	P, R
35	Number of permits issued for entry into specially protected areas and sites of special scientific interest in the AAT	P
36	Numbers of species protected at various levels of conservation status	R
37	Species and numbers of species killed, taken or interfered with, or disturbed in the Antarctic and the sub-Antarctic for scientific research	C
38	Mean sea level for the Antarctic region	C
39	Average chlorophyll concentrations for the Southern Ocean across latitude bands 40–50°S, 50–60°S, 60°S–continent	C
40	Average sea-surface temperatures in latitude bands 40–50°S, 50–60°S, 60°S–continent	C

Table 3.3 (*cont.*)

SIMR indicator number	Indicator title	Indicator type[a]
41	Average sea-surface salinity in latitude bands: 40–50°S, 50–60°S, 60°S–continent	C
43	Fast-ice thickness at Davis and Mawson	C
46	Annual tourist ship visits and tourist numbers	P
48	Station and ship person-days	P
49	Medical consultations per 1000 person-years	C
50	Volume of waste water discharged from AAT stations	P
51	Biological oxygen demand of waste water discharged at AAT stations	P
52	Suspended solids content of waste water discharged at AAT stations	P
54	Amount of waste incinerated at AAT stations	P, R
56	Monthly fuel usage of the generator sets and boilers	P
57	Monthly incinerator fuel usage of AAT stations	P
58	Monthly total of fuel used by vehicles at AAT stations	P
59	Monthly electricity usage at AAT stations	P
60	Total helicopter hours	P

SIMR, system for indicator management and reporting; AAT, Australian Antartic Territory.
[a] Indicator types: C, condition; P, pressure; R, response.

Ecological processes (such as eutrophication) and pressures of exotic pests are considered. Urban ecosystems are also described.

By way of contrast, there is relatively little about biology or nature in Japan's SoE report. A section on the status of wildlife is more about the number of endangered species than anything else. Some of the biological data given in this report are based over time. For example, there are the yearly trends in swan and goose populations and population levels of several species of whales.

One other example is the New Zealand SoE report. The chapter on biological diversity considers the sources, status and completeness of the information. There is some information on the degradation of ecosystems and effects of introduced species.

Overall, the SoE reports during the 1990s have tended to focus on the current status of species, communities and ecosystems. This, together with the information on pressures and responses, does provide a baseline for future SoE reports of some kind. The ecological content of SoE reports needs to be considered alongside the purpose of such reports. If, on one hand, SoE reports are provided as general accounts of pressures on the environment and responses of society, then perhaps many of the existing styles of report are appropriate. On the other hand, SoE reports could be used more in the sense of reporting on progress, having established management procedures. In this case, ecological monitoring data would have a very central role in the reports.

3.9   Indicators and indices of environmental quality and sustainability in state of the environment reports

A detailed discussion of biological indicators and indices is given in Ch. 5. This section serves as a general introduction to the concept of indicators and indices with particular reference to SoE reporting.

The use of the terms index (or indices) and indicator are now very commonly used in SoE reports. Indeed, there have been considerable efforts by some government departments to determine what are the most useful and most acceptable indices and indicators for environmental reporting: that is, useful in the sense of operational aspects of SoE reporting and acceptable in the sense of what is acceptable to interested parties. Unfortunately, the term indicator has come to be used in a very generic sense and has almost replaced ecological variable or ecological parameter.

The use of the terms indicator and index in the context of environmental management and SoE reporting have become rather complex and confused. Indeed, in some reports the terms indicator and index are used interchangeably. Some definitions of indicator and index follow.

Indicator

An indicator is a parameter, or value derived from parameters, that describes the state of a phenomenon/environment/area with a significance extending beyond that usually associated with a parameter value (OECD, 1994).

An indicator can also be a quantitative measure (i.e. distance from a goal, target, threshold, benchmark) against which some aspects of policy performance can be assessed.

In biology, an indicator is a selected variable (population level, species level, etc.) that indicates or tells us something by way of its condition, behaviour or presence (or absence). These indicators may be used to indicate some condition of the environment. The biological indicators that tend to be used in monitoring include those species which have narrow environmental tolerances such as some lichen species, salmon and mayflies.

Examples of environmental indicators include levels of ozone concentrations (indicating overall concentration of photochemical oxidants) and total suspended particulate concentration (indicating the quality of air). Examples of biological or ecological indicators include the relative abundance of species of lichens (indicating levels of SO_2 in the air), number of salmon in a river (indicating water quality) and numbers of earthworms in the soil (indicating soil quality).

In this book, the term 'biological indicator' is used in the sense of a level of biological organization that by its presence, absence, condition or behaviour indicates something about the condition of the environment (see also Slocombe, 1992). An analogy is the indicator (flashing light) on a car. When the light starts flashing there is a change, that is the indicator light has changed from not being on to that of flashing off and on. In other words, an electrical component of the car has changed its behaviour. Indicators are used to indicate a future change in the direction of the car (or are supposed to do so).

The term indicator is widely used in SoE reporting and in monitoring biological diversity. (On p. 18 there is an example of indicator variables at different levels of biological organization.) However, the use here is slightly different. Noss (1990), for example, defined indicators as 'measurable surrogates for environmental endpoints such as biodiversity that are assumed to be of value to the public'. He went on to suggest that indicators should have some of the following characteristics:

- be sufficiently sensitive to provide an early warning of change
- be distributed over a broad geographical area
- be capable of providing a continuous assessment over a wide range of stress
- be relatively independent of sample size
- be easy and cost effective to measure
- be able to differentiate between natural and human-induced stress.

This use of the term 'indicator' is broadly different to the use in the sense of indicator species. I suggest that it would be useful to use the term 'biological variable' in the sense that Noss has used instead of indictors.

Index

The OECD (1994) defined 'index' as a set of aggregated or weighted parameters or indicators. This is confusing because of the use of both terms, index and indicator.

An index, as used here, is a number or score derived or transformed from quantitative data in an objective manner (see also Slocombe, 1992). The flashing light indicator on a car was mentioned above as an analogy to a biological indicator. An index associated with cars could be the subjective assessment of passenger comfort over long journeys expressed within a scale or score of 0–10. Still with cars, a running cost composite index could be derived for cars based on five-year periods of fuel consumption, maintenance, insurance, deterioration, etc. By way of comparison, in the field of economics, a price index (such as the retail price index) is a weighted average of prices of consumer goods and services produced in the economy measured over time.

A classification of environmental indices has been devised by Mallett (1999) to help in the understanding of the nature of an index (Table 3.4). He described a range of indices from a 'simple percentage' to 'composite' indices. In general, he identified three broad categories:

- simple ratios (e.g. the BTO Common Bird Census index; p. 288)
- scoring systems (e.g. the Trent biotic 'index' – more correctly a score; p. 174)
- composite indices such as the water quality index (WQI), where concentrations of various contaminants are scored and then added together.

Two of the examples given in Table 3.4 are composite indices. Composite indices are where two or more values are combined. Often the values are weighted. A composite index can be a very simple way of presenting a lot of complex information. However, the calculation of a composite index is a subjective exercise and some information can be lost. Is it realistic or meaningful to express the condition of the environment via one composite index?

Table 3.4 *Types of environmental index*

Type	Description	Example
1a	Simple percentage	$\nu_i = \frac{x_i}{x_t} \times 100$
1b	Simple ratio: temporal	$\nu_i = \frac{x_i}{x_b}$
1c	Simple ratio: reference value	$\nu_i = \frac{x_i}{x_b}$
1d	Simple ratio: other	$\nu_i = \frac{x_i}{y_i}$
2	Simple function	$\nu_i = f(x_i)$
3	Biotic score	Macro-invertebrate community index; index of biotic integrity
4	Composite function	$D = \frac{A+B+C}{13} \times 100w$
5	Composite: specific issue	Greenhouse gas index
6	Composite: general issue	Air quality index, environmental quality index, sustainability index

As with indices of the state of the economy and indices of social conditions, there have been attempts to establish indices of the state of the environment. These indices of the state of the economy are made widely known and typically are reported daily in the news media. There is a demand for daily news about the state of the economy and, for individuals, the state of their investments. Why not daily or weekly reporting of the state of the environment?

3.10 Examples of indicators and the use of indicators in state of the environment reports

In New Zealand, the Ministry for the Environment established an programme in 1996 for the purpose of identifying a national system of environmental performance indicators (the publication of the New Zealand SoE took place in 1997). A number of areas were selected including land, air, freshwater, climate change, ozone, marine environment, terrestrial and freshwater biological diversity, waste, transport, energy and landscape values. Examples of the indicators that have been confirmed are shown in Table 3.5.

3.11 Indices and monitoring for the state of the environment and sustainability

For many decades, indices have been used in programmes to monitor pollution, changes in biotic communities and so-called 'environmental standards' or 'quality of the environment'. Environmental indices include those that are based on physical and chemical parameters, those based on biological parameters and those based on perceived aesthetic qualities of the environment.

Environmental indices that have been used to assess and monitor environmental quality include not only those used in relation to pollution but also those used to assess and monitor landscapes and aesthetics. There are many ways of assessing landscape quality and a number of schemes incorporating scoring systems have been developed (e.g. Ribe, 1986; Tips & Savasdisara, 1986). These scores of landscape quality can be used in the sense of an index and thus be used to monitor changes. Other environmental indices include those for noise, radioactivity, air quality and water quality (see Thomas (1972) for a review of environmental indices).

Comparing indices

A wide array of environmental indices can form the basis of SoE reports. How, then, can indices of soil quality and biodiversity be compared? How to compare apples and pears? Normalization can be used. A widely accepted

Table 3.5 *Summary of confirmed environmental performance indicators derived by the New Zealand Ministry for the Environment*

Area	Stage 1: ready to implement	Stage 2: further development required
Air	Particulate matter (PM$_{10}$) Carbon monoxide Nitrogen dioxide Sulphur dioxide Ground level ozone	Benzene Particulate matter (PM$_{2.5}$) Lichen diversity/ coverage Visibility
Freshwater	Dissolved oxygen Ammonia Temperature Clarity Trophic state index Percentage of population with good water supply Periphyton (effects of slime on bathing)	Occurrence of native fish: Giant Kokopu, Red-finned Bully Macroinvertebrates (insects in rivers) Periphyton (effects of slime in rivers) Riparian condition Wetland condition and extent Groundwater: nitrates, abstraction quantity Water abstraction
Land	Changes in areas susceptible to hill country erosion Percentage change in area of slip at selected sites	Change in area susceptible to high-country degradation Acidity or alkalinity of soil Organic matter Change in area susceptible to agricultural impacts Change in area susceptible to reduction in soil health Bulk density of soil pH soil test Organic carbon

method of normalization has proved popular in these circumstances, particularly where the quality of wildlife, air, noise, dust, recreation and other environmental variables are being assessed. That is, interpretation of the environmental variable can be achieved by converting measured variables into normalized numbers or an index by means of a uniform scale. A hypothetical example is shown in Fig. 3.2.

Kreisel (1984) proposed this method, that is the use of empirically derived indices, for describing spatially and temporally varying environmental quality of a metropolitan area (the city of Dortmund). Environmental variables were selected from three categories (air pollution, noise pollution and recreational facilities as one example of human-created environments) and each variable was then normalized on a 0 to 1 scale. To derive composite indices, the normalized variables or indicators of environmental quality were weighted, using a method based on the Delphi technique and then aggregated. The results were then converted to isopleths plotted onto a map of Dortmund (Fig. 3.3). The Delphi technique (Linstone & Turoff, 1975; Sackman, 1975) as used by Kreisel for making numerical comparisons and weighting is worth noting in some more detail. The technique was developed in the 1960s by Helmer at the Rand Corporation as one of the many techniques used in decision analysis, probability estimates and long-range forecasts. The technique is sometimes used as a means for promoting a consensus of the views expressed by a panel of experts or a committee. Basically, Delphi aims to encourage independent and unbiased assessments from each individual.

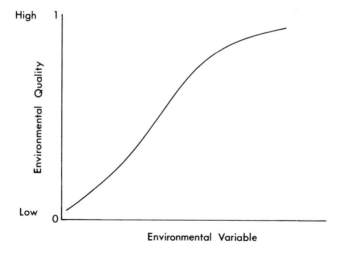

Fig. 3.2 Normalization or conversion of a hypothetical environmental variable (such as SO_2 levels or species richness) to levels of environmental quality (scored from 0 to 1).

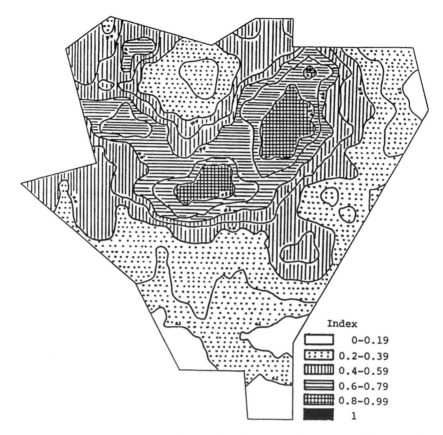

Index

☐ 0-0.19
⫶⫶⫶⫶ 0.2-0.39
▥▥▥ 0.4-0.59
≣ 0.6-0.79
▦▦ 0.8-0.99
■ 1

Fig. 3.3 One examples of the use of isopleths plotted onto a map of Dortmund; an index for air pollution. (From Kreisel, 1984.)

In brief, the technique uses a questionnaire that is circulated amongst a panel of experts who are not aware of the identity of fellow members of the panel. There are several stages.

1. A questionnaire addressing the environmental issues is designed by a small team.
2. The questionnaire is circulated amongst the panel by post.
3. The replies are analysed then redistributed stating the median and interquartile range of replies. Individual panel members are asked to reconsider their answers and those falling outside the interquartile range are invited to state their reasons. Reasons may include lack of knowledge or, by way of contrast, some specialist knowledge unknown to other members of the panel.
4. Analysis of the replies from stage 2 are recirculated, together with the reasons put forward in support of the 'extreme' positions. In the light of

this information, the panel members are asked to reconsider their original replies.

5. Additional stages of circulation and reiteration are undertaken when it is felt necessary to do so.

With the use of computer programs, the process can be reduced from several weeks to a few hours. The reliability and validity of this and other kinds of technique for decision analysis is very much dependent on the composition of the panel members and the design of the questionnaire. Although Delphi has been widely used for environmental assessment and for monitoring programmes (Linstone & Turoff, 1975; Richey *et al.*, 1985), this technique, like the many others, does have its critics (Sackman, 1975).

Wildlife indices

In 1971, the MITRE Corporation in the USA prepared a report on environmental indices as a background report for the Council on Environmental Quality. The report considered a total of 112 indices (together with detailed calculations and equations) ranging from indices for odours in the home to fish kills and wildlife indices. The MITRE report suggested two types of wildlife index: one for endangered species (those species that are approaching extinction) and one for 'troubled' species (species that have been greatly reduced in numbers but have some populations still thriving).

The calculation of these wildlife indices takes into account a subjective weighting of the position of a species in a food chain, with endangered carnivores scoring 5 and rodents 1. This is because, it was argued, changes in the groups higher in the food chain may indicate effects that have occurred lower down. However, this argument does not apply to outbreaks of canine distemper or to pressures on top carnivores from human hunters or poachers.

Some wildlife indices have been developed as a basis for assessment of conservation status. The 'threat number' (Perring & Farrell, 1983) was devised to assess the conservation status of endangered flora in the UK. Although not specifically devised for monitoring the status of plants, threat numbers could be used with other data for monitoring the extent of degradation in natural and semi-natural areas. Wildlife indices and species evaluations in general (Spellerberg, 1992) have an important role in assessing the status of a species and focusing attention on the conservation needs of those species.

Composite environmental and sustainability indices

It has been argued that a composite or comprehensive index (an index with a single value) could provide a useful and concise summary about the state of, and the changes in the state of, the environment. It could also be argued that this type of index is an oversimplification and misleading.

Considerable research on composite indices was carried out in the 1970s. For example, Inhaber (1976) in Canada constructed an environmental quality index (EQI). This index consisted of four main subindices for air, water, land and miscellaneous aspects such as pesticide and radioactive levels.

An overall EQI was proposed by Parker (1991). This was constructed so that changes in various aspects of the environment could be recorded relative to values in a 'base' year. In the following year, the UK Labour party published an EQI for the UK. Hope and Parker (1995) proposed an alternative EQI based on Parker's earlier research: the Hope and Parker index ranges from 0 (best) to 100 (pressure in an arbitrary reference year). The variables used include air-, water- and land-quality measurements.

Sustainability is a concept that continues to feature in most if not all SoE reports. Perhaps it is not surprising, therefore, that there are composite indices for measuring levels of sustainability and progress towards sustainability. Monitoring sustainability (or monitoring how 'green' a country's activities are) includes some elements of ecological monitoring along with the monitoring of social and economic states.

A task force of the World Economic Forum (2001) with collaborators (Yale University Center for Environmental Law and Policy, Columbia University Center for International Earth Science Information Network) has recently published an environmental sustainability index (ESI) to provide measures of environmental performance (a report is available on-line at http://www.ciesin.columbia.edu/indicators?ESI).

The purpose of the ESI is to benchmark environmental performance, identify comparative environmental results that are above or below expectations, iden-tify 'best practice' options and investigate interactions between environmental and economic performance.

The ESI is derived from the state of systems (air, water, ecosystems, etc.), the stresses on systems (pollution), human vulnerability (loss of food resources), social and institutional capacity to cope with environmental challenges and global stewardship.

The full list of the variables is shown in Table 3.6. In *The Economist* (27 January 2001), this ESI is described as 'a detailed assessment of dozens of variables that influence the environmental health of economies'. This is an interesting but perhaps not surprising view from this publication.

The key findings included that environmental sustainability is measurable and that, although important, economic conditions are not a fundamental policy constraint. At that time (early 2001), league tables were published.

Table 3.6 *Variables, indicators and components of the World Economic Forum environmental sustainability index*

Component	Indicators	Variables
1. Environmental systems	Air quality, water quality, biological diversity, terrestrial systems	Urban SO_2 and NO_2 concentration; internal renewable water per capita; water inflow from other countries; dissolved oxygen concentration; suspended solids; percentage mammals threatened; percentage of breeding birds threatened; soil degradation; percentage area of land affected by humans
2. Reducing stresses	Reducing air pollution; reducing water stress; reducing ecosystem stress; reducing waste and consumption pressures; reducing population pressures	Emissions per populated land area of nitrogen oxides and SO_2; coal consumption; vehicles per populated land area; rate of fertilizer consumption; pesticide use; industrial organic pollutants; change in forest cover; radioactive waste; percentage change in population
3. Reducing human vulnerability	Basic human sustenance; environmental health	Daily per capita calorie supply as a percentage of total requirements; percentage of population with access to improved drinking water; death rate from intestinal infectious diseases; in those under five year old mortality rate
4. Social and institutional capacity	Science/technology; capacity for debate, regulation and management; private sector responsiveness; environmental information; eco-efficiency; reducing public choice distortions	Research and development scientists and engineers per million population; scientific and technical articles per million population; IUCN member organizations per million population; degree to which environmental regulators promote

Table 3.6 (*cont.*)

Component	Indicators	Variables
		innovation; percentage of land area under protected status; number of sectorial EIA guidelines; number of ISO 14001 certified companies per million dollars GDP; Dow Jones sustainability group index membership; memebership of World Business Council for Sustainable Development; environmental strategies and action plans; energy efficiency; extent of renewable energy production; price of premium petrol; subsidies for energy or material usage; reducing corruption
5. Global stewardship	International commitment; global-scale funding/ participation; protecting international commons	Number of memberships in environmental intergovernmental organizations; percentage of CITES reporting requirements met; compliance with environmental agreements; Global Environmental Facility participation; FSC accredited forest area; ecological footprint deficit and historical cumulative CO_2 emissions

GDP, gross domestic product; all other acronyms given on p. xiv–xvii.

The relevance of league tables is questionable. Far more important is how the ESI is derived via the 22 indicators and 67 variables and how such an index can be used as a measure of overall progress towards environmental sustainability.

There are five core components (environmental systems to global steward-ship) and 22 indicators of environmental sustainability (air quality to protecting international commons). Each indicator is calculated from a number of measur-able variables. For example, air quality has been assigned the variables of urban SO_2 concentration, urban NO_2 concentration and urban total suspended part-iculate concentration. In summary, the variable scores are averaged to obtain the 22 indicator scores. The ESI is calculated by taking the average of the 22 indicators and then converting this into a standard normal percentile.

The living planet index

The state of very small samples of the world's species is used as the basis for the living planet index (Box 3.3). This is maintained by the WWF (together with UNEP, WCMC, the Centre for Sustainability Studies, Mexico and Redefining Progress). This index is an indicator of the state of the world's natural ecosystems. It is a measure of the natural goods and services provided by nature and is calculated from the averages of three subindices of natural wealth: the forest species population index, the freshwater species population index and the marine species population index.

The population data for all species used in the index were assembled by UNEP-WCMC. Each subindex is measured by the average population trend over time for a sample of over 730 animal species. For example, the forest index includes 319 species populations. The marine index includes 217 species populations.

The three ecosystem indices are calculated on a regional basis and are expressed in simple graph forms (there is baseline of 100 for 1970). The graphs show the decline in the state of the species, and over 30 years (1970–2000) there has been an overall decline of about 37%. Confidence limits have not been calculated for the index because of uncertainties within the underlying popula-tion data (Loh, 2002).

Ecological footprints

How much of nature's resources does it take to provide humans with food and materials (services) and provide a sink for waste materials (CO_2, physical waste)? How much area of land, freshwater and sea is required to support each country at the current standard of living? Clearly there are many differences around the world but in industrialized areas the amount of nature's resources being used to provide a city with nature's services and sinks is much larger in area than the city itself. That is, cities have impacts on nature and we can think of these impacts as 'human footprints'. Ecological footprints are composite indices of the state of sustainability.

Box 3.3   The *Living Planet Report 2002*: the living planet index and the ecological footprint

The living planet index

The living planet index is the average of three ecosystem population indices that monitor the changes over time in populations of animal species in forest, freshwater and marine ecosystems' respectively (see the Figure).

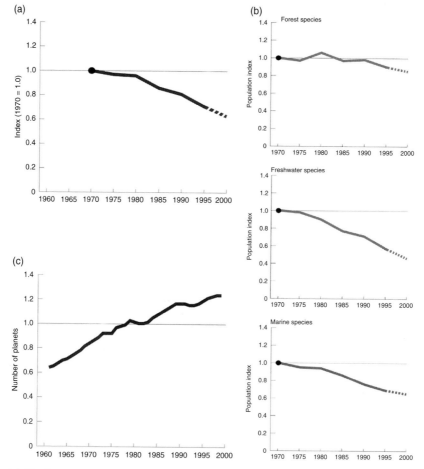

(a) The living planet index 1970–2000.

(b) Population indices 1970–2000 for forest, freshwater and marine species.

(c) The world ecological footprint 1961–1999.

Each ecosystem index measures the average population trend over time of a sample of animal species. The forest index included 319 species populations, which show an average decline of about 12% from 1970 to 1999. The freshwater index included 194 species populations, which fell, on average, by about 50% between 1970 and 1999. The marine index included 217 species populations, which declined by about 35% on average over the same period.

The three ecosystems indices are calculated on a regional basis. The forest species population index is the average of separate trends in temperate and tropical forests. The freshwater species population index combines average trends from six continents, and the marine species population index is based on trends in six regional oceans. In all three ecosystem types, the most severe declines have been in the southern temperate or tropical regions of the world.

The world ecological footprint

The ecological footprint is a measure of human use of renewable natural resources expressed as the number of planets, where one planet equals the total biologically productive capacity of the Earth in any one year. The most current national accounts methodology is used.

Six production area types are used for human activities:

croplands
grazing land
forests
fishing grounds
urban
energy land used to offset CO_2 production.

A nation's consumption is calculated by adding imports to, and subtracting exports from, domestic production (net consumption = domestic production + imports − exports). Domestic production is adjusted for production waste and, in the case of crops, the amount of seeds necessary for growing the crops.

This balance is calculated for over 200 items such as cereals, timber, fish-meal and cotton. These resource uses are translated into 'global hectares' by dividing the total amount consumed in each category by its global average productivity, or yield. Biomass yields are taken from Food and Agricultural Organization statistics.

Cropland, forests and grasslands vary in productivity. In order to produce a single measure (the 'global hectare'), the calculations normalize production areas across nations to account for differences in land and sea productivity.

'Equivalence factors' relate the average primary biomass productivity of different types of land to the global average primary biomass productivity for a given year. A hectare of land with world-average productivity has an equivalence factor of 1.0. For example, every hectare of pasture is assigned an equivalence factor of 0.47, since on average pasture is about half as productive as the average hectare on the Earth's surface.

(From Loh (2000) with permission of Jonathan Loh.)

In 1996, Mathis Wackernagel and William Rees (Fig. 3.4) published their book *Our Ecological Footprint: Reducing Human Impact on the Earth.* They defined the ecological footprint as a measure of the 'load' imposed by a given population on nature and suggested that it represents the land area necessary to sustain current levels of resource consumption and waste discharge by that population. Another definition of an ecological footprint is 'an estimate of the land area necessary to sustain current levels of resource consumption and waste assimilation for a given population' (Bicknell *et al.*, 1998). Perhaps 'land' could be interpreted as being the area of the environment including aquatic habitats or resources or perhaps the term 'land-equivalent' could be used.

As one example, Wackernagel and Rees (1996) provided a calculation of the Netherlands' ecological footprint that was 15 times larger than the Dutch territory. They go on to suggest that our ecological footprint keeps growing while our per capita 'earthshares' continue to shrink.

The objective in writing their book (see also Wackernagel & Yount, 1998) was to make the case that humans have no choice but to reduce our ecological footprint. Since 1996, the concept of the ecological footprint has been widely used in many countries. There has also been much debate in the literature about the strengths and weaknesses of the concept.

The method used to calculate ecological footprints has been evolving and almost certainly there will continue to be research on the methods. There is, of course, no agreed method but this cannot be used as an excuse to reject the concept of ecological footprints.

In the *Living Planet Report*, the ecological footprint has been calculated for many countries. The method used for calculation of the ecological footprints in the *Living Planet Report 2002* is outlined in Box 3.3. The living planet index alongside the ecological footprint provides the basis for the *Living Planet Report 2002* (Box 3.3). Together these are indicators of global sustainability. In this report, the world ecological footprint is described as having increased by 50% between 1970 and 1997 (an increase of about 1.5% per year). At the same time the living planet index declined by about 33%.

In volume 32 (2000) of the journal *Ecological Economics*, there are several papers that provide commentaries on ecological footprints. The fact that the ecological footprint does not represent the full range of impacts of humans on the earth is often put forward as a major limitation. Other comments point to its simplicity

(a)

Fig. 3.4 Mathis Wackernagal (a) and William Rees (b), authors of *Our Ecological Footprint: Reducing Human Impact on Earth.*

(which some see as an advantage) and to the perception that it is neither ecological nor economic (Opschoor, 2000). Rapport (2000) argued strongly that the ecological footprint does not provide sufficient calculus for portraying the relationship between humans and nature. As well as describing its limitations, Rees (2000) argued that the strengths include the fact that it captures Odum's (1971) suggestion that 'great cities are planned and grow without any regard for the fact that they are parasites on the countryside which somehow must supply food, water, air and degrade huge quantities of waste'. Rees also suggested that the concept builds on the critical importance of natural capital to economic development and also recognizes the importance of the second law of thermodynamics for human affairs.

(b)

Whatever the criticism of the concept, there seems no doubt that human use of nature is increasing at a rate that is not sustainable. The ecological footprint is simple and it seems strange that such a concept is rejected on the basis of its simplicity. What are the alternatives to indicators of sustainability? Perhaps the critics of the ecological footprint (as an indicator of sustainability) are frightened by the simple message. Some might say, 'ecological footprints: so what?' Perhaps the discussion about ecological footprints needs to move on from methods to applications. How can this indicator (along with others) be used to help to achieve environmental sustainability? That is where ecological monitoring has a role.

Ecological rucksacks

The concept of the ecological rucksack was introduced in 1993 by Schmidt-Bleek (see Schmidt-Bleek, 1994). This concept and the concept of material input per unit service (the mass of materials needed to produce a unit of service) is put into a global context in the book *Factor 4* written by von Weizsaecker *et al.* (1997). Doubling wealth while halving the resource use is the basis for the book.

There is an ecological rucksack (index of environmental stress) associated with each and every product and this can be calculated to indicate the weight of all materials that have been displaced in the production of a product or service. That is, an ecological rucksack equals the mass (in kilograms) of those natural resources that were moved from their original place to produce the product or service. A kilogram of metal obtained from mining may have required tonnes of ores to be processed. For example, a gold ring may originate from three tonnes of rock.

The Takeda Foundation in Japan promotes the creation and application of engineering intellect and knowledge. The Takeda World Environmental Award for 2001 was awarded to Friedrick Schmidt-Bleek (Box 3.4) and Ernst von Weizsaecker for the development and promotion of the concepts of ecological rucksacks and material input per unit service.

Ecological rucksacks are considered by Schmidt-Bleek to be broad measures of environmental stress. All products or services carry with them ecological rucksacks of all the materials displaced or processed during the life cycle of the product or service.

Another way of considering an ecological rucksack might be that all goods and services have been produced at some environmental cost. Whether you purchase a car or buy a holiday, it is a fact that there were environmental costs incurred. These environmental costs could, at their simplest, be based on the energy required to produce that product. Such an index of environmental costs could have a useful application. For example, the purchase price of cut flowers or tomatoes or fresh fish could be exactly the same in two different countries. However, there could be considerable differences in energy costs if in one country the produce was obtained locally and in the other country the produce was flown halfway around the world. It is now commonplace to have energy ratings put on household appliances (referring to the amount of energy used to function). Should the energy consumption or a measure of the ecological rucksack be included with the price of all goods and services?

Box 3.4   Takeda World Environmental Award for 2001

The 2001 Award winner was Schmidt-Bleek for the '**proposal, promotion and implementation of the MIPS [material input per unit service] and ecological rucksack concept**' [together with Ernst Ulrich von Weizsaecker. 'I developed the MIPS and ecological rucksack concept to make sure that we can produce wealth for all the people on this planet and still live in peace with nature', said Bio Schmidt-Bleek, the President of the Factor 10 Institute at Carnoules, Provence and recent winner of the Takeda World Environment Award. 'Unfortunately, current economic and environmental policies will not get us to a sustainable future', he continued and explained that the major problem today is that we reward those who waste natural resources through old-fashioned fiscal policies and punish those who hire people for work. He added, 'there are already today a number of forward looking companies who increase the resource productivity of their products and services. They go in this direction because they want to be still in business in 10 or 20 years in a world that simply does not have enough resources to globalize the present western life style for 8 or 10 billion people. Today, many tons of non-renewable nature are wasted just to produce a single PC'. In Schmidt-Bleek's language, personal computers carry an enormous 'ecological rucksack'.

Schmidt-Bleek's concept calls for considerable economic, social and technical innovation to satisfy people's needs with much less input of natural resources than today – at least a factor 10 – for generating the same, or even better, value or utility as output. This relationship, the MIPS, is used as a design principle and to measure and compare the 'ecological price' of all goods, infrastructures and services.

The Takeda Foundation in Japan promotes the creation and application of engineering intellect and knowledge.

3.12   Conclusions

Monitoring ecological change has a significant role to play in SoE reporting. Where indices or indicators are used, there is a need to have indicators and indices that show broad progress. These need to be linked to complimentary data that play a more diagnostic role in alerting us to conditions that are not progressing as expected.

On one hand we need reliable and sometimes quite technical data. On the other hand, there is a need for a greater understanding of the application of indicators and indices and their role in SoE reporting. There are many options that have been proposed and while there is much debate about the usefulness, or lack of usefulness, in each case, the main purpose appears to be lost in academic arguments. Perhaps there is a need for simple, clear and readily understandable indicators and indices but supplemented by more technical and detailed data.

All the indices discussed in this chapter are showing the same trend. The fact is that whatever index is used it will show that human impact on nature and use of nature is not sustainable. There is a valuable role for SoE reporting but it has yet to become popularized and become embedded in the minds of us all.

4

Biological spatial scales in ecological monitoring

4.1   Introduction

Ecological monitoring is undertaken over time and at different biological and spatial scales. The relevance of time to ecological monitoring is made clear throughout this book. In addition to time scales, there are biological and spatial scales that must be considered. The levels of biological organization can be considered in at least three different ways:

- ecological: individual organisms to ecosystems
- taxonomic: from species to species assemblages
- genetic: from molecules to chromosomes.

The 1992 Convention on Biological Diversity (Appendix) reinforced the need to remind ourselves that biological diversity occurs at all levels of organization and that biological diversity does not mean only diversity of species. Ecological monitoring programmes have been undertaken at many different levels of biological organization as the examples in this chapter demonstrate.

Land is used for many purposes, and conflicts between different land uses are becoming more common and more intense. The greatest diversity of land uses is possibly to be found in the industrialized countries, where there is a desperate need for good information on land use and land cover as a basis for planning and resource management. Mapping and classification of land at many different spatial scales has evolved over many years to become a more highly sophisticated science (see, for example, the methods described by Bailey (2002)). There are many land-use recording and monitoring programmes (both national and regional) for many types of feature (ecological communities, land use, soil, water, land capability).

Whereas land-use capability and physiography are 'static' and classification requires only occasional surveillance, land-use and ecological characteristics are continually changing at different temporal scales. Consequently, there is a need for a sophisticated method for monitoring (at different spatial scales). The basis depends on the aims of the monitoring, whether it be for identification of water-resource planning or the identification of ecologically sensitive areas.

There are, however, few examples of long-term land-use monitoring programmes and this can, in part, be attributed to a lack of logistical and financial support for long-term programmes, which leads to lack of uniformity in the variables being measured and lack of coordinated databases.

The aim of this chapter is to introduce and emphasize the relevance of biological scales and the relevance of spatial scales to ecological monitoring.

4.2   Biological scales and ecological monitoring

Monitoring biological diversity

From about the time leading up to the 1992 Convention on Biological Diversity, there has been much research and considerable conservation efforts directed at assessment and monitoring of various levels of biological diversity. The convention identified priorities for inventory and monitoring, including:

- habitats and ecosystems
  - those with high diversity, large numbers of endemic species or threatened species
  - those required by migratory species
  - those areas of social, economic, cultural or scientific importance
  - those areas representative, unique or associated with key evolutionary processes
- species and communities
  - threatened, wild relatives of domesticated or cultivated species
  - of medicinal, agricultural or other economic importance
  - of social, scientific or cultural importance
  - of importance for research.

Conservation, management, assessment and monitoring of various levels of biological diversity has become an imperative. Often the science is brought under pressure by political pressure. There is often limited time and there may be a limited number of trained personnel. This sometimes results in innovative methods. For example, the development of programmes for state of the environment reporting has possibly contributed to the widespread use of so-called indicator taxa.

The UNEP has drawn up a list of general priorities for the types of biological diversity information that should be collected and this is a good example of the range in biological scale. The key parameters for monitoring biological diversity at the country level (UNEP, 1993) are:

- monitoring genetic diversity
- species monitoring
- habitat monitoring
- protected area monitoring.

Indicator taxa (particularly species and biotic communities) have become useful because often conservation decisions are required for large geographical scales. For example, Pearson and Carroll (1998) used birds, butterflies, tiger beetles, mean annual precipitation and spatial statistical models to investigate the applicability of indicators of species richness for conservation planning on a continental scale. They found that spatial models are optimal for these analyses because species data typically are not spatially independent.

The use of target organisms at the species level has come under considerable scrutiny, particularly where one species is used to reflect conditions amongst other species and at the biotic community level of organization. Overall, there seems to be consensus that indicators at the species level do have a role in ecological monitoring but that they must be considered with caution. Carignan and Villard (2002) concluded in their critical review of indicator species for monitoring the state of the environment that indicator species are useful if:

- many species representing various taxa and life histories are include in the monitoring programme
- their selection is based on sound quantitative databases
- caution is applied when interpreting population trends.

Carignan and Villard (2002) also suggested that there is less debate on the validity of indicators at higher level (biological communities, forest stand, landscape level) than there is at the species level. Certainly, biological communities appear to be promising biological indicators. For example, Tscharntke *et al.* (1998) assessed communities of trap-nesting bees and wasps and their natural enemies as indicators of ecological change or habitat quality (Box 4.1). They found that communities of trap-nesting bees and wasps were large enough to examine community structure and interactions but small enough to be easily managed.

Noss (1990, 1999) provided a useful reminder of the fact that biological diversity can be assessed and monitored both at different levels of biological organization and in terms of composition, structure and function. He referred to 'indicator variables' for inventory, monitoring and assessment at four spatial levels of organization (Table 4.1, Fig. 4.1). I suggest that these could be called 'variables' to avoid confusion with 'species indicators' or 'community indicators', which are used specifically to indicate some state or condition or change in state or condition. The LTER network (p. 51) has provided an important basis and indeed essential information for research on long-term and large-scale relationships between biological diversity and ecosystem functioning (e.g. Symstad *et al.*, 2003). It has been suggested that long-term studies of planned and accidental changes in species richness and in species composition may affect ecosystem functioning and that it may vary over space and in time. This is just one of the many valuable outcomes of the LTER network research (Hobbie *et al.*, 2003).

Box 4.1   Use of trap-nesting bees and wasps and their natural enemies as biological indicators of community structure and interactions

Most biological diversity monitoring relies on indirect methods and consequently six approaches have been suggested:

- species richness of indicator taxa or a suite of indicators from unrelated taxa or representatives from different trophic levels
- abiotic habitat characteristics such as edaphic factors or microclimate
- predictions of species richness based on richness or higher taxa
- extrapolations from one focal group to larger groups
- calculation of species richness using the number of easily separated 'morphospecies' (Oliver & Beattie, 1996)
- small, manageable communities with information on both species richness and ecological interactions.

The study of trap-nesting bees and wasps was undertaken in agricultural and horticultural areas of Germany. The traps were internodes of the common reed *Phragmites australis*. The exposure of standardized traps is an experimental approach. The authors found that species richness and abundance of bees (not wasps) was closely related to plant species richness of the habitat. Also, with increasing isolation of fragmented habitats, both species richness of natural enemies and percentage mortality declined significantly.

Several characteristics make small communities of bees and wasps suitable as indicators including:

- aculeate Hymenoptera can be used as sensitive indicators of habitat quality or environmental change

- the sensitivity of this approach relies not only on the presence/absence of data but also on ecological interactions
- population or community monitoring is supplemented with species interactions
- dissections of nests make it possible to reconstruct events during the past vegetation period
- all species used as indicators with trap nests at least reproduce in the 'target' habitat
- communities of trap-nesting bees and wasps are large enough to be suitable for the examination of community structure and interactions but also small enough to be easily managed.

(a)

(b)

Close up of traps.

(c)

Traps in a field.

A bee on a trap.

(With permission from Tscharntke, T., Gathmann, A. & Steffan-Dewenter, I. (1998). Journal of Applied Ecology, 35, 708–709.)

Typically, threatened, rare and endangered species are used as indicator species and consequently the focus of conservation projects or programmes is often on the taxa and not on what is supposedly being indicated (e.g. habitat degradation, exploitation, etc.). However, Bonn *et al.* (2002) have asked the

Table 4.1 *Suggested indictor variables for inventory, monitoring and assessing terrestrial biological diversity at four levels of organization; composition, structure and function are also included*

Level of organization	Indicators			Inventory and monitoring tools
	Composition	Structure	Function	
Regional landscape	Identity, distribution, richness and proportions of patch (habitat) types and multipatch landscape types; collective patterns of species distributions (richness, endemism)	Heterogeneity; connectivity; spatial linkage; patchiness; porosity; contrast; grain size; fragmentation; configuration; juxtaposition; patch size frequency distribution	Disturbance processes (areal extent, frequency or return interval; rotation period; predictability; intensity; severity; seasonality); nutrient cycling rates; energy flow rates; patch persistence and turnover rates; rates of erosion and geomorphic and hydrologic processes; human land-use trends	Aerial photographs (satellite and conventional aircraft) and other remote sensing data; geographic information system technology; time series analysis; spatial statistics; mathematical indices (of pattern, heterogeneity, connectivity, layering, diversity, edge, morphology, autocorrelation, fractal dimension)
Community–ecosystem	Identity, relative abundance, frequency, richness, evenness and diversity of species and guilds; proportions of endemic, exotic, threatened and endangered species; dominance–diversity curves; life-form proportions;	Substrate and soil variables; slope and aspect; vegetation biomass and physiognomy; foliage density and layering; horizontal patchiness; canopy openness and gap proportions; abundance, density, and distribution of	Biomass and resource productivity; herbivory, parasitism and predation rates; colonization and local extinction rates; patch dynamics (fine-scale disturbance processes), nutrient cycling rates;	Aerial photographs and other remote sensing data; ground-level photo stations; time series analysis; physical habitat measures and resource inventories; habitat suitability indices (multispecies); observations, censuses and

119

Table 4.1 (cont.)

Level of organization	Indicators			
	Composition	Structure	Function	Inventory and monitoring tools
	similarity coefficients; C4:C3 plant species ratios	key physical features (e.g. cliffs, outcrops, sinks) and structural elements (snags, down logs); water and resource (e.g. mast) availability; snow cover	human intrusion rates and intensities	inventories, captures and other sampling methodologies; mathematical indices (e.g. of diversity, heterogeneity, layering dispersion, biotic integrity)
Population–species	Absolute or relative abundance; frequency; importance or cover value; biomass; density	Dispersion (microdistribution); range (macrodistribution); population structure (sex ratio, age ratio); habitat variables (see community–ecosystem structure, above); within-individual morphological variability	Demographic processes (fertility, recruitment rate, survivorship, mortality); metapopulation dynamics; population genetics (see below); population fluctuations; physiology; life history; phenology; growth rate (of individuals); acclimation; adaptation	Censuses (observations, counts, captures, signs, radio-tracking; remote sensing; habitat suitability index; species–habitat modelling; population viability analysis
Genetic	Allelic diversity; presence of particular rare alleles, deleterious recessives or karyotypic variants	Census and effective population size; heterozygosity; chromosomal or phenotypic polymorphism; generation overlap; heritability	Inbreeding depression; outbreeding rate; rate of genetic drift; gene flow; mutation rate; selection intensity	Electrophoresis; karyotypic analysis; DNA sequencing; offspring–parent regression; sibling analysis; morphological analysis

From Noss (1990) with permission.

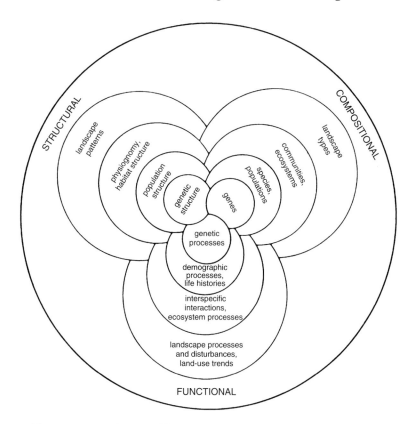

Fig. 4.1 Suggested composition and structural and functional biological diversity shown as interconnected spheres, each encompassing multiple levels. Noss suggested that this concept may assist with selection of indicators. (With permission from Noss, 1990.)

question, 'are threatened and endemic species good indicators of patterns of biodiversity on a national scale?' They suggest that, although nationally threatened and endemic species are important conservation targets, reserve networks that focus solely on these species may not be sufficient to conserve species overall.

In 1995, Pearson suggested seven criteria to meet the requirements of efficiency, comparison and communication. These are listed in order of priority for monitoring.

1. Potential economic importance
2. Higher taxa occupy a breadth of habitats and broad geographic range
3. Patterns observed in the indicator taxon are reflected in other related and unrelated taxa

4.   Biology and general life history well understood
5.   Populations are readily survey and manipulated
6.   Taxonomically well-known and stable
7.   Specialization of each population within a narrow habitat.

Pearson (1995) also noted that scientific robustness and efficiency of data gath-ering are only as useful as their ability to be communicated to non-scientists (see also Ch. 7). In 2003, David Pearson (personal communication) continued to uphold these arguments and noted that the scientific rigour needed for choosing and testing bioindicators has been generally ignored. He suggested that bio-indicators have not been used wisely among many who have tried to adapt this concept to individual situations. 'What the criteria are and how many are needed should be flexible. What is essential is that there are unambiguous criteria by which all can assess that the bioindicator is appropriate or useful for that situation. I think these two steps are mandatory for anyone attempting to deduce broad generalizations from a few selective taxa.'

4.3   Monitoring aquatic species and the effects of pollution

In the 1980s, there was a field experiment undertaken to look at the effects of acidification on lake species. This provided the basis for monitoring species populations. The effects of eight years' gradual, manipulated acidifica-tion on a small lake ecosystem in Ontario was undertaken by Schindler and his colleagues (1985). In that experiment, which commenced with a two-year base-line survey, the pH was slowly decreased from 6.8 to 5.0 and there were dramatic effects on the lake ecosystem, which was studied over a 10-year period. The selection of variables and processes chosen for that monitoring programme (Table 4.2) is still of interest today because it is unusual to have such a wide range of variables and processes being employed.

The baseline survey was undertaken from 1974 to 1976. In 1976, there were few distinguishable chemical changes and little biological change. In 1977, the relative abundance of chrysophycean species (phytoplankton) declined slightly. The phytoplankton production, biomass and chlorophyll was within limits of natural variation for lakes in the area. In 1978, several 'key' organisms in the lake's food web were severely affected and primary production was slightly higher than in any previous year. In the next year, some algae species formed highly visible thick mats in littoral areas. In 1980, the condition of the lake trout had declined; phytoplankton biomass had increased relative to the level in the previous year, and an acidophilic diatom, previously rare in the lake, appeared in large numbers. The trout continued to be affected; spawning behaviour

Table 4.2 *Biological parameters used to monitor a lake ecosystem during acidification over a period of seven years*

1.	Species composition of epilimnion algae groups
2.	Phytoplankton productivity, biomass and diversity (Simpson's index)
3.	Epilimnion chlorophyll levels
4.	Density of dipterans emerging each year
5.	Percentage composition biomass of zooplankton groups
6.	'Condition' of trout
7.	Population levels of Crayfish (*Orconectes virilis*) and Sculpin (*Cottus cognatus*)
8.	Size class structure of Minnows (*Pimephales promelas*) and Pearl Dace (*Semotilus margarita*)

From Schindler *et al.*, 1985.

changed in 1982 and in 1983 their condition was very poor and there was evidence of cannibalism amongst the trout.

Many previous studies of acidification have relied largely on pH measurements and abundance of more common species of fish only. The report by Schindler *et al.* (1985) suggested that these are not sensitive, reliable indicators of early damage caused by acidification. For example, twice weekly pH measurements did not reveal the disappearance of 80% of the alkalinity from the lake in the first year of acidification. They also suggested that most large fish are not sensitive indicators of early stages of acidification damage. The damage at lower trophic levels would, it was predicted, cause almost complete extinction of the trout within a decade. If trout are to be monitored, then it needs to be more than just population size that is measured; other parameters include age–class structure and yield (yield is that part of production utilized by the consumer at a higher trophic level, or by humans). Results from research in other Canadian lakes led Oglesby (1977) to conclude that fish yield has the advantage over traditional biotic and abiotic parameters partly because fish yield relates to socio-economic terms. This is perhaps a reminder that the objective of monitoring may not be solely for ecological reasons but there may be economic considerations as well.

In Schindler's (1985) study, it was found that, although the variables of phytoplankton production, species diversity and species richness were relatively insensitive to acidification, there were noticeable changes in species assemblages. For example, shifts from a large chrysophycean community to one where chlorophycean, cyanophycean and peridinean species were often dominant in the phytoplankton. These shifts are characteristics of eutrophication.

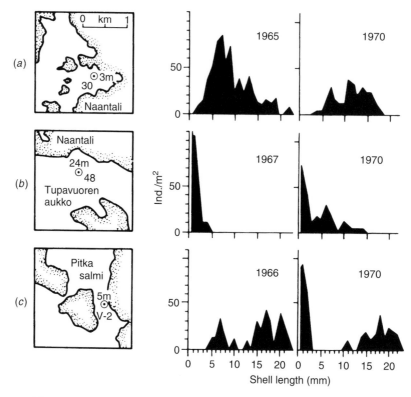

Fig. 4.2 Size–class distribution in populations of the mollusc *Macoma baltica* at three polluted sites off southwest Finland. (*a*) Decreasing population, no recruitment; (*b*) expanding population, intense recolonization prior to 1967; (*c*) aging population in 1966, newly started recruitment in 1969–1970. (Redrawn from Leppakoski, 1979.)

The variables of species richness, species composition, species diversity and even biomass, when employed in isolation, would seem to have limited use in monitoring changes in ecosystems. Even age–class structure or size–class structure has limitations despite the interesting observations that have been made (Fig. 4.2). The interpretation of this kind of data is difficult without information on abiotic variables and other biotic information. After all, the species in Fig. 4.2 are part of a community and a selection of community variables is best employed in monitoring ecosystems. However, to be realistic, collections of large amounts of data may be costly and not easily justified unless, as suggested above, there are benefits to balance the costs. If ecosystem monitoring is to be done in detail, then perhaps the community being sampled should be considered with particular care. Leppakoski (1979), for example, had suggested that large amounts of macrobenthos can be collected at low cost and is a suitable community that can be used in monitoring effects of pollution in aquatic environments.

4.4   Monitoring large mammals in large-scale landscapes

Distance sampling techniques have often been used to monitor mammals in tropical forests. A computer package by the name of TRANSECT (Laake *et al.*, 1994) has been particularly useful in this respect. However, as so often happens in ecological monitoring, the methods used are more often determined by immediate events and conditions rather than ecological monitoring theory. Large-mammal monitoring is perhaps the most challenging of the Smithsonian Institution's efforts for many reasons (Major Boddicker, personal communication). These include the behaviour choices that the mammals introduce, the lack of sightings, the replication of observations of the same individuals and the large numbers of few individuals.

Boddicker *et al.* (2002) developed ecological monitoring protocols for large mammals in the lowland tropical forests of Amazonian Peru and later in the rainforest of Gabon. These led to the introduction of an occurrence index and an abundance index (Box 4.2). Particularly valuable has been the development of unconventional methods to meet the demands of biological and spatial scales.

Many people attempting to monitor mammals in tropical forests have referred to the difficulties posed by poor visibility and by the fact that some large species cannot be approached for safety reasons. Furthermore, there may be huge organizational challenges, such as those described by Walsh & White (1999) in their account of developing a monitoring programme for elephants in the Gamba complex of southwestern Gabon. They also draw attention for the need to improve statistical methods as well as the need to pay more attention to the design phase of surveys.

Experience of monitoring mammals populations in African forests has led Plumptre (2000) to make the following seven recommendations:

- pilot studies are necessary
- obtain data on decay rates of signs
- obtain at least 100 sightings of groups
- if possible and if the budget allows, survey regularly
- record the number of individuals seen in any group
- obtain measures of production rate of signs so that the associated coefficient of variation is small
- consider the use of a variety of methods that could be used.

Box 4.2   Monitoring large mammals within an adaptive framework

This monitoring project, in the lowland tropical forests of Amazonian Peru, was designed to assess monitoring protocols for large mammals and note changes in populations brought about by development of natural gas resources (Boddicker *et al.*, 2002). Mammal assessment sites were based on four vegetation monitoring plots, each 1 ha in size, established by UNESCO under Man and the Biosphere (see also Fig. 2.1). The following methods were used:

- direct observations of large mammals while walking transects
- identification of mammal vocalizations
- scent-post surveys
- signs of large mammals
- live trapping along transects.

An occurrence index was developed from the five methods of collecting data. This provided a confirmed species list based on accumulated evidence from various methods. When the accumulated points of evidence reached a threshold, it was concluded that the species was present at the site. Each mammal event was assigned a value based on a point system that reflected the quality of the evidence. Unambiguous evidence was given 10 points: species collected, species observed. High-quality evidence was given 5 points: bone, hair, identification by local people, tracks, vocalizations and odours. Low-quality evidence was given 4 points: beds, nests, etc., faeces and scats, signs of feeding. The index was calculated from the sum of the accumulated points for each species.

An abundance index (based on the data for the occurrence index) was obtained by multiplying evidence values from the occurrence index by the number of independent observations of that type of evidence.

The indices used in this ecological monitoring of large mammals helps to provide a better understanding of the ecology of these mammals. The design is simple and is appropriate for the vast landscape in which the work was undertaken. Unconventional variables have been used, such as olfactory stimulants and vocalizations. Such variables seem appropriate for large mammals monitoring in large-scale landscapes. Similar work has continued in the rainforest of Gabon (Major Boddicker, personal communication).

Construction of drilling site, San Martin, Peru.

Cloud Forest, looking west From the drilling site.

Red Brocket Deer in a stream below the site.

Brazilian tapir tracks, San Martin.

(Information and photographs kindly supplied by Major Boddicker.)

Having drawn attention to the associated errors brought about by rates of production and decomposition of dung or nests (indirect signs of animals), Andrew Plumptre (personal communication) believes that census methods and monitoring of primates needs a re-assessment. He feels that suspect material has crept into the literature.

4.5   Monitoring woodland and forest communities

A review and assessment of long-term monitoring of woodland nature reserves by Peterken & Backmeroff (1988) not only resulted in a set of useful rules for ecosystem monitoring (Table 7.2, p. 238) but also confirmed the scarcity of woodland and other vegetation plots used for monitoring. Relatively few permanent vegetation plot studies have been established in central Europe or in Scandinavia. The longest and most thorough studies seem to have been undertaken on dune and other coastal vegetation in the Netherlands, but other countries such as Sweden have recognized the value of long-term monitoring of plant-community changes especially in relation to detecting effects of pollution on ecosystems. A Register of Permanent Vegetation Plots in the UK compiled in the 1980s showed that, even at that time, there was a diverse range of vegetation being used for ecological monitoring (Hill & Radford, 1986; Table 4.3).

Peterken and colleagues (e.g. Peterken & Backmeroff, 1988; Peterken & Mountford, 1998) have described the history of five studies of changes in the structure and species composition of unmanaged woodland nature reserves in the UK, some based on transects and some on permanent quadrats. The two oldest transects were Lady Park Wood (Gwent and Gloucestershire) established in 1944 and Denny Wood (New Forest, Hampshire) established in 1956.

Table 4.3 *Types of vegetation monitored by permanent plots in the UK; these represent 63 projects being carried out by 27 people*

Type	Vegetation
Lichens	–
Coastal sites	Shingle heathland, shingle coast, dune grassland, saltmarsh, serpentine quarry
Calcareous grassland and scrub	Railway and roadside verges, chalk grassland, Juniper scrub (*Juniperus communis*), scrub on limestone, calcareous flushes
Acid and neutral grassland and scrub	Railway verges, acid grass, neutral grass, upland grass
Heathland, moorland and birch scrub	Birch (*Betula* spp.) on bog and moor, blanket bog, heathland and heather moor
Coniferous woodland and plantation forestry	Sitka Spruce (*Picea abies*), Scots Pine (*Pinus sylvestris*), pine on bog, amenity trees
Broadleaved woodland	Acid oakwood, lowland coppice with standards, mixed woodland, calcareous ashwood, birch on bog and on moor
Aquatic vegetation	Freshwater loch

From Hill & Radford, 1986.

Denny Wood (Fig. 4.3), is an ancient, mixed deciduous wood-pasture dominated by Beech (*Fagus sylvatica*), Pedunculate Oak (*Quercus robur*) and Holly (*Ilex aquifolium*), with canopy trees ranging in age from about 70 to 300 years. In the 1980s, concern was expressed about the effects of acid precipitation on woodland in this part of England. Occasional measurements of size–class distributions, growth rates and mortality data from Denny Wood had shown that Beech had regenerated much more than oak and that many of the large Beech but not oak had not died. In an assessment of these trends in the mid 1980s, Manners & Edwards (1986) saw no reason to implicate atmospheric pollution and concluded that periodic drought (especially in 1976) aided by attacks of honey fungus has been largely responsible for a disintegrating wood. In the absence of more detailed data and synoptic recording of abiotic factors, it is reasonable to draw only general conclusions.

In more recent research (Mountford et al., 1999) changes that have taken place over 40 years have been assessed. These reports are based on a transect in unenclosed forest that was recently rediscovered and also a portion of Denny Wood that lies within the Denny enclosure. The latter is an area where, since 1956, all individual trees, shrubs and saplings growing within a 1000 m by 20 m transect have been periodically mapped and measured. The authors of this assessment of Denny Wood draw attention to the way that single natural

Fig. 4.3 Denny Wood in the New Forest, southern England. This is an acid mixed woodland (oak, Beech, Holly) that was designated an ecological reserve by the Forestry Commission in 1952. The post in the foreground of the lower photograph marks the corner of a permanent quadrat. (Photograph by B. Lockyer.)

events have played a major role in the ecology of the wood. The drought in 1976 and storms of 1987 and 1990, for example, generated large volumes of dead wood.

George Peterken (personal communication) believes that unmanaged woods such as Denny Wood play an important reference or baseline for comparison with managed woods. Simple measurements can generate a wealth of information on woodland ecology. A more detailed monitoring programme could include nutrient cycling measurements, because the sensitivity of this process to physical and chemical perturbations has been well demonstrated in other studies. With over three decades of records already available and at a time when effects of pollution on ecosystems need more careful monitoring, there has long been justification for the Denny transect and other woodland monitoring studies.

In retrospect, George Peterken (personal communication) suggests that the problem is that much monitoring is undertaken with vague criteria, such as whether or not a site of special scientific interest is in 'favourable' condition. Monitoring can be based more on political than scientific grounds. Furthermore, he suggests that it is all too often the case that individuals keep the monitoring going and that there is no clearing house or 'ecological records office' to coordinate the monitoring.

The political need for improved forest-monitoring data is ever increasing. Most if not all of the products of the 1992 Earth Summit have had implications for forest management. Following that Earth Summit, Canada convened an international seminar of experts on sustainable development of forests (boreal and temperate). This seminar was held in 1993 in Montreal and subsequently there was an initiative to establish and implement agreed criteria and indicators for sustainable forest management. The Montreal Process is the working group on criteria and indicators for the conservation and sustainable management of temperate and boreal forests. It was established in Switzerland in 1994.

Various levels of biological diversity are used as variables for monitoring biological diversity in forests. This is one of the outcomes of the Montreal Process and is a good example of the use of a range of levels of biological diversity for ecological monitoring. The criteria and indicators for the conservation and sustainable management of temperate and boreal forests are shown in Table 4.4.

The Montreal Process has been taken further in Europe through other initiatives. Independent forest certification programmes are also emerging. For example, there is the International Forest Stewardship Council, which in the UK has manifest itself as the Woodland Assurance Standard. These are audit

Table 4.4 *The Montreal Process criteria and indicators for the conservation and sustainable management of boreal and temperate forests: only criterion 1 is given in detail; the other criteria are mentioned as headings only*

Criterion	Indicators
1. Conservation of biological diversity	Ecosystem diversity: extent of forest type relative to total forest area; extent of forest type and age class or successional stage; extent of area by forest type in protected area categories; extent of areas by forest type in protected area by age class or successional stage; fragmentation
	Species diversity: the number of forest-dependent species; the status (threatened, rare, vulnerable, endangered or extinct) of forest-dependent species
	Genetic diversity: number of forest-dependent species that occupy a small portion of their former range; population levels of representative species from diverse habitats monitored across their range
2. Maintenance of productive capacity of forest ecosystems	
3. Maintenance of forest ecosystem health and vitality	
4. Conservation and maintenance of soil and water resources	
5. Maintenance of forest contribution to global carbon cycles	
6. Maintenance and enhancement of long-term multiple socio-economic benefits to meet the needs of societies	

systems that could be seen as a form of monitoring, but more importantly they put a duty on forest managers to undertake ecological monitoring.

4.6   Monitoring marine ecosystems

Of all ecosystems, estuarine ecosystems have probably been most severely damaged over the longest timescale. This is because so many estuaries throughout the world have been centres of habitation; consequently, high population densities occur around them, with the resulting impacts of pollution in its various forms. A major part of the estuarine ecosystem is the coastal wetlands, which, because of its high productivity, is particularly rich in wildlife. This has meant that the anthropogenic impacts on wetlands have resulted in great losses of wildlife.

The physical nature and geography of coastal ecosystems presents difficulties for the collection of data for long-term monitoring programmes. This is because the tides, and in some locations the inhospitable terrain, make it very difficult to establish permanent sites and make it difficult to sample or record the fauna and flora.

As with other ecosystems, there is often pressure to undertake ecological assessment and ecological mapping or marine ecosystems in as reliable and rapid way as possible. One approach is to use surrogates, that is, ecological variables that correlate strongly with the number of species. This approach has been tested by Olsgard *et al.* (2003), who tested two polychaete surrogates, one for higher taxa and one for indicator groups. The aim was to assist with the prediction of species richness in marine soft-bottom communities (Fig. 4.4). They found that the group Terebellida was particularly suited as an indicator group. This was because the group contains long-lived large species that are relatively easy to sort from the sediment and they are well defined taxonomically.

Methods for monitoring the long-term ecological effects of industrial impacts on estuaries has progressed in a very exciting way from simple ground surveys (sometimes in conjunction with aerial photography) to very sophisticated digital remote sensing. These improvements and developments in monitoring method-ology have been brought about, in part, by the need to be more cost effective, in part by the need to be more precise and in part by the advances in remote sensing technology.

One early example of these developments emerged from long-term moni-toring studies of the effects of a major refinery (one of the largest in Europe) and an oil-fired electricity-generating station on salt marshes in Southampton Water on the south coast of England (Shears, 1989). The marshes (Fig. 4.5) had been affected by industrial effluent since the 1950s and damage occurred up until 1971. By then an area of marsh about 1000 m by 600 m close to the outfalls had been affected (Dicks & Hartley, 1982). At that time, the refinery commenced a programme of enhanced effluent control and subsequently there was extensive recovery of the salt marshes. The salt marsh species *Salicornia* and *Sueda* were primary colonizers (Fig. 4.6), later to be followed by *Spartina* spp. (which helps to stabilize the mud with its extensive roots systems). Spread of

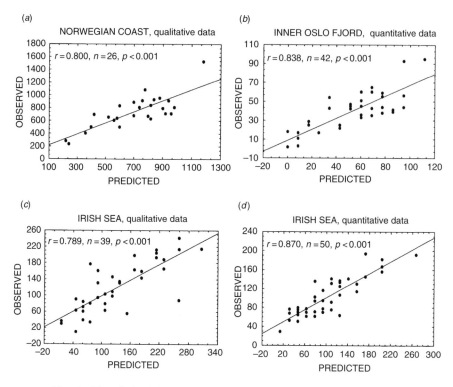

Fig. 4.4 The relationships between the total number of benthic species (annelids, molluscs, crustaceans and echinoderms) observed and expected to occur in different collecting sites. This is based on the proportion (mean ratio) of terebellidad polychaetes in the study areas. (From Olsgard *et al.* (2003) with permission of Frode Olsgard and Kluwer.)

Spartina was also given some assistance by a refinery-funded plant-transplanting programme to areas where natural colonization was slow.

Monitoring commenced with a twice-yearly ground-vegetation mapping technique that used a simple subjective scale of abundance of salt-marsh plant density (Dicks, 1976, 1989). At that time, the objective of the monitoring was to describe the damage to the salt-marsh community and to identify the cause of the damage. Following the improvements in effluent management from 1970 onwards, a broader monitoring programme was established to include littoral and benthic animals. The abundance categories for salt-marsh plants were also revised to include the following for each species: abundant (majority of plants less than 50 cm apart and often close to each other), common (individual plants 50–100 cm apart), rare (individual plants more than 100 cm apart).

In general terms, this monitoring, which largely involved twice-yearly ground vegetation mapping, was successful and produced correlations between the spatial distribution of the plants and proximity to effluent streams.

Fig. 4.5 Fawley salt marsh on the edge of Southampton Water. One view looks towards a refinery and the other towards an oil-fired electricity-generating station. (Photographs by B. Lockyer.)

Extending the surveys to animal groups was valuable particularly as some oligochaete worms were found to be very sensitive to effluent levels and thus good pollution indicators. The information has been used by the refinery in its management of effluents and has led not only to a decrease in effluent levels but

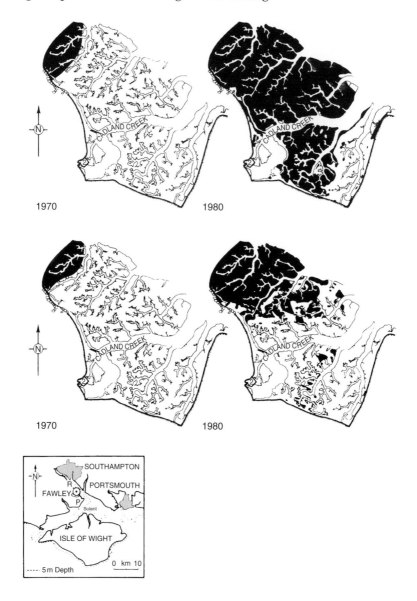

Fig. 4.6 Fawley salt marsh. Above, area of marsh (black) covered by *Salicornia* spp. in 1970 and 1980. Below, the same but for *Spartina* spp. R, refinery; P, electricity-generating station. (Redrawn from Dicks & Iball, 1981.)

also to a better understanding of ecological monitoring techniques necessary to assess the impacts of a large refinery.

However, these ground-vegetation surveys are time consuming and difficult to carry out. Shears (1989), therefore, explored the use of remote sensing techniques for monitoring these salt marshes. He used airborne thematic mapper remote sensing data in conjunction with field spectroscopy and plant surveys to monitor the effects of the refinery effluent on the salt marshes. The objective was to detect differences in the vegetation community and to map the distribution of the main salt-marsh plant species. The ground surveys were based on 10 parallel transects, seated at 100 m intervals, across the salt marsh. Two transects were in an area unaffected by effluents; two were in an area that had fully recovered and the remaining six were in areas still affected by the effluent. On each of the transects, 10 sample points were chosen at random for collection of ground spectral data and plant biomass. Ground spectra were collected to help to interpret the remote sensing imagery and were obtained using a NERC EPFS Spectron SES90 spectroradiometer. Plant biomass and species composition were also recorded from a $0.5\,m^2$ quadrat at each sampling point.

Computer image-processing techniques were used to classify the marsh. An unsupervised classification, CLUSTER, where the computer automatically divides the data into spectrally distinct classes was found to give the most accurate results when compared with the ground survey; the resulting images could then be used to distinguish between several ecological zones on the salt marsh. Although the previous ground surveys by Dicks and his colleagues had identified different areas at different stages of recovery, it was only later that Shears was able to show possible uses of airborne remote sensing and thematic mapping imagery to detect those parts of the salt marshes dominated by *Spartina* and *Salicornia* spp. and thus provide an effective ecological monitoring technique.

Remote sensing using both aircraft and satellites has become established as a very sophisticated science with many applications. For example, for large-scale applications of coastal monitoring, the satellite imagery from LANDSAT and SPOT gave advantages in terms of both costs and time. Remote sensing techniques and the use of GIS, which allow the merging of databases and mapping, have provided an even more sophisticated basis for monitoring the effects of impacts on coastal ecosystems. For example, Jensen *et al.* (1990), in their assessment of environmental sensitivity indices, suggested that although information on oil-sensitive taxa on maps of these indices can be useful, there is a limit to the amount of information that can be placed on a single map. Therefore, they suggested that information on the spatial distribution of oil-sensitive taxa was more effectively stored in a GIS database; individual files being created for each taxonomic group. A basic concept of creating databases using remote sensing and GIS technology is shown conceptually in Fig. 4.7.

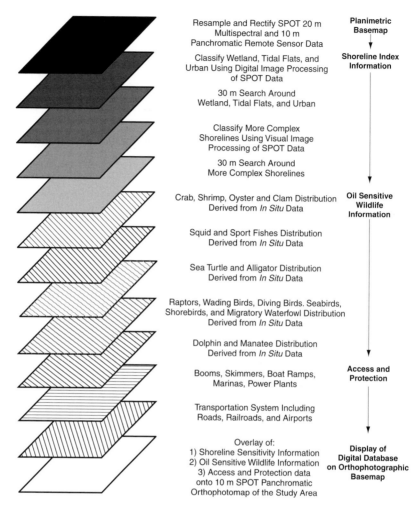

Resample and Rectify SPOT 20 m
Multispectral and 10 m
Panchromatic Remote Sensor Data

Planimetric Basemap

Classify Wetland, Tidal Flats, and
Urban Using Digital Image Processing
of SPOT Data

Shoreline Index Information

30 m Search Around
Wetland, Tidal Flats, and Urban

Classify More Complex
Shorelines Using Visual Image
Processing of SPOT Data

30 m Search Around
More Complex Shorelines

Crab, Shrimp, Oyster and Clam Distribution
Derived from *In Situ* Data

Oil Sensitive Wildlife Information

Squid and Sport Fishes Distribution
Derived from *In Situ* Data

Sea Turtle and Alligator Distribution
Derived from *In Situ* Data

Raptors, Wading Birds, Diving Birds. Seabirds,
Shorebirds, and Migratory Waterfowl Distribution
Derived from *In Situ* Data

Dolphin and Manatee Distribution
Derived from *In Situ* Data

Booms, Skimmers, Boat Ramps,
Marinas, Power Plants

Access and Protection

Transportation System Including
Roads, Railroads, and Airports

Overlay of:
1) Shoreline Sensitivity Information
2) Oil Sensitive Wildlife Information
3) Access and Protection data
onto 10 m SPOT Panchromatic
Orthophotomap of the Study Area

Display of Digital Database on Orthophotographic Basemap

Fig. 4.7 Conceptual arrangement for environmental sensitivity mapping with remote sensing and geographical information systems. This can be achieved if various kinds of information can be registered on a common planimetric base map and then interrogated. (From Jensen *et al.* (1990); original figure kindly provided by John R. Jensen.)

4.7   Monitoring pollution across countries

The acidification of inland waters throughout northern Europe and in Scandinavia has been well documented and is one example of monitoring at the scale of countries. Despite widespread public interest in 'acid rain' and its effects, we should perhaps remind ourselves that low-pH precipitation is not a new phenomenon. Palaeolimnology provides convincing evidence that acidification of the atmosphere and rain began in the early 1800s (Stevenson *et al.*, 2002). The

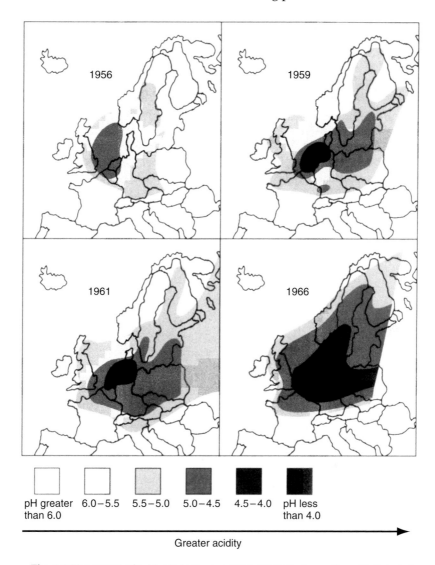

Fig. 4.8 The spread of acid rain in Europe 1956–1966. (Redrawn from Ehrlich *et al.* (1977) after Oden (1971).)

Industrial Revolution (mid eighteenth century) was a major contributor to low-pH precipitation. Acidity of precipitation has increased about four-fold in northeastern USA since 1900. Records of 'acid rain' in Europe have been available now for more than 40 years (Fig. 4.8). The broad effects of diminished pH levels of inland waters is now well known, usually leading to diminished fish size and species richness. Despite some very interesting retrospective monitoring of acidification of inland water (example on p. 123), much more research is required and an understanding of the biological processes involved can be advanced by the establishment of monitoring programmes and by way of experiments.

4.8   Spatial scales and ecological monitoring

It has long been recognized that inventory and monitoring of different levels of biological diversity need to be undertaken at different spatial scales. Scales range from small study plots through local to regional and global (e.g. Sandhu et al., 1998). In the *Global Biodiversity Assessment* report (Heywood, 1995), there are examples of the scales that could be applied to inventory and monitoring biological diversity (see Section 7, of that report). The scales include:

1.   Site or local level: mining and manufacturing industries that have environmental policies may use maps of specific areas to determine optimal locations of activities to avoid impacts on nature.
2.   Subnational level: government agencies may use information on protected areas.
3.   National level: ministries in many countries monitor population changes of economically important species.
4.   Regional and continental: agencies use information on species distribution to prioritize where to allocate conservation funding.
5.   Global level: international agencies use information from wild species to monitor compliance with treaties such as CITES.

Ecological monitoring is commonly undertaken at a local level in relation to specific local activities. However, there has been an increase in the number of ecological monitoring programmes at larger spatial scales, such as regionally and countrywide. Beyond that scale, there are examples of ecological monitoring at very large spatial scales covering oceans and entire continents. Several issues arise when considering how ecological monitoring can best be achieved at different spatial cases.

Spatial uncertainty. How is it best to deal with the errors involved in collection and analysis of spatial data?

Objectives. The objectives will almost certainly be different at different spatial scales.

From local to national. Can local data become part of nationwide data? Can the same variables and methods be used at local, regional and national levels?

Collaboration, communication and liaison issues. There may well be a need to ensure that ecological monitoring at different spatial scale is managed in such a way that the managers of the programmes are able to share information about standards and protocols.

Processes used. Different processes will be needed at different spatial scales.

There is uncertainty in all aspects of science including ecological science. There is spatial uncertainty in ecology at three levels (see Hunsaker *et al.*, 2001): uncertainty in the analysis of data, uncertainty in the ecological processes and, finally, uncertainty in the boundary conditions. In any ecological monitoring programme, such levels of uncertainty must be recognized and acknowledged at all levels.

Assessments of some environmental monitoring programmes have drawn attention to some of these issues. For example, in 1990 the US National Research Council published detailed assessments of marine environmental monitoring programmes (Bernstein & Weisberg, 2003) and the recommendations drew attention to the need for greater regionalization of efforts, greater standardization of methods and the need for centralized data management.

The fact that the objectives of ecological monitoring programmes may be different at different spatial scales does not mean that the different programmes cannot be related. Indeed, there would seem to be good reasons of efficiency to ensure that there were links between monitoring programmes at different scales.

The importance of scale problems in reporting landscape patterns and in reporting landscape change has often centred around the grain or the spatial resolution. In one report (O'Neil *et al.*, 1996), there is the suggestion that the grain should be at least two to five times smaller than the features of interest. Clearly, the dimension of the sample unit are important, and analysis of the effects of different spatial scales on changes over time has been the subject of much research since the 1980s when Wiens (1989) put forward an hypothesis to explain the influence of scale.

In one study, the semi-arid Texas grasslands have been used to assess the influence of spatial scale on temporal patterns (Fuhlendorf & Smeins, 1996) using data collected over a 45-year period. They found that many processes were contributing to variations at both scales and that different processes are important at each of the different scales. Large-scale sites were characterized by low variation between units and high variation within units. At the small scale (quadrat), they found that variation between units was high.

4.9   Monitoring land-use change

Changes in land use have been dominated by agriculture throughout the world and it is estimated that about 9 000 000 km^2 of the Earth's surface has been converted into permanent croplands. It is relatively easy to estimate, in broad terms, a global figure such as this; however, until recently, it has been difficult to determine accurately on a countrywide basis the extent and nature of change in a major land use such as agriculture. Some of these difficulties were initially overcome by the US Geological Survey with LANDSAT data coupled with aerial photography and ground surveys. As long ago as 1975, mapping and monitoring land use and land cover was undertaken at a scale of 1:25 000 with the aim of producing baseline data every few years.

In the UK, data on land use come from widely varied sources and historically there has been no one comprehensive source. Agriculture is the dominant land use in the UK and perhaps not surprisingly comprehensive records of agricultural land have long been provided by the detailed agricultural statistics produced by the government. All of these inventories and monitoring programmes have been based on different land-use classifications and different data storage and retrieval systems. Some have been map based, some simply recording total area. Therefore, it has until recently been difficult to develop comprehensive national land-use statistics or land-use maps, let alone comprehensive monitoring programmes.

In the UK, national land-use classification schemes have long been discussed and in the 1970s there were some published reports. Later, a feasibility study of methods and costs was commissioned and the National Land Use Stock Survey was published in 1985. Sample sites across the UK were surveyed at periodic intervals (1978, 1984, 1990) and then in 2000 a report was published (Haines-Young et al., 2000) to mark the launch of a very significant survey of the UK countryside, the Countryside Survey 2000. This was sponsored by several agencies: DEFRA, NERC, the National Assembly for Wales, the Scottish Executive, the Environment Agency, Scottish Natural Heritage and the Countryside Council for Wales.

The survey was based on both detailed field data (random samples of 1 km^2 grid squares) and satellite imagery. The Countryside Information System (CIS), a software package that was developed for analysis and presentation of data for the Countryside Survey 2000, continues to be developed further. There are versions that provide access to datasets based on 10 km^2 units in addition to the 1 km^2 units.

The Countryside Survey 2000 is an important milestone and the most comprehensive survey of the UK countryside and it will be a baseline for future environmental monitoring and land-use change. A *Land Cover Map 2000* was one of the major components of the Countryside Survey 2000. Several projects are now beginning to be coordinated and standardized. These include as CIS, a biological diversity network and a multi-agency geographic information system for the countryside. The Countryside Survey 2000 website has an HTML version of the report *Accounting for Nature*.

Monitoring the quality or the state of the landscape

Many environmental impacts occur at a large scale and much of environmental management is directed at large-scale changes. These impacts include changes in agricultural practices, afforestation, urbanization and drainage of wetlands. It has been argued, therefore, that there are very good reasons for monitoring environmental quality at the landscape scale. But what landscape variable can be used in such monitoring programmes? In a review of this topic, O'Neil *et al.* (1997) drew attention to a number of potential variables. They suggested, for example, that biological integrity could be assessed on the basis of:

- total change in land cover
- frequency distribution of patch size of natural vegetation
- spatial configuration of the patches
- frequency distribution of distances between patches
- loss of linear landscape features
- length of forest edges.

The quality of the landscapes, in general, could be monitored using any of these variables. Similarly, the state of watersheds could be monitored using the extent and change in natural vegetation and also the extent of changes along riparian zones. Landscape resilience could be monitored using the distances between patches. They argued that the resilience of landscapes has a threshold which can be identified by the extent of the number of patches of natural vegetation. Above a certain threshold, the landscape may be able to recover, but below that threshold, there may be unmanageable deterioration such as erosion and desertification (see also Box 5.1, p. 154).

In an interesting final note, O'Neil *et al.* (1997) suggested that the levels of efficiency and accessibility of routine monitoring of the weather might apply to monitoring the quality of the landscape. That suggestion is surely very timely and important. It seems strange that so much attention is given to the quality of the weather but not to the quality of the landscape. Perhaps in the not too distant future, we will see regular reports on the quality of the landscapes and, more importantly, such monitoring will be linked to environmental management.

Land classes and classification

Hierarchical classifications have commonly been incorporated in land classifications. One example of a hierarchical system developed by the US Geological Survey made use of two levels of data collection based on information from aerial photography and satellite imagery (Table 4.5). Level 1 had nine land classes of which five were biological: tundra, 'barren' land, wetland, forests and

Table 4.5 *The US Geological Survey land-use and land-cover classification for use with remote sensing*

Level I	Level II
1. Urban or built-up land	11. Residential
	12. Commercial and services
	13. Industrial
	14. Transportation, communication and utilities
	15. Industrial and commercial complexes
	16. Mixed urban or built-up land
	17. Other urban or built-up land
2. Agricultural land	21. Cropland and pasture
	22. Orchards, groves, vineyards, nurseries and ornamental horticultural areas
	23. Confined feeding operations
	24. Other agricultural land
3. Rangeland	31. Herbaceous rangeland
	32. Shrub–brushland, rangeland
	33. Mixed rangeland
4. Forest land	41. Deciduous forest land
	42. Evergreen forest land
	43. Mixed forest land
5. Water	51. Streams and canals
	52. Lakes
	53. Reservoirs
	54. Bays and estuaries
6. Wetland	61. Forested wetland
	62. Non-forested wetland
7. Barren land	71. Dry salt flats
	72. Beaches
	73. Sandy areas other than beaches
	74. Bare exposed rock
	75. Strip mines, quarries and gravel pits
	76. Transitional areas
	77. Mixed barren land
8. Tundra	81. Shrub and brush tundra
	82. Herbaceous tundra
	83. Bare-ground tundra
	84. Wet tundra
	85. Mixed tundra
9. Perennial snow or ice	91. Perennial snowfields
	92. Glaciers

rangelands. The nature of the data collected at each level in the hierarchy depends, in part, on the method of data recording. Data from satellites could, for example, identify areas of tundra (high-latitude or high-altitude treeless regions dominated by low shrubs, lichens, mosses and sedges); aerial photography could identify categories at the next level such as bare-ground tundra, wet tundra or herbaceous tundra. Ground surveys could extend the hierarchy to another level to identify species associations within wet tundra or herbaceous tundra.

Land classification schemes can be used as a basis for evaluating the potential of land for one or more uses or they can be used to tell us what is present. Evaluation is a form of classification and may be undertaken in relation to identification of the potential of land for many purposes such as mineral resources, agriculture or forestry, production of wood for fuel, or for settlement (e.g. Bailey, 2002). Classification schemes used to identify what is there might be directed at soil types, physiographic features, aquatic habitats or woodland types. Land classification has to be based on a suitable choice of land units or regions and the choice of those land units or regions will depend on the objectives of the land classification. For example, a hypothetical woodland could be classified as a mixed deciduous woodland or as a recreational area or both, depending on the woodland's attributes and use.

Data and monitoring land-use change

The purpose of land classification, whether it be for monitoring change or as a basis for assessment of an area's potential, will determine the methods used for data collection. Anyone looking for data on land use may be surprised at the difficulties in many countries in obtaining published accurate data. This difficulty exists despite the contrary impression given by widely quoted figures on land uses and the publication of various land-use statistics.

Land-use data collection, analysis and classification is a vast subject and here it is possible only to summarize the more important aspects as a background to descriptions of some land-use and landscape monitoring programmes.

The frequency of data collection will also dictate certain methods; that is, some land attributes are continuously changing and, therefore, need repeated monitoring whereas some attributes such as soil type are static and would need only occasional data collection.

It is also useful to distinguish between land surveys used to map spatial patterns and other surveys that are designed to measure extent of land use in each class or category. It is also necessary to distinguish between land mapping that is designed to provide information on what is there now and information on schemes designed to assess potential value or potential use of the land (capability maps and surveys).

Mapping of ecological variables, using a range of techniques, is popularly called ecological mapping but that is a simplification. Ecological mapping is indeed concerned with mapping of organisms, assemblages and habitats (whether it be from the ground, air or space) but it can also be concerned with mapping fungi on a forest floor or microbes in the soil. Ecological mapping became a sophisticated science in the Netherlands as long ago as 1970, where the identification and monitoring of areas for conservation and ecological importance were undertaken (Kalkhoven et al., 1976). Ecological mapping of coastal zones to enable the identification of ecologically sensitive areas was also one of the early developments. That information has been used for clean-up operations after oil spills so that areas in greatest need of protection were given priority.

Sources of data for monitoring land use and land cover can be either secondary (e.g. Ordnance Survey maps) or primary and collected specifically for the purposes of monitoring. Primary data for monitoring land-use and land-type change has previously been obtained from ground surveys, aerial photography and satellite imagery. Ground surveys or inspection of the land on foot potentially allow as much land to be examined in as much detail as time and facilities permit. Such systematic recording has considerable advantages in terms of accuracy but may be costly in terms of time and effort. Aerial photography has long been used to study changes in landscapes and land uses, and this method has been well researched.

In practice, aerial photography requires careful interpretation and by itself cannot provide detailed information about ecological attributes such as the interior structure of a woodland or the species composition of grassland. Nevertheless the technique is suitable for less-structurally complex vegetation such as coastal dune vegetation. Remote sensing techniques with satellite imagery have greatly improved the accuracy of recording and monitoring land-use change and have widened the application of landscape and land-use monitoring; nevertheless, the success of satellite imagery is dependent on acquisition of good ground reference data.

In most cases, the extent of land being classified and monitored would be so great that comprehensive coverage at regular intervals for monitoring would be impractical. In the absence of comprehensive coverage, there has to be a method of data collection by sampling units, of which there are three broad methods: area, line and point sampling. Line sampling simply uses points along a line between two locations as a basis for recording land uses and land cover, a technique used in surveys of area but not spatial patterns. Point and area sampling are similar in practice but points have a location and no area. In both methods, land use and land cover are recorded at a point or in a sample area.

The importance of giving some care to the choice of number of sampling units and the resolution of spatial sample units can be illustrated with reference to ecological features such as woodlands. There are many shapes and sizes of wood-lands scattered throughout the countryside, some as small as 1 km^2. Depending on the frequency and distance between points used for sampling, some wood-lands may fall between points or the proportion of land sampled could fail to record certain types of woodland in some localities. Similarly, the spatial distribu-tion of rare plant species or rare animals in a community may not be detected by the sampling methods adopted. Land-cover surveys may, therefore, require addi-tional and more detailed surveys in some parts of the region being considered.

Spatial distribution data have traditionally been presented in printed map form only, but this becomes cumbersome to handle especially when several categories of information (such as soil characteristics, water characteristics, current land uses, land capability) are being analysed. Recent advances in the sophistication of computers and GIS software has greatly influenced the way data is handled, stored, analysed and presented. Computerized databases are becoming a more and more important and integral part of biological monitor-ing and so too are computers that store spatial and descriptive data about objects and the relationship of those objects to other objects on the Earth's surface. Those computers and the computer programs are the basis of GIS and they provide a powerful and exciting way of analysing environmental impact assessments, of manipulating data and of answering questions such as what effect will a land-use change have on other land uses.

The use of GIS enables interaction of databases and information that tradi-tionally have been presented in map form. For example, land features such as biological communities or habitats can be portrayed in the form of digital terrain models. Additional information can be added to the model as required; for example, it is possible to follow the impact of planting a conifer plantation or constructing a reservoir. The flexibility and powerful capacity of GIS have brought about dramatic changes in the way land-use data are monitored and GIS has also become a great asset for ecologists.

The technique has now been available for over 30 years and applications continue to be seen in ecological monitoring. As long ago as 1972, the LANDSAT series of satellites with systematic and regular coverage of the Earth's surface played a role in providing a source of data for use in analysis of land-use mapping and for monitoring programmes. The two principal sensors, the multi-spectral scanner and the thematic mapper, were used widely for monitoring changes in medium- to large-scale features.

Later came SPOT (Systeme Probatoire d'Observation de la Terre; based on satellites placed in orbit in the mid 1980s) for use in monitoring land-use change. In the USA, both National Aeronautics and Space Administration (NASA) and the National Academy of Sciences used remote sensing with satellites on a global scale, particularly in connection with the International Geosphere–Biosphere Programme (IGBP). About this time, there were many applications of satellite technology for monitoring environmental changes, particularly broad-scale changes. Vujakovic's (1987) work, for example, used satellite imagery for studying the use of buffer zones in relation to conserving the ecological integrity of protected areas in Botswana. In this study, classification of semi-arid vegetation based on LANDSAT MSS Band 5 was achieved with relatively high accuracy for that time.

The use of GIS in nature conservation and monitoring has been well documented since the early 1990s. The application of GIS in ecological monitoring has ranged from continents to groups of small islands. For example in 1999, the Small Island Ecosystem Group was established by the US Geological Survey and the French National Scientific Research Centre to identify a network of small islands managed as nature reserves as potential sites for long-term ecological monitoring (Françoise Gourmelon, personal communication). The islands are located in protected areas of the Mediterranean, north Atlantic and Pacific Oceans. It is expected that this monitoring will contribute to the research on the effects of global climate change as well as having important applications for monitoring the effects of ecological management. The French component concerns islands in the Mediterrean and Atlantic where data are collected every six years.

Box 4.3   An assessment of satellite imagery as a tool for ecological monitoring: a case study from India

This study took place in the Western Ghats Hill range of peninsular India. The region has relatively low-lying hills and the natural vegetation (characterized by west coast tropical evergreen forest) has been extensively modified by human impacts. The purpose of the research was to:

- determine if it was possible to identify (in a landscape of a few tens of square kilometres) a small number of landscape element types or patches belonging to different types or distinctive sets of species
- determine whether or not these landscape elements could be identified with an acceptable low level of error on the basis of satellite imagery
- determine whether or not there are distinctive species assemblages within the landscape element types.

The research had to cope with contrasting spatial scales; from small patches of less than a square kilometre to a landscape scale. Judicious sampling was, therefore, necessary. It was found that landscape element types could be identified accurately on the basis of supervised classification (see below) and that these elements, coupled with satellite imagery, could be used to organize an ecological monitoring programme. The authors concluded that unsupervised classifications are not appropriate for ecological monitoring.

Supervised classification

A satellite image is a collection of data that provides information of the reflectance of various points on the ground at different wavelengths. In order to extract information from this image, the analyst attempts to sort these individual pixels with reflectance information into a finite number of classes (classification). There are two ways of doing this. Supervised classification is where the analyst controls the process and unsupervised is more computer automated. In supervised classification, the analyst identifies field locations for the major land cover categories. This information is used to 'train' the classifier and thus classify the image. Ground truthing tends to produce more reliable classifications.

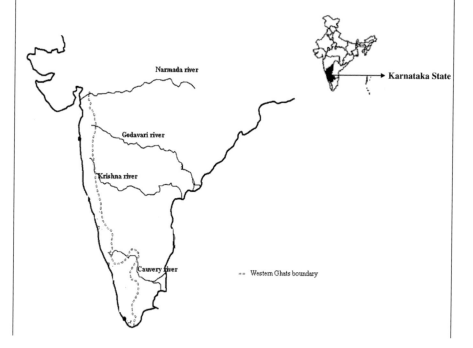

(a)

(b)

Examples of the landscape element types in the Karnataka region of the Western Ghats of India.

(With permission from Nagendra and Gadgil, 1999.)

Ecological monitoring and GIS have also been used for research on continental landscapes. In 1999, Nagendra and Gadgil reported an assessment of the use of satellite imagery as a tool for monitoring species diversity in the Karnataka region of the Western Ghats of India (Box 4.3). There, the landscape was mapped on the basis of seven types. This was on the basis of field identification as well as supervised and unsupervised classification of satellite imagery. They reported that unsupervised classification of satellite imagery did not permit classification of landscape types with a high enough level of accuracy.

Despite the applications of GIS at different spatial scales, it appears that GIS software may have limited programme facilities for large-scale projects. The Khoros Research Group at the University of New Mexico is one group currently addressing some of the problems associated with large-scale projects (see http://www.khoral.com/).

5

Biological indicators and indices

5.1   Introduction

The subject of indicators and indices (and difference between them) was introduced in Ch. 3 in the context of environmental reporting. The terms indicator and index are used here in the same way as they are used by Slocombe (1992). The purpose of this chapter is to provide examples of biological indicators (such as indicator species) and biological indices (such as water quality indices).

Biological indicators range from single organisms through populations and species to biological communities. Biological indices are derived from attributes of biological communities (such as abundance and sensitivity to pollution). Physical disturbance and changes in environmental variables such as temperature or salinity result in change in variables such as the species composition of the biotic community. Such community changes can, therefore, usefully be monitored to assess current states of the environment and also in a predictive sense in relation to environmental assessments (see Ch. 10).

The use of species indicators and community indices has grown rapidly, particularly in relation to monitoring water quality. There has also been much detailed research on the use of biological indicators for the detection of pollution and specific pollutants. Biological variables and indicator species used for monitoring of pollutants are many and varied, ranging from cells, tissues and organs to whole organisms (Table 5.1) including unicellular organisms, plants and animals.

5.2   Plant and animal biological indicators

Presence and absence

Earlier (p. 22) there was reference to Alister Hardy's description of plankton monitoring. In that same very readable book (Hardy, 1956) there is

Table 5.1 *Examples of some reports from the 'popular' and 'scientific' literature concerning the use of biological indicators*

Taxa	Reference
Microbial communities	McCormick *et al.* (1991)
'Lower' plants	Burton (1986)
Lichens	Skye (1979)
'Weeds'	Holzner (1982)
'Higher' plants	Manning & Feder (1980)
Coelenterates	Hanna & Muir (1990)
Aquatic invertebrates	Thomas *et al.* (1973)
Terrestrial invertebrates	Allred (1975)
Earthworms	Gooneratne *et al.* (1999)
Bees	Samiullah (1986)
Fish	Gruber Diamond (1988)
Snakes	Bauerle (1975)
Birds	Ratcliffe (1980)
Marine birds	Batty (1989)
Mammals (raccoons)	Gaines *et al.* (2002)
Mammals (deer)	Sawicka-Kapusta (1979)

an example of how the presence and absence of a plankton species can be an indicator of environmental condition. Hardy described the distribution of arrow worms (*Sagitta setosa* and *Saggita elegans*) and noted that coastal water can be distinguished from the more oceanic water by the presence of these species, the former being found in coastal water and the latter in oceanic water.

In general, the presence of every plant and animal (and its condition and behaviour) is potentially an indication of the conditions under which it is existing or existed previously (see p. 19 for comments on indicators of past conditions). Examples are the occurrence of Common Stinging Nettles (*Urtica dioica*), which is an indication of possible high levels of nitrogen in the soil, and the appearance of Rosebay Willow Herb (*Chamaenerion angustifolium*), which indicates that the soil may have been disturbed.

The presence or absence of patches of plants and/or bare soil can be used as indicators of the 'state of the health' of the landscape. Such variables have been used as indicators of rangeland health in North America (Whitford *et al.*, 1998). Similar variables have been used to try and achieve a balance between conservation and sustainable use of rangelands in Australia (Box 5.1).

Throughout history, different cultures have known that the presence of certain species, especially plant species, indicated certain conditions (or that certain conditions were required for growth of certain plants). For example West Africans have long recognized good soil by the presence of the Gau tree (*Acacia albida*), the Gaya Grass (*Andropogon gayanus*) and also the Roan Antelope (*Hippotragus equinus*). Plant indicators have had an interesting application in the location of over 70 different minerals. For example, the presence of a species of basil (*Ocimum homblei*) in Zimbabwe is an indication of a high copper content in the soil. The use of indicator plants has also been widely used in prospecting for gold. The presence of many species of animals can be used as indicators of certain environmental condition, such as high organic load. The insect group Ephemeroptera (mayflies), which have aquatic larvae, contains species that, with very few exceptions, are intolerant to organic enrichment, and so they have been incorporated into programmes monitoring water quality.

In practice, the use of the presence or absence of organisms as indicators of environmental conditions requires some caution. Indicator species should be tested under different spatial and temporal conditions. One method used to assess

---

Box 5.1   Measuring the 'state of the health' of a rangeland in Australia: achieving a balance between conservation and sustainable use

This case study from Australian rangeland sites uses remote-based techniques and indicators of landscape function. The 1992 Biological Diversity Convention refers to both conservation and sustainable use (and equitable use). The challenge lies in being able to ensure that nature is conserved but is also used in a sustainable manner. In the Australian rangelands, there have been attempts to ensure a balance between production and conservation. The balance can easily alter in either direction. What are needed are some indicators that can be used to support a robust and efficient ecological monitoring of the condition of various rangeland sites.

John Ludwig and colleagues have employed landscape function indicators together with ground and remote-based techniques. The aim has been to determine what to monitor and where. For example, they had to identify and monitor changes caused by grazing livestock and soil erosion. They also needed to identify where to monitor because it would be impractical to monitor all locations.

It has been shown in both the USA and in Australia that simple vegetation patch attributes and soil attributes can be used as indicators of the 'state of the health' of the landscape.

   Landscapes with a high cover of perennial plant patches capture water and nutrients in sediments. Thus the cover of vegetation patches or conversely the extent of bare soil can be used as indicators of landscape function.

   Ground-based transects combined with remote-based videography were used to monitor the patches. A theoretical sigmoid response curve was used to interpret where to measure patch indicators. Sites located near the inflection point will be sensitive to changes in the indicator being monitored.

Ground-based photograph of the landscape (left) and aerial-based videography (right) of a cattle watering point (a) and inside a cattle exclosure (b).

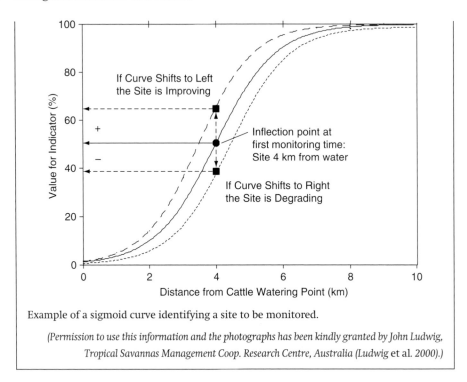

Example of a sigmoid curve identifying a site to be monitored.

(Permission to use this information and the photographs has been kindly granted by John Ludwig, Tropical Savannas Management Coop. Research Centre, Australia (Ludwig et al. 2000).)

or quantify the 'value' of a biological indicator is the indicator value, which was developed by Dufrene & Legendre (1997). This method is based on the specificity of an organism to the environment and also its fidelity. Thus those species with a high specificity and a high fidelity are considered to have a high indicator value. This approach was used by McGeoch *et al.* (2002) in their study of dung beetles as biological indicators. They found that sets of species were reliable indicators across the habitat. They concluded that the indicator value method is an effective tool for assessing biological indicators.

Biological indicators must also be treated with caution because of the variability inherent between populations. Genetic differentiation amongst some plant species, for example, can lead to some populations being commonly associated with alkaline soils whereas in other parts of the range of the species, populations may tolerate slightly acid conditions.

Behaviour and physiology

The behaviour and physiology of some animals have been useful in monitoring the quality of the environment. At a simple level, and an example familiar to us all, is the dawn chorus. A survey of those species of birds taking part in the dawn chorus at different locations in an urban area will soon show that fewer bird species take part in the chorus near disturbed and polluted areas: a phenomenon that may remind us of Carson's *Silent Spring* (1962; Table 2.1, p. 32). Still with birds, the behaviour of a caged canary succumbing to dangerous gases was used for many years by miners to detect methane; in that instance the canary was a biological early warning system (BEWS).

The behaviour and respiratory physiology of several aquatic organisms including fish have successfully been used to monitor water quality. For example, the respiratory and cardiac activity of various species of trout (*Salmo*) and other fish species have been used in sophisticated, automated BEWS. In 1988, Diamond and colleagues reviewed research and development of BEWS; although they supported the use of fish and other organisms to monitor water quality and to act as early warning systems, they suggested that the additional use of chemical monitoring for basic water-quality parameters could help to identify causes of changes in water quality. In other words, they supported the combined use of biological and chemical monitoring.

One particularly interesting fish species once proposed as a 'bio-sensor' was the Elephant-nosed Mormyrid (*Gnathonemus petersi*). This species navigates and communicates by emitting sequences of pulses from an electric organ on its tail. So sensitive and reliable is this behaviour that a research and development company (Aztec Environmental Control Ltd) once devised a sophisticated pollution-monitoring system based on the computer-controlled detection of the pulses emitted by these fish. However, not all individual fish will exhibit the same response and, therefore, the use of this fish as an indicator has to be based on the mean results from a group of fish.

Microevolution

One classic example of an indicator of the extent of pollution has been the spread of melanic forms of the Peppered Moth (*Biston betularia*) throughout polluted areas of the UK and Europe. Industrial melanism occurs when dark or melanic forms of an organism such as an insect become relatively more common than other forms, sometimes as a result of predation pressures (predators are less able to detect dark forms against dark or polluted surfaces). Particularly striking was the spread but then later decline of melanic forms after the Clean Air Act in England and Wales. This decline in melanism coincided with a period of increasing species richness of lichens on trees (Brakefield, 1990).

5.3   Biological community indicators

Populations of animals and plants occur in communities and, therefore, the species indicator concept can be extended to communities of indicator species. Different soils (for example serpentine soils, chalk soils and acid soils) all support indicator plant communities. The characteristic flora of serpentine soils, which are low in calcium and high in magnesium, is a good example of a plant indicator community. In North American serpentine soils, the vegetation is usually sparse and stunted and includes narrowly endemic species (e.g. *Quercus durata*, *Ceanothus jepsoni*, *Garrya congdoni* and *Cupressus sargentii*), which make a striking contrast with vegetation on other nearby soils. Another example is heathlands, which are found on oligotrophic, acid soils; the indicator heathland plants are the low-growing, dwarf ericoid shrubs. Use of plant associations as an indicator of soil chemical composition and even depth of ground water was undertaken in the USSR in the 1960s (Khudyakov, 1965) where water of 'best quality' was discovered under associations with dominance of *Alopecurus pratensis*, *Agropyron pectiniforme* and *Stipa capillata*.

The ecological characteristics of ground vegetation has long been used as an indicator of soil characteristics. This has led to several methods being developed, particularly indicator scales. For example the Ellenberg system has a list of indicator values (e.g. Hill *et al.*, 2000; Dzwonko, 2001) that can be used to monitor ecological and environmental change. There has been considerable research to show that the values are correlated with measurable environmental variables. The Ellenberg indicator values have been used and assessed throughout Europe. One study (Dzwonko, 2001) showed that the mean Ellenberg indicator values were good predictors of the environmental variables in ancient woodlands (Carpathian foothills) but poorer indicators in the recent woodlands.

Hill *et al.* (2000) have questioned how to extend the applications of the Ellenberg indicators to other geographical areas. They concluded that an advantage of the Ellenberg system is that it is universal and, therefore, can be used as a basis for making comparisons of different communities on scales that are ecologically meaningful.

5.4   Indicators of pollution

It has long been known that heavy metals and organochlorides penetrate ecosystems and, as a result, some organisms will accumulate pollutants in varying amounts. For example, in England where there were very polluted parts of the river Thames, the mollusc *Anodonta* spp. was once found to have 20 times the level of cadmium compared with the same species from the river Test near Southampton Water (Leatherland & Burton, 1974). Although bioaccumulation occurs in a wide variety of taxonomic groups, it does not necessarily follow that the source of pollution is near those organisms in which the pollutants have accumulated. An extreme but worrying example was the discovery of DDE (a derivative of DDT) in bodies of penguins in Antarctica, thousands of miles from any use of agrochemicals. Nevertheless, a wide range of species have been used as pollution indicator species (Table 5.1) and many more species, especially marine species, have recently been tested for their use as indicator species.

A recent departure from the conventional concept of species or species assemblages as indicators of pollution is the thought that DNA of an entire biological community could be used. Following the successful decoding of the human genome, some scientists are now looking not only to other species but also to biological communities. Is it possible that genetic changes in populations could ever be used as early indicators of pollution?

There are many uses of the term indicator species and this varied use provides a basis for distinguishing between the types of indicators.

- *Sentinels.* Sensitive organisms can be introduced into the environment, for example, as early warning devices (canaries in coal mines) or to delimit the effect of an effluent.
- *Detectors.* Some species occur naturally in the area of interest and may show a measurable response to environmental change (e.g. changes in behaviour, mortality, age–class structure, etc.).
- *Exploiters.* The presence of some species indicates the probability of disturbance or pollution. They are often abundant in polluted areas because of lack of competition from eliminated species.
- *Accumulators.* Some organisms take up and accumulate chemicals in measurable quantities.
- *Bioassay organisms.* Selected organisms can be used as laboratory reagents to detect the presence and/or concentration of pollutants, or to rank pollutants in order of toxicity.

This book is mainly concerned with changes in ecological systems and methods of detecting and monitoring changes rather than the use of organisms to detect levels and extent of pollution. The latter topic, which makes use of accumulator types of indicator, is verging on another large area, environmental toxicology and ecotoxicology. In this section, we consider mainly the detector type of indicator with some reference to the exploiter and accumulator type.

Detectors and exploiters

Desirable properties of indicator species for use as detectors and exploiters in connection with monitoring pollution include:

- the organism should have narrow tolerances to environmental variables, for example stenothermal, stenohaline as opposed to eurythermal and euryhaline
- the organism should be sedentary or have a limited dispersal
- the organism should be easy to sample and, therefore, presumably common would be an advantage
- accumulation of pollutants should occur without killing the organism (that is unless mortality is used as the variable)
- preferably the organism should be long lived so that different age classes can be sampled.

As plants are sedentary and many are easy to sample, it is not surprising that they have had wide use as detector and exploiter types of indicator. Responses of plant indicator species may take the form of changes in distribution or of tissue damage. Symptoms of chronic injury include premature senescence and bronzing or chlorosis. However, results should be interpreted carefully because some injuries thought to be caused by pollutants may have been caused by disease, insects or environmental stress such as drought.

Algae have commonly been used as indicators of pollution. One study (Stevenson *et al.*, 2002) assessed the role of algae as early warning indicators of ecological damage in the Everglades in the USA. Experimental research has shown cause and effect relationships between phosphate enrichment and a decrease in calcareous algae assemblages. The research by Stevenson *et al.* (2002) showed that calcareous algae assemblages are less abundant in regions with high phosphate concentrations in the water or sediments. They concluded that the challenge of ecological assessment of wetlands may be reduced by using biological indicators (and the sound scientific approach found in the protocols of risk assessment).

Lichens and mosses are long-lived organisms and their sensitivity to airborne pollutants is well known. In the case of lichens, pollutants such as SO_2 affect the algae component of the lichen and consequently the symbiotic relationship between alga and fungus breaks down. The use of lichens as bioindicators has been so well documented over the last 130 years that there are now many 'easy to follow guides' for use of lichens in monitoring air pollution. In addition to use as indicators, lichens have ben used in the development of an index of atmospheric purity (p. 170).

It is the sensitivity of lichens as well as their long lifespan that makes them useful as indicator species, especially for SO_2 levels. For example, in the 1970s, Hawksworth & Rose (1976) developed a method of estimating mean winter levels of atmospheric SO_2 based on the presence of certain indicator lichen taxa. Using a scale of 0 to 10 (highest and lowest concentrations of SO_2), it was

Table 5.2 *Increase in lichen abundance in northwest London based on a survey of 29 sites; all species were previously extinct or very rare*

Lichen species	Sites now present	Percentage of 29 sites	Mean (range) of abundance[a]
Evernia prunastri	8	27.6	1.6 (1–4)
Hypogymnia physodes	16	55.2	1.8 (1–3)
Parmelia caperata	1	3.4	1
Parmelia subaurifera	7	24.1	1.7 (1–4)
Parmelia sulcata	13	44.8	1.8 (1–5)
Usnea subfloridana	2	6.9	2 and 4

[a] Abundance is scored on a 1–5 scale with 5 maximum.
From Rose & Hawksworth, 1981.

possible to prepare maps with zones of pollution based on the lichen species composition in each zone. One limitation of lichens as indicator species is that they are rather slow to respond to changes in levels of SO_2, that is the effects take place over years rather than in weeks or days.

In North America and in Europe, the use of lichens as detector, exploiter and also accumulator types of indicator of air pollution has been extensively described. Although many questions remain unanswered about the physiological basis of effects of SO_2 on lichens and there may be confounding factors such as microclimate, droughts and the buffering effects caused by the substrata, lichens have been used successfully to monitor air pollution either as an accumulator indicator (see the report of work by Pilegaard (1978) below) or as a detector indicator. There are many studies that describe changes in lichen ecology as a result of increasing acidity of the substratum and, conversely, an increase in abundance of lichens with decreasing pollutants. Rose & Hawksworth (1981), for instance, showed that several species of lichen that were extinct or very rare in Greater London around 1970 had later extended their ranges considerably and had become more abundant (Table 5.2). From about 1962 onwards, the levels of SO_2 recorded at six London recording stations decreased and this change in level of pollutants may have been the main reason for the lichen recolonization.

Changes in the epiphytic cover on trees, with special note being made of the lichens, was used by Cook *et al.* (1990) in a study of melanic frequencies in the Peppered Moth. Indeed, the method used (a method previously developed by Lees *et al.* (1973)), and their method of data collection could usefully be part of a monitoring programme at a long-term ecological monitoring site or permanent vegetation plots. In their study, epiphytic cover at 1.5 m from ground level on oaks was scored as a fraction of the circumference occupied by bare bark, pleuro-coccoid algae, crustose lichens, foliose lichens and bryophytes. The height of 1.5 m is a fairly standard height and is used in other studies such as population studies of trees, where diameter of the tree at 1.5 m would be recorded. High species richness and cover by lichens was found at some sites well away from sources of pollution, but as might be expected there was a lack of uniformity in the results.

There have been some applications of plant indicator species in the detection of ancient woodlands, classification of woodlands and in the aging of hedges. Prior to recording the distribution and status of ancient woodlands and subsequent surveillance, there has to be a process of woodland classification and ancient woodland identification. The different rates at which various flowering plants and ferns are able to colonize new woodlands is the basis of ancient woodland indicators (Bunce, 1982) and it is possible, for example, to identify a number of woodland vascular plant species that are strongly associated with ancient woodland. Lists of typical indicator species of ancient woodlands and recent secondary woodlands have now been established in surveys that, when completed, could provide baseline information for future monitoring programmes. One example of part of a list of woodland vascular plants associated with ancient woodland in the south of England is shown in Table 5.3.

Historical recording of hedgerows became popular in the UK following Hooper's (1970) publication of a method for dating hedgerows. That method has since contributed to some landscape assessments and surveillance of those landscapes. Hooper's method was very simple and was based on the idea that a 27.4 m (30 yard) length of hedge contains approximately one woody species for each 100 years of its existence (Pollard et al., 1974). Willmot's (1980) thorough research on dating hedges using woody species showed that less of the variation in number of woody species in hedges was caused by the age of the hedge than had previously been implied. Previous management of the hedge, its location (whether next to a road or between fields) and other variables could all contribute to the number of woody species found. In addition, the fact that some hedges are remnants of woodlands or have been part of a woodland would affect the number of woody species. Consequently, Willmot (1980) concluded that the number of woody species as an indicator of hedge age should only be used for dating groups of hedges when a local relationship has been established empirically between age and number of species.

The use of whole live animals as detector indicators of air pollution is rare and this is perhaps because there are few suitable sedentary terrestrial species. One interesting suggestion for monitoring and mapping air pollution involved the use of an oribatid mite, common in some orchards (Andre et al., 1982).

Table 5.3 *Ancient woodland vascular plants in woodland in the south of England: the following 36 species (from a list of 100 species) are thought to be most indicative of a long continuity and are, therefore, good indicators of an ancient woodland*

Species	Common name
Adoxa moschatellina	Moschatel, Townhall Clock
Allium ursinum	Wild Garlic
Anemone nemorosa	Wood Anemone
Blechnum spicant	Hard-fern
Carex laevigata	Smooth Sedge
Carex strigosa	Loose-spiked Wood Sedge
Convallaria majalis	Lily-of-the-valley
Daphne laureola	Spurge Laurel
Dryopteris carthusiana	Narrow Buckler-fern
Dryopteris pseudomas	–
Epipactis purpurata	Violet Helloborine
Euphorbia amygdaloides	Wood Spurge
Galium adoratum	Sweet Woodruff
Helleborus viridis	Green Hellebore
Hordelymus europeaus	Wood Barley
Hypericum androsaemum	Tutsan
Lathrea squamaria	Toothwort
Luzula forsteri	Forster's Woodrush
Luzula sylvatica	Greater Woodrush
Melampyrum pratense	Common Cow-wheat
Milium effusum	Wood Millet
Oxalis acetosella	Wood Sorrel
Paris quadrifolia	Herb Paris
Platanthera chlorantha	Greater Butterfly Orchid
Polygonatum multiflorum	Solomon's Seal
Polystichum setiferum	Soft Shield-fern
Populus tremula	Aspen
Quercus petraea	Sessile Oak
Ranunculus auricomus	Goldilocks
Sanicula europaea	Sanicle
Sorbus torminalis	Wild Service Tree
Thelypteris oreopteris	Mountain Fern
Ulmus glabra	Wych Elm
Vaccinium myrtillus	Bilberry
Veronica montana	Wood Speedwell
Viola reichenbachiana	Pale Food Violet

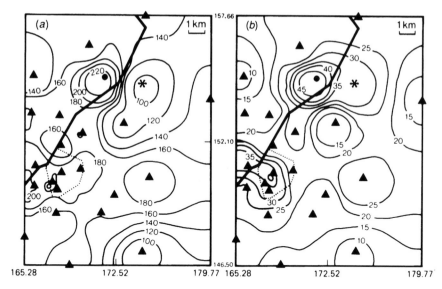

Fig. 5.1 The use of the oribatid mite *Humerobates rostrolamellatus* as a detection indicator of SO$_2$ pollution. (a) The SO$_2$ mean contour level map (μg/m^3). (b) Percentage mortality contour map for the mite. The heavy line represents the Willebroek Canal and the dotted line represents the pentagonal inner loop of Brussels. The stations are indicated by triangles except sites 7 and 8, which are, respectively, represented by a circle and an asterisk. Lambert coordinates (i.e. true directions at each point) are specified in the margin. (Redrawn from Andre *et al.*, 1982.)

The environmental biology of the mite *Humerobates rostrolamellatus* had been well studied and this species is known to be sensitive to SO$_2$. The response time of this species to air pollution was shorter than that for lichens and the doubts about lichens because of confounding factors such as drought do not apply to these mites. The method used by Andre *et al.* (1982) was simple; adult mites collected from orchards in the country were placed in vials that were then left at 24 pollution-monitoring sites in Brussels. The mites were retrieved after a few weeks of exposure and the mortality was then used to construct the map in Fig. 5.1. Two centres of high mortality can be seen, one in the centre of Brussels and one near an industrial park on the Willebroek Canal. The use of this mite as a bioindicator seemed quite convincing although further research such as on synergistic effects needs to be undertaken.

The use of detector and exploiter indicators for monitoring water quality has been described for many years but in very early monitoring studies it was realized that single invertebrate species have limited use as indicators. For example, although the absence of certain species from formerly clean water suggests pollution, single species of organisms such as *Tubifex tubifex* or *Chironomus tentans* cannot be used as indicators of pollution unless relative abundance is recorded. Although the move away from reliance on single species to communities began many decades ago, ecological studies supporting community biological indicators are relatively recent.

It is possible to undertake biological and ecological monitoring in retrospect (see p. 19). In the 1980s, one study used diatoms to indicate previous acidification of two Scottish lochs. This work was described by Flower & Battarbee (1983). There is now a growing number of long-term ecological studies of acidification of lakes and it was of interest, therefore, that in the 1980s, those authors were able to use diatom analysis of core samples to show declines in pH during the past 130 years.

It is the change in the relative abundance of various species that form the basis of community indicators. In the 1970s for example, Patrick (1972), in her account of aquatic communities as indicators of pollution in North American natural streams, described how species richness at different trophic levels in many streams remained fairly similar throughout the year but species composition changed greatly. Therefore, in natural streams, there is no dominance by a single species: these circumstances change when pollution occurs and the change follows a pattern not unexpected, with exploiter type indicators becoming the dominant species. The first effects of an increase in organic load are to cause some species such as diatoms to become more common. The species composition of protozoans changes from well balanced to almost complete dominance by ciliates. Similarly, there is a shift in the species composition of herbivores and carnivores and some of these are represented by large populations. It is these kinds of change reflected in the indicator species that have been particularly useful in monitoring water pollution by way of community indicators.

Accumulators

The effects of air pollutants on plants have been well researched. Lichens have been used both as detector indicators and as accumulator indicators. Pilegaard's (1978) research in an industrial area of Denmark was a good example of how lichens have been used as accumulator indicators to monitor airborne metals and SO_2. Using samples of the epiphytic lichen *Lecanora conizaeoides*, which is tolerant of SO_2, it was possible to show that concentrations of all nine heavy metals found in *L. conizaeoides* have step gradients, with highest values nearest to the industrial area.

Hair, shells, bones and internal organs such as liver, kidneys and muscle from a wide range of animal indicator species have been used in biological monitoring of environmental contaminants.

The use of deer antlers is one example of animal material being used as a bioindicator of environmental pollution. That deer antlers can be used

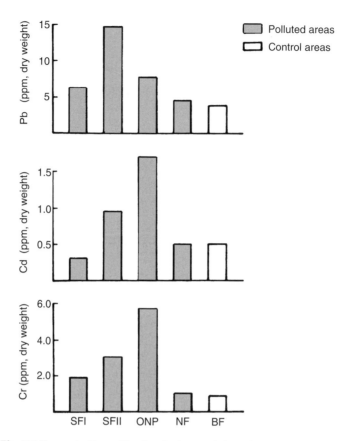

Fig. 5.2 Concentrations of lead, cadmium and chromium in Roe Deer antlers. SFI, Silesian forests 1938–1950; SFII, Silesian forests 1951–1973; ONP, Ojcow National Park; NF, Niepolomice Forest; BF, Bialowieza Forest. (Redrawn from Sawicka-Kapusta, 1979.)

successfully as bioindicators of environmental pollution was confirmed in a study of Roe Deer in Polish forest by Sawicka-Kapusta (1979). For example, data for three metals from five groups of animals, one of which was a control, show very clearly that industrial pollutants do accumulate in these deer antlers (Fig. 5.2). There seemed no doubt that deer antlers could be used as indicators of environmental pollution especially as the antlers represent about 130 days of growth and can be sampled annually without harm to the animals.

There are many collections of deer antlers throughout the world but without any baseline research those antlers could not reliably be used for monitoring environmental pollutants. However, research by Jones & Samiullah (1985) established that samples taken from deer of known ages can be used to monitor physiological and environmental metal concentrations despite possible variations in the concentrations of metals in different parts of the antlers. That research has provided a baseline for later studies involving the use of antlers in collections representing a timescale of several hundred years.

Although many animal species have been used as accumulator indicators for monitoring levels of pollution in water, there are a number of advantages and disadvantages compared with chemical monitoring. First, assuming careful selection of the indicator organism for its known ability to accumulate particular pollutants under certain conditions, one advantage is that higher accumulations of the pollutant in the organism compared with concentration in the environment facilitate easier and cheaper analysis. Second, the temporal variation in the level of chemical pollutants in water can be overcome by the use of indicator organisms, which allow a time-averaged index of pollutant availability. Third, the use of indicator organisms enables availability of pollutants to be measured directly, thus avoiding the need to try to detect pollution levels from speciation of metals (occurrence of a metal in different forms). The disadvantage of using accumulator indicator species in the detection of pollutants arises from the fact that a number of biotic and abiotic variables may affect the rate at which the pollutant is accumulated. Clearly both laboratory and field tests need to be undertaken so that the effects, if any, of extraneous parameters can be identified. The extraneous parameters can be grouped under three general headings. First, abiotic parameters such as water temperature levels, salinity levels, pH and other parameters may obscure or alter the availability of pollutants, either directly or via the behaviour of the organism. Second, biotic factors such as age, size, sex, stage in the sexual cycle, growth and diet, body lipids and behaviour may all affect the rate of uptake and rate of accumulation of metals or organochlorines. Third, there may be synergistic effects resulting from the presence of more than one pollutant.

5.5   Status of biological indicators in monitoring programmes

There has been much research on the nature of biological indicators but a disappointing amount of research on the role of biological indicators in pollution-monitoring programmes. The reason for this can be attributed mainly to the apparently cheaper methods of pollution detection made possible by sophisticated technology. Machines may be more reliable than biological organisms and it has to be admitted that care has to be taken when it comes to interpreting the physiology, behaviour and ecology of biological indicators. Cause and effect relations are never easy to confirm without good research.

Nevertheless, it is one thing to detect levels of pollutants and another thing to monitor the effects on organisms and ecosystems. All too often the effects of pollution on ecosystems have been obvious and there has been much time and effort spent on trying to establish the cause. There are many biological indicators that can be used successfully as effective warning systems and as cheap and reliable components of long-term pollution-monitoring programmes. To this end, much use has been made of establishing baseline information on the conditions of rivers and then using information from aquatic communities for ecological monitoring. For example, there is the River Invertebrate Prediction and Classification System (RIVPACS). This was a software package developed in the UK to assess the quality or condition of rivers (Wright *et al.*, 2000). The overall approach is to compare the observed fauna (aquatic macroinvertebrates) with the expected or 'target' fauna. The difference between the observed and expected fauna can then be used as the basis of an index of quality of river sites. Use of RIVPACS or similar programmes now occurs in Australia (Wells *et al.*, 2002) and other countries. (A general introduction to RIVPACS is given on p. 311.)

The Environmental Monitoring and Assessment Programme and its indicators

The EMAP is a long-term monitoring programme which was established by the EPA to determine the condition and trends in the environment. Critical to this programme was the process leading to the selection of indicators. A conceptual outline for the indicators is shown in Fig. 5.3. The response indicators are the primary measurement endpoints for EMAP. These quantify the overall conditions of ecosystems. The exposure indicators are measures of exposure to pathogens, acidity, GMOs, etc. The habitat indicators refer to the conditions of habitats necessary to support biological populations. Stressor indicators include pollutant, management and natural-process indicators.The stressor indicators reflect activities or perturbations that result in change in habitat condition.

The selection of EMAP indicators was based on 11 criteria, not all of which were ecological (Hunsaker *et al.*, 1990):

- they correlate with changes in processes
- they are appropriate for regional monitoring
- they can integrate effects over time and space
- they are unambiguous
- they can be quantified
- they can be related to the overall structure and function of ecosystems
- they are responsive to stressors of concern or to management strategies
- they are measured in a standardized manner
- they should have a low measurement error
- there should be an historical database
- they should be low cost.

Indicators identified for the EMAP for lakes and streams have been grouped into the three broad categories of response indicators, exposure and habitat

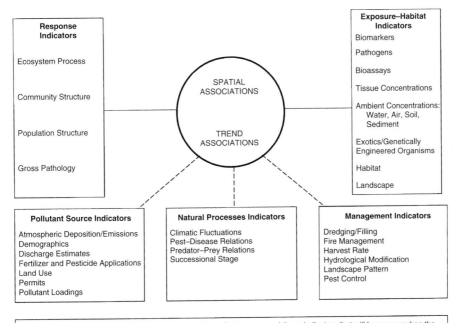

Diagram of EMAP conceptual indicator strategy showing indicator types and those indicators that will be measured on the EMAP sampling frame. The circle indicates that analyses is by statistical association.

Fig. 5.3 A conceptual outline for the indicators used in the US Environmental Monitoring and Assessment Programme, showing the types of indicators used. The circle indicates that analysis is by statistical association. (Adapted from Hunsaker *et al.*, 1990.)

indicators and stressor indicators (Hughes *et al.*, 1992). The response indicators come from measurements that describe the biological conditions of organisms, populations, biological communities and ecosystems. Examples of response indicators include fish assemblages, trophic state and zooplankton assemblages.

Exposure and habitat indicators are used to determine the natural range of ecological conditions. They include water quality, physical habitat structure and chemical contaminants in fish.

The stressor indicators describe characteristics or processes that are the sources of pollution or perturbations. They may include natural processes such as drought and floods but they also include pollutant loadings and introduced species.

5.6   Biological indices

Biological indices are single numbers that summarize the characteristics of biological communities. These indices can be used as a simple way of reporting environmental quality. Monitoring programmes can generate very large amounts of data, so large that there may be considerable demands on laboratory time for sorting of the material, let alone analysis of the data.

Biological monitoring programmes are commonly part of an interdisciplinary management programme and a further problem may arise in that it is sometimes difficult to summarize the results in a manner that is understandable to non-biologists. Sometimes, indices are based on relatively complicated calculations and the resulting index may seem distant from the biological attributes of the community being examined. Summarizing data and communication issues may, however, be addressed by using one of several biotic indices that are based on biological attributes such as sensitivity to certain kinds of pollution.

Lichens as indicators and an index of atmospheric purity

Lichens can be used as indicators of pollution. The presence or absence of certain species and the species assemblage can tell us something about the quality of the air. One index of atmospheric purity is based on lichens. A common example is given by:

$$IAP = \sum_{n}^{1} \frac{(Q \times f)}{10}$$

where IAP is the index of atmospheric purity, n is the number of species at the site, f is the frequency (cover) of the species and Q is the mean number of other species growing with the species in the area.

The IAP method has been used in many studies of lichens and there are many variations of the method (Herzig et al., 1989).

The MITRE report (see p. 101) suggested two types of wildlife index, one for endangered species (those species that are approaching extinction) and one for 'troubled' species (species that have been greatly reduced in numbers but some populations still thrive). The calculation of these wildlife indices took into account a subjective weighting of the position of the species in a food chain, with endangered carnivores scoring 5 and rodents 1. This was because, it was argued at the time, changes in the groups higher in the food chain indicate effects that have occurred lower down.

Many other wildlife indices have been developed and one old example is the threat number (Perring & Farrell, 1983), which was devised to assess the conservation status of endangered flora in the UK. Although not specifically devised for monitoring the status of plants, threat numbers could be used with other data in monitoring the extent of degradation in natural and semi-natural areas. Wildlife indices and species evaluations in general (Spellerberg, 1992) have an important role in assessing the status of a species and focusing attention on the conservation needs of those species.

5.7   Indices of water quality

A few examples of the wide range of indices used in assessing and monitoring water quality are described here. Other examples are given in Ch. 9. The examples described here represent almost 100 years of research and they have been developed in several countries.

The saprobian system

Developed in continental Europe in the early 1900s, the saprobian system or saprobian index was developed for assessing the degree of organic pollution in water. It was received with wide approval but the index had its critics. Saprobes are organisms of decomposition and decay and the saprobian system was based on a series of such groups, each of which is associated with different stages of oxidation in organically enriched water. The distribution of aquatic organisms is determined by many factors but in particular by the level of organic matter in the water and by the level of dissolved oxygen. The classification expressed the degree of independence (of the groups) of decomposing organic nutrients. That is, saprobity is the state of the water quality with respect to the content of decaying organic material, but it probably reflects oxygen tolerances as well, especially in the mesosaprobic zone (Fig. 5.4). The saprobic zones and the saprobic index are calculated by first allocating groups of organisms to each of the zones. The index can be calculated by several equations; one example is Pantle and Buck's index:

$$\frac{\sum(h_3 S_i)}{\sum h} = \frac{38}{21}$$

Where S_i is the individual saprobic index for each species and h is the relative abundance according to a scale of estimation (1 rare, 3 frequent and 5 abundant). In the calculation, each species is given a number according to the group to which it belongs; lowest value for a xenosaprobic species then the next highest value for an oligosaprobic species and so on. An example of the calculation for Pantle and Buck's index is shown in Table 5.4, in which oligosaprobic organisms are rated 1 and polysaprobic organisms 4. Pantle and Buck's index gives a value of 1.0 to 1.5 for slight organic pollution and 3.5 to 4.0 for heavy pollution.

In Europe, the oligosaprobic zone (clean, unpolluted water) is characterized by organisms such as mayfly larvae, mosses such as *Fontinalis antipyretica* and

Table 5.4 *Calculation of Pantle and Buck's index*

Species	Saprobic rating	S_i	Abundance (No.)	h	hS_i
Tubifex tubifex	p	4	2	1	4
Dendrocoelum lacteum	β	2	5	3	6
Asellus aquaticus	α	3	10	3	9
Gammarus pulex	o/β	1 (or 2)	30	5	5
Baetis rhodani	o/β	1 (or 2)	20	5	5
Erpobdella octoculata	α	3	2	1	3
Limnaea stagnalis	β	2	10	3	6
Totals				21	38

p, polysaprobity; β, Beta-saprobity; α, alpha-saprobity; o, oligosaprobity. Inserting the values of 21 for n and 38 for h_3S_i in the Pantle and Buck's equation gives an index value of 1.81. See text and Fig. 5.4. From Hellawell (1978).

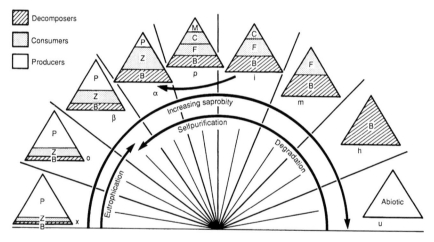

Fig. 5.4 Structure of a saprobic community. x, xenosaprobity; o, oligosaprobity; β, beta-mesosaprobity; α, alpha-mesosaprobity; p, polysaprobity; i, isosaprobity; m, metasaprobity; h, hypersaprobity; u, ultrasaprobity; B, bacterial; F, colourless flagellates; C, ciliates; M, mixotrophic algae and flagellates; Z, zooplankton and other consumers; P, phytoplankton and other producers. (Redrawn from Sladecek (1979) with permission of John Wiley & Sons Ltd.)

some flat worms such *Planaria gonocephala*. In the α-mesosaprobic zone (water some distance from a pollution source that is poorly oxygenated), characteristic organisms include exploiter indicator species such as chironomids, *Tubifex* spp. and larvae of some insects such as Alder Fly (*Sialis lutaria*).

The saprobic method has been criticized for various reasons, mainly because the classification of organisms is not absolute and because some organisms may be in one or more zones. Some people have suggested that products of pollution are many and varied and are both physical and chemical. A given organism might be intolerant to some kinds of pollution but not others or may be less sensitive to some forms of pollution. This casts some doubt on the usefulness of some biotic indices.

Whereas the saprobic system can be applied to all freshwater systems and all organisms, many biotic indices have previously been restricted in their use to running water and aquatic macro-invertebrates. There was much debate, therefore, about the uses of the so-called 'British biotic indices' (and other indices based on these) and the saprobic indices. In particular, some scientists believed that the biotic indices had limited use because of their use in running water. For example, Sladecek (1973), who gave strong support to the saprobic system at that time, described the limitations of the biotic indices and gave a fairly convincing demonstration that there is no substantial difference amongst the procedures and results obtained from either the biotic indices or the saprobian system. More recently, however, there have been adaptations of these biotic indices so that they can be used for aquatic systems other than running water.

The Neville Williams index

There have been many biological studies of polluted rivers and lakes and, therefore, the effects of organic pollution in particular have been well documented for many years. The early studies led to the development of a number of indices based largely on the effects of organic pollution on species composition. For example, the Neville Williams index was based on a subjective classification of taxonomic groups of invertebrates into those that are tolerant, intolerant and indifferent to pollution.

This was an index that received wide use in the earlier attempts to monitor water quality. For aquatic communities, tolerant groups would include leeches, some molluscs and some oligochaetes. Intolerant groups include Ephemeroptera, Trichoptera and Plecoptera. A similar, simple dichotomy could be assembled for other groups such as lichens (in relation to airborne pollution), grasses (in relation to trampling) and birds (in relation to vegetation structure and physical disturbance). In New Zealand, in the 1970s, Dacre & Scott (1973) used the numerical ratio of Ephemeroptera to Oligochaeta as a basis for an index. However, such an index is probably too simplistic.

The Neville Williams index is calculated as the ratio of the percentage of tolerant organisms to the percentage of intolerant organisms. Values greater than 1 indicate polluted or disturbed conditions while values less than 1 would indicate conditions relatively free of pollution or disturbance.

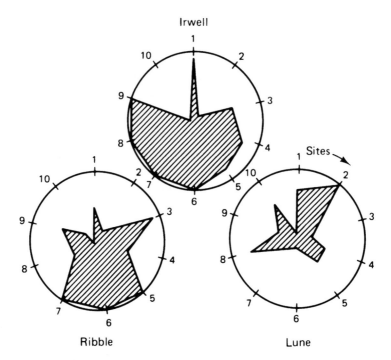

Fig. 5.5 Clock diagrams comparing the Neville Williams biotic index for water quality in three rivers in England (the Irwell, Ribble and Lune). The Neville Williams index is plotted radially. The smaller the central shaded area, the less polluted the water. (Redrawn from Williams & Dussart, 1976.)

These and other indices can usefully be expressed in the manner suggested over 30 years ago by Williams & Dussart (1976), that is in the form of clock diagrams where the extent of shading indicates the degree of disturbance or pollution (Fig. 5.5).

The Trent biotic index (score)

In England in 1964, Woodiwiss described a biotic index (or score) that was easy to calculate and simple to understand. That biotic index, which had its origins in the saprobian system, was based on the concept of aquatic indicator species and was weighted by a number of certain defined groups in relation to their sensitivity to organic pollution (Table 5.5). Qualitative samples are taken with a hand net and where possible shallow riffles are used as sampling points. A sampling time of 10 minutes was suggested by Woodiwiss.

In the example provided in Table 5.5, we can see how this biotic index (now known as the Trent biotic index, because of its introduction in England by the former Trent River Board in the late 1950s) is used. The lowest score of 0 indicates gross organic pollution and the top score of 15 indicates completely unpolluted conditions. The number of 'groups' (defined in Table 5.5) indicates which column to use. The row to be used is indicated by the presence or absence of certain indicator species. For example, in a sample there may be between 11 and 15 taxonomic groups present and we then look down the appropriate column. In our hypothetical example, there are no plecopterans but there is one species of Ephemeroptera. The biotic index would, therefore, be 7. The number 7 by itself is really meaningless because it is possible to arrive at the value of 7 in several different ways; clearly the index value has to be qualified by reference to the various indicator groups.

Overall, this index is derived from two effects, the reduction in community richness and the progressive loss of certain detector indicator groups (chosen on a subjective basis) and weighted by the number of defined 'groups'. The index does not require a standardized sampling procedure, which overcomes some of the effects of sampling problems. In its original form, this biotic index had a score ranging from 0 to 10, but in the expanded form (Table 5.5) the scores range from 0 to 15, which increases 'the sensitivity range' of the index towards the 'clean' end. There is a subjective choice of taxonomic groups and so, therefore, different sets of groups could be chosen for different river systems if it was considered that different types of river could not be compared using one form of the index. The main objective of this index is to monitor changes over time and not to compare index values between river systems; even so, changes in the index value tell us little about the magnitude of the change in the biotic community. This index does not make use of changes in proportional species composition and such changes could occur as a result of changes in water quality.

The Trent biotic index went through various modifications. For example Chandler (1970) decribed the addition of a semi-quantitative component (or weighting). That was done in order to provide a scoring system (rather than an index) that would be dependent on relative abundance of groups present. Numbers of each macro-invertebrate group are recorded from five-minute sampling periods. In Table 5.6, it can be seen that each abundant intolerant species would gain a high score but an abundant tolerant species would attract a very low score. In this system, a score is allocated to each taxonomic group and the score for the sample is then derived by summing the individual group scores. The Chandler biotic score can be modified to give the average Chandler biotic score by dividing it by the number of species, species groups or genera used in the calculation. Similarly, the Biological Monitoring Working Party (BMWP) score (see below) divided by the number of taxa gives an average score per taxon (ASPT).

Compared with the Trent biotic index, the Chandler index quantifies sensitivity to organic pollution of macro-invertebrates by producing more detailed

Table 5.5 *The Trent biotic index in its expanded form*

Organisms in order of tendency to be absent as the degree of pollution increases		Biotic indices for 2 total numbers of groups present[a]									
Indicators	Species richness	0–1	2–5	6–10	11–15	16–20	21–25	26–30	31–35	36–40	41–45
Plecoptera nymphs present (stoneflies)	More than one species	–	7	8	9	10	11	12	13	14	15
	One species only	–	6	7	8	9	10	11	12	13	14
Ephemeroptera nymphs (mayflies)	More than one species[b]	–	6	7	8	9	10	11	12	13	14
	One species only[b]	–	5	6	7	8	9	10	11	12	13
Trichoptera larvae present (caddis flies)	More than one species[c]	–	5	6	7	8	9	10	11	12	13
	One species only[c]	4	4	5	6	7	8	9	10	11	12
Gammarus present (crustacean)	All above species absent	3	4	5	6	7	8	9	10	11	12
Asellus present (crustacean)	All above species absent	2	3	4	5	6	7	8	9	10	11
Tubificid worms and/or Red Chironomid larvae present	All above species absent	1	2	3	4	5	6	7	8	9	10
All above types absent	Some organisms such as *Eristalis tenax* not requiring dissolved oxygen may be present	0	1	2	–	–	–	–	–	–	–

Clean

Increasing pollution →

Polluted

[a] The term group refers to the limit of identification without resorting to excessive keying-out. The groups are each known species of platyhelminthes (flatworms); *Baetis rhodani* (mayfly); annelids (worms excluding genus *Nais*); each family of Trichoptera (caddis flies); Genus *Nais* (worms); each species of Neuroptera larvae (Alder Fly); each known species of Hirodinea (leeches); Family Chironomidae (midge larvae except *Chironomus thumni*); each known species of Mollusca (snails); *Chironomus thumni* (Blood Worms); each known species of Crustacea (Hog Louse, shrimps); Family Simulidae (Black Fly larvae); each known species of Plecoptera (stonefly); each known species of other fly larvae; each known genus of Ephemeroptera (mayfly, excluding *Baetis rhodani*) each known species of Coleoptera (beetles and beetle larvae); each known species of Hydracarina (water mites).

[b] *Baetis rhodani* (Ephemeroptera) not counted.

[c] *Baetis rhodani* is counted in this section for the purpose of classification.

Modified from Hawkes (1979) after Woodiwiss (1964).

Table 5.6 *The Chandler biotic index*

Indicators	Points scored for increasing abundance				
	P	F	C	A	V
Each species of *Planaria alpina*	90	94	98	99	100
Each species of Taenopterygidae	90	94	98	99	100
Each species of Perlidae, Perlodidae	90	94	98	99	100
Each species of Isoperlidae, Chloroperlidae	90	94	98	99	100
Each species of Leuctridae, Capniidae	84	89	94	97	98
Each species of Nemouridae (excluding *Amphinemura*)	84	89	94	97	98
Each species of Ephemeroptera (excluding *Baetis*)	79	84	90	94	97
Each species of cased caddis, Megaloptera	75	80	86	91	94
Each species of *Ancylus*	70	75	82	87	91
Rhyacophila (Trichoptera)	65	70	77	83	88
Genera of *Dicranota, Limnophora*	60	65	72	78	84
Genera of *Simulium*	56	61	67	73	75
Genera of Coleoptera, Nematoda	51	55	61	66	72
Amphinemura (Plecoptera)	47	50	54	58	63
Baetis (Ephemeroptera)	44	46	48	50	52
Gammarus	40	40	40	40	40
Each species of uncased caddis (excluding *Rhyacophila*)	38	36	35	33	31
Each species of Tricladida (excluding *P. alpina*)	35	33	31	29	25
Genera of Hydracarina	32	30	28	25	21
Each species of Mollusca (excluding *Ancylus*)	30	28	25	22	18
Chironomids (excluding *C. riparius*)	28	25	21	18	15
Each species of *Glossiphonia*	26	23	20	16	13
Each species of *Axellus*	25	22	18	14	10
Each species of Leech (excluding *Glossiphonia*)	24	20	16	12	8
Haemopsis sp.	23	19	15	10	7
Tubifex sp.	22	18	13	12	9
Chironomus riparius	21	17	12	7	4
Nais sp.	20	16	10	6	2
Each species of air-breathing species	19	15	9	5	1
No animal life			0		

Levels of abundance in the scoring system is given by the numbers sampled in a five minute sample: P, present (1–2); F, few (3–10); C, common (11–50); A, abundant (51–100); V, very abundant ($\geq$100).
Modified from Hawkes, 1979.

lists of species that have been ranked on their sensitivity to diminishing water quality. It is a subjective ranking and could be modified according to conditions and needs. In theory, there is no upper limit for Chandler biotic scores but values would rarely exceed 3000 and polluted streams have values up to about 300.

In the 1970s, the use of qualitative indices such as the Trent biotic index was compared with quantitative indices (Murphy, 1978) in an analysis of water quality from three rivers in Wales (UK). Figure 5.6 shows the temporal and spatial variation in the indices and it is clear that marked temporal variations in the Shannon–Weiner and Margalef indices mask any spatial patterns that may exist. Murphy (1978) concluded that, for relatively unpolluted rivers, qualitative indices give a more consistent spatial discrimination between sites and that the average Chandler biotic score would seem to be least affected by wide ranges in physical conditions.

The BMWP is a straightforward type of index (score) that was developed in the UK and refined over several years. This score was calculated on the basis of selected invertebrate families incorporated into a system that can then be used to assess the biological condition of a river. The BMWP score system (Table 5.7) has three simple steps: first, all macro-invertebrate families present at the site are listed; second, scores are allocated to those families; and, third, the scores for all the families are summed to give a cumulative site score. Simple and easy to calculate, the BMWP score system was developed for use in national river pollution surveys but was also developed in response to economic and logistic constraints.

In freshwater ecology, biotic indices have often been used for detecting pollution. An abundance–biomass comparison (ABC) method was proposed by Warwick (1986) for detecting pollution effects on marine macrobenthic communities. This is based on the differences between distribution of biomass among species and distribution of numbers of individuals among species. These difference can be shown by K-dominance plots (see Lambshead et al., 1983). When the community is approaching equilibrium, the biomass becomes increasingly dominated by one of a few species, each represented by a few individuals. When plotted as K-dominance curves, 'numerical diversity' is greater than 'biomass diversity'. When under severe stress, benthic communities become increasingly dominated by one of a few species and few larger species are present. Under those conditions, 'numerical diversity' is lower than 'biomass diversity'.

In the Netherlands, Meire and Dereu (1990) have explored the use of the ABC index for application in detecting pollution effects in intertidal areas (see also Warwick, 1986). They suggested that ABC values could be converted into an

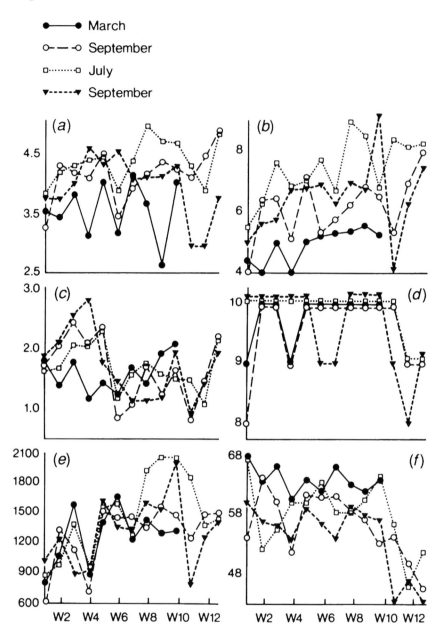

Fig. 5.6 Seasonal differences in various indices for the river Wye.
(a) Shannon–Weiner index; (b) Margalef index; (c) Menhinick index; (d) Trent biotic index; (e) Chandler biotic score; (f) average Chandler biotic score. W1 to W13 are the sampling sites from upstream to downstream. There is an input of sewage between sites W10 and W11. (Redrawn from Murphy, 1978.)

Table 5.7 *Allocation of biological scores (using the Biological Monitoring Working Party Scores)*

Families	Score
Siphlonuridae, Heptageniidae, Leptophlebidae, Ephemerellidae, Potamanthidae, Ephemeridae	10
Taeniopterygidae, Leucridae, Capniidae, Periodidae, Periidae	10
Chloroperlidae	10
Aphelocheiridae	10
Phryganeidae, Molannidae, Beraeidae, Odontoceridae, Leptoceridae, Gooeridae, Lepidostomatidae, Brachycentridae, Sericostomatidae	10
Astacidae	8
Lestidae, Agriidase, Gomphidae, Cordulegasteridae, Aeshnidae	8
Corduliidae, Libellulidae	8
Psychomyiidae, Philopotamidae	8
Caenidae	7
Nemouridae	7
Rhyacophilidae, Polycentropodidae, Limnephilidae	7
Neritidae, Viviparidae, Ancylidae	6
Hydroptilidae	6
Unionidae	6
Corophiidae, Gammaridae	6
Platycnemididae, Coenagriidae	6
Mesovelidae, Hydrometridae, Gerridae, Nepidae, Naucoridae, Notonectidae, Pleidae, Corixidae	5
Haliplidae, Hygrobidae, Dytiscidae, Gyrinidae, Hydrophilidae, Clambidae, Helodidae, Dryopidea, Elminthidae, Chrysomelidae, Curculionidae	5
Hydropsychidae	5
Tipulidae, Simuliidae	5
Planariidae, Dendrocoelidae	5
Baetidae	
Sialidae	4
Piscicolidae	
Valvatidae, Hydrobiidae, Lymnaeidae, Physidae, Planorbidae, Sphaeriidae	3
Glossiphoniidae, Hirudidae, Erpobdellidae	3
Asellidae	3
Chironomidae	2
Oligochaeta (whole class)	1

From National Water Council, 1981.

index and proposed, therefore, the average of the difference between cumulative biomass and abundance:

$$\text{ABC index} = \frac{B_i - A_i}{n}$$

where B_i is the percentage dominance of species i (ranked from highest to lowest biomass) and A_i the percentage dominance of species i (ranked from the most to the least abundant species) and n is the total number of species.

This ABC index is negative in heavily stressed conditions and positive for conditions where there is no pollution. Meire and colleagues (e.g. Meire & Dereu, 1990) have used this index in several studies and they are now exploring the possibility of its use as one of the indices to described the ecological status of waterways within the European Water Framework Directive.

As was the case throughout the UK and continental Europe, the latter half of the twentieth century saw the development and application of many biotic indices for use in connection with water quality assessment in North America. For example, in the late 1970s in Wisconsin, Hilsenhoff (1977, 1987) developed a biotic index of stream organic pollution that was based on arthropods only. This index was a measure of organic load and nutrient pollution. Since that time, many biotic indices have been developed on the basis of Hilsenhoff's biotic index.

Later, it was recognized that there was a need for the development and use of protocols in relation to the evaluation of the condition of water quality. This came about partly because of the pressure to have cost-effective biological assessment of streams and rivers. One example is the widely used rapid bioasessment protocols (RBPs) for use in biological assessments of streams and rivers (Barbour et al., 1999).

Many indices have been developed in North America in the context of measuring biological integrity. For example, there is a benthic index of biotic integrity and one example of this has been in connection with environmental monitoring in Chesapeake Bay (Llanso et al., 2003). Most of these indices of biotic integrity have been developed for water resources (e.g. Simon, 1999) but some have been developed for other systems or taxa such as bird communities (O'Connell et al., 1998).

Many of the freshwater biological indices are based on invertebrates. One departure from this has been the use of fish communities (Karr, 1981, 1987; Angermeier & Karr, 1986). This index is based on 12 criteria including the number of species, the species richness and species composition of certain groups of fish, and the proportion of omnivores etc. A value is given for each of these criteria and the sum of the values gives the index. An advocate of biological variables for ecological monitoring (as opposed to physical and chemical variables), Karr is critical of the use of indicator species. Quite rightly, he draws attention to the fact that the cause of declining indicator species may be difficult to determine. He also suggests that the use of species diversity indices is invalid because of the interdependence of species richness and relative abundance. Instead, he proposed an ecological guild approach because of its integrative ecological perspective. He concludes, however, that adoption of guilds as a simplistic reflection of environmental quality could lead to problems similar to those seen in the 1960s when there excessive dependence on a number of quantitative indices of species diversity (Karr, 1987).

The macro-invertebrate community index in New Zealand

Pollution assessment of water bodies can be based on biological indicators. The macro-invertebrate community index (MCI) and the quantitative version was proposed in New Zealand by Stark in 1985 to assess organic load in stony riffles of streams and rivers. This index is based on pollution tolerances of macro-invertebrates and is calculated on the basis of previously allocated scores (between 1 and 10). Based on the BMWP scoring system, the MCI uses mostly genera (but also some higher taxonomic groups). Although developed for stony–riffle sites, it has subsequently been used widely for other sites such as muddy stream bottoms. Stark (1993, 1998) proposed a further version of the MCI that used a five-point scale of coded abundances (rare, 1–4 animals; common, 5–19 animals; abundant, 20–99 animals; very abundant, 100–499; extremely abundant, greater than 500). The MCI has proved to be a very popular index and it is now widely used amongst regional councils in New Zealand. An assessment of the advantages and limitations of the MCI can be found in Stark (1998) and Boothroyd & Stark (2000).

The MCI, the quantitative version and the Stark version are calculated using the equations below and the taxon scores listed in Table 5.8. For the MCI, the calculation is based on presence or absence of the taxa.

Taxon scores based on the MCI were previously included in the Stream Health Monitoring and Assessment Kit (see Biggs et al., 2002), which had been designed for use by farming communities (Boothroyd & Stark, 2000). This kit can also be used by schools and community groups. Different levels of monitoring intensity can be undertaken. For example, its biological score for level two monitoring is an average score based on ten stone samples.

SIGNAL and Australian rivers

In Australia, Chessman (1995, 2003) adapted a version of the BMWP system previously used in the UK. Called the 'stream invertebrate grade number average level' (SIGNAL), this biotic index is the equivalent of the New Zealand MCI. The SIGNAL system was originally developed in eastern Australia but has

Table 5.8 *Taxon scores for use in calculation of the macro-invertebrate community index and its quantitative and Stark versions*

Taxon	Score
Ephemeroptera	
Ameletopsis	10
Arachnocolus	8
Atalophlebioides	9
Austroclima	9
Coloburiscus	9
Deleatidium	8
Ichthybotus	8
Isothraulus	8
Mauiulus	5
Neozephlebia	7
Nesameletus	9
Oniscigaster	10
Rallidens	9
Siphlaenigma	9
Zephlebia	7
Plecoptera	
Acroperla	5
Austroperla	9
Cristaperla	8
Halticoperla	8
Megaleptoperla	9
Spaniocerca	8
Spaniocercoides	8
Stenoperla	10
Taraperla	7
Zelandobius	5
Zelandoperla	10
Megaloptera	
Archichauliodes	7
Mecoptera	
Nannochorista	7
Odonata	
Aeshna	5
Antipodochlora	6

Table 5.8 (*cont.*)

Taxon	Score
Odonata (cont.)	
Austrolestes	6
Hemicordulia	5
Procordulia	6
Xanthocnemis	5
Hemiptera	
Diaprepocoris	5
Microvelia	5
Sigara	5
Coleoptera	
Antiporus	5
Berosus	5
Dytiscidae	5
Elmidae	6
Hydraenidae	8
Hydrophilidae	5
Liodessus	5
Ptilodactylidae	8
Rhantus	5
Scirtidae	8
Staphylinidae	5
Mollusca	
Ferrissia	3
Gyraulus	3
Latia	3
Lymnaeidae	3
Melanopsis	3
Diptera	
Aphrophila	5
Austrosimulium	3
Calopsectra	4
Ceratopogonidae	3
Chironomus	1
Cryptochironomus	3
Culex	3
Empididae	3
Ephydridae	4
Eriopterini	9

Table 5.8 (*cont.*)

Taxon	Score
Diptera (cont.)	
Harrisius	6
Hexatomini	5
Limonia	6
Lobodiamesa	3
Maoridiamesa	3
Mischoderus	4
Molophilus	5
Muscidae	3
Neocurupira	7
Nothodixa	5
Orthocladiinae	2
Parochlus	8
Paradixa	4
Paralimnophila	6
Paucispinigera	6
Peritheates	7
Podonominae	8
Polypedilum	3
Psychodidae	1
Sciomyzidae	3
Stratiomyidae	5
Syrphidae	1
Tabanidae	3
Tanypodinae	5
Tanytarsini	3
Physa	3
Glyptophysa	5
Potamopyrgus	4
Sphaeriidae	3
Tanytarsus	3
Zelandotipula	6
Oligochaeta	1
Hirudinea	3
Trichoptera	
Aoteapsyche	4
Beraeoptera	8
Confluens	5
Costachorema	7

Table 5.8 (*cont.*)

Taxon	Score
Trichoptera (cont.)	
Ecnomidae/*Zelandoptila*	8
Edpercivalia	9
Helicopsyche	10
Hudsonema	6
Hydrobiosella	9
Hydrobiosis	5
Hydrochorema	9
Kokiria	9
Neurochorema	6
Oeconesidae	9
Olinga	9
Orthopsyche	9
Oxyethira	2
Paroxyethira	2
Philorheithrus	8
Plectrocnemia	8
Polyplectropus	8
Psilochorema	8
Pycnocentrella	9
Pycnocentria	7
Pycnocentrodes	5
Rakiura	10
Tiphobiosis	6
Triplectides	5
Zelolessica	10
Lepidoptera	
Hygraula	4
Collembola	6
Acari	5
Crustacea	
Amphipoda	5
Cladocera	5
Copepoda	5
Isopoda	5
Ostracoda	3
Paranephrops	5
Paratya	5
Tanaidacea	4

Table 5.8 (*cont.*)

Taxon	Score
Platyhelminthes	3
Nematoda	3
Nematomorpha	3
Nemertea	3
Cnidaria	
Hydra	3

From Stark (1998) with modifications.

now been revised for use throughout Australia and it has a wide range of applications in Australia. For example, it has been included the Australian River Assessment System, which is a derivative of the British RIVPACS (see p. 311).

Originally, grade numbers were set subjectively for 11 families. Later, the grading system was refined and extended. The SIGNAL score for macro-invertebrate samples is calculated by averaging the pollution sensitivity grade numbers of the families present. These may range from 10 (most sensitive) to 1 (most tolerant).

A new version of SIGNAL has been published by Chessman (2003). For SIGNAL2, the grades have been derived from 171 families, 6 chironomid subfamilies and 33 higher taxa (macro-invertebrate orders, subclasses and phyla).

5.8   The development of biotic indices

This brief account of a few biotic indices may give the impression that biotic indices are simple and easy to develop. Nothing could be further from reality. In other words, this brief account does not take into consideration the many years of surveys and research that have led finally to the development and application of these biotic indices.

Much could be said about the advantages and disadvantages of using biotic indices as a basis for monitoring change in ecological systems. For example, the use of these biotic indices provides a quick, easy and cheap method of classifying certain characteristics of polluted water on a linear scale and also provides a basis for monitoring effects of pollution, but only organic pollution. That is, the biotic indices described above would not be suitable, without modification, for monitoring effects of certain kinds of industrial pollution where heavy metals were predominant.

Time-consuming identification to species level can be avoided with the use of these indices. However, not all the taxonomic groups used for calculating the index may be suitable as detectors of change or be sensitive to all kinds of pollutant. Clearly, there needs to be a judicious selection of the 'groups'. The species diversity or numbers of individuals is not recorded and in reality these biotic indices tell us little about community structure. One other potential disadvantage is that, depending on the sampling methods employed, natural changes in relative abundance of groups may be difficult to detect. For example, even in the case of a most basic sampling method, the kicking technique for sampling stream-bottom fauna, duration and extent of kicking can result in a less than representative sample. Some applications of biotic indices with reference to water pollution are described in later chapters.

In pollution-monitoring programmes, biotic indices cannot usually be interpreted in a useful manner without information on chemical changes in the water. These indices do, however, provide an important first step in monitoring the effects of disturbance and pollution.

Most of the indices described here are based on the same ecological principles (i.e. the species richness and the extent to which an organism can tolerate exposure to various physical and chemical factors).

Although the concept of an indicator species is simple, it is no simple task to transform that concept into practice. Once an idea for an index has been discussed, there then comes the need to consider the baseline data or reference condition, which taxa to sample, which methods of sampling to use, the timing of sampling and the choice of sampling equipment. Plans on paper all too often turn out to be too ambitious or not cost effective (often too much sampling is undertaken) when the pilot study has been completed. Results from the pilot study can provide a very useful basis on which to modify the definitive monitoring programme.

Temporal and spatial patterns of distribution of various taxa being sampled may demand that preliminary studies be undertaken for at least one year, leading to a further time period required for analysis and interpretation of data. Several years of modified sampling may then be required before it is possible to distinguish natural changes from those changes brought about by human activities.

Development and use of biotic indices requires an interdisciplinary approach and, therefore, good communication among those disciplines (biological, economic, social, etc.) is necessary. That need for good communication is important at both the research and development stage and at the interpretation stage, particularly when ecological concepts may have to be considered along with the planning and developments undertaken by local authorities and by industry.

It is not surprising that a great deal of work has been done to establish and put into practice protocols. The RBPs for use in the biological assessment of streams is one example (Barbour *et al.*, 1999). The techniques for the RBPs are based on evaluation of water quality (physiochemical variables), habitat characteristics and analysis of periphyton, benthic, macro-invertebrates and fish. There is emphasis on the need for reference conditions in relation to the interpretation of the data. Finally, the principles underlying the RBPs are well worth noting not only because they continue to be relevant today but also because they have been seen to be relevant throughout the development of indices for assessment and monitoring. These principles are (Barbour *et al.*, 1999):

- cost effective
- scientifically valid
- provisions for multiple site investigations
- rapid processing of results for management decisions
- science easily translated
- environmental impacts of the methods are minimal.

6

Diversity and similarity indices

6.1 Introduction

For many years, environmental variables and indices have been used to monitor pollution, changes in biotic communities and so-called 'environmental standards' or 'quality of the environment'. Environmental indices include those that are based on physical and chemical parameters, those based on biological parameters and also those based on perceived aesthetic qualities of the environment. The aim of this chapter is to provide an introduction to a selection of indices, their calculation, uses and limitations. Applications of some of these indices in monitoring programmes are described in later chapters.

6.2 Number of species, species composition, abundance and diversity

Perhaps the simplest variable that could be used in biological and ecological monitoring is the number of species. In addition to that there is the species composition and also the proportional abundance of species. The difference between the number of species, species composition and proportional abundance of species in a hypothetical community may be illustrated as shown in Fig. 6.1. Sample 1 or the 'baseline' contains four species. Samples 2, 4 and 5 also contain four species (the species richness is the same for each sample) but there are differences in species composition and total biomass. In sample 2, there are also four species but the species composition has changed. There is a difference between the relative abundance of each group of species in samples 1 and 4. The species richness of sample 3 is greater than found in the other samples. Sample 5 has a higher biomass but the species composition and proportional abundance of the species is the same as in sample 1.

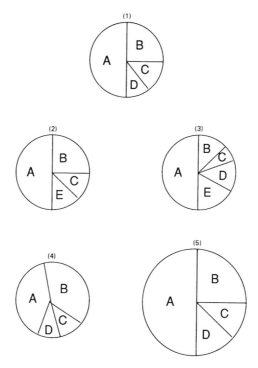

Fig. 6.1 Biotic community variables. Species indicated by letters A–E, abundance by the area of sector and biomass by area of the circle. Comparisons are made with 1. In 2, the species composition has changed; in 3, the species richness has increased; in 4, the relative abundance has changed; in 5, the biomass has increased. (After Hellawell, 1977.)

6.3   Species richness

Species richness or species abundance is the total number of species present. Data from a study on the effects of people trampling on a lawn provide us with information (Table 6.1) with which to examine changes in species richness and also species composition. Abundance of the various plant species has been expressed as percentage cover in the sample areas. There is no doubt that the 'low-trampling' site has a greater number of grassland species (higher species richness) than the 'high-trampling' site. Both the species richness and the species composition in these samples could provide a basis for monitoring the effects of physical impacts on lawns. It is also useful to note that the high-trampling site is dominated by two species, *Agrostis tenuis* and *Poa annua* and that there is a marked difference in the abundance of *P. annua*. The species diversity index in Table 6.1 is discussed below.

A subsample of a community from a collection of night-flying moths caught with a light-trap provides us with a further example with which to explore

Table 6.1 *'Subsample' of data from a study of the effect of people trampling on grassland*

	Species abundance (mean percentage cover per metre quadrat (x))	
	Site one 'high' trampling	Site two 'low' trampling
Species		
Agrostis tenuis (Common Bent-grass)	49	46.7
Poa annua (Annual Meadow Grass)	45	3.5
Festuca rubra (Red Fescue)	4	3.7
Plantago coronopus (Buck's Horn Plantain)	0	5.0
Plantago lanceolata (Ribwort)	0.5	1.3
Anthoxanthum odoratum (Sweet Vernal Grass)	1.0	0
Bellis perennis (Daisy)	0.5	11.3
Trifolium repens (White Clover)	0	21.2
Taraxacum officinale (Common Dandelion)	0	2.0
Prunella vulgaris (Self Heal)	0	0.5
Leontodon autumnalis (Autumnal Hawkbit)	0	3.8
Hypochaeris radicata	0	1.2
Summary data		
Number of species	6	11
Total of x	100	100
Species diversity calculation[a]	$\frac{(100)^2}{49^2 + 45^2 \ldots}$	$\frac{(100)^2}{46.7^2 + 3.5^2 + \ldots}$
Species diversity	2.25	3.5

[a] Hill's Species diversity index is given by $(\sum x)^2 / \sum x^2$.

From Hill, 1973.

various features that might usefully be used in monitoring. Looking at Table 6.2, we see that the total number of species of moths (species richness) for each month varies between five and ten and was highest in July. The number of individuals for each species caught each month is used as a measure of abundance. Although total numbers of individuals of each species is an obvious way of expressing abundance or population size, alternative methods have been adopted for other taxonomic groups, such as percentage cover for grassland species (Table 6.2).

6.4   Species richness indices

Species richness can be expressed as the number of species in a sample or habitat, or it can be expressed more usefully as species richness per unit area.

Table 6.2 *Species richness and species diversity of a 'subsample' of night-flying moths collected with a light-trap at monthly intervals*

Family and species	Month			
	May	June	July	August
Amphipyrinae				
Apamea monoglypha (Dark Arches)	0	7	149	87
Apamea secalis (Common Rustic)	0	0	2	85
Meristis trigrammica (Treble Lines)	47	96	2	0
Arctiinae				
Spilosoma lutea (Buff Ermine)	4	44	31	0
Ennominae				
Biston betularis (Peppered Moth)	22	36	92	5
Hadeninae				
Leucania impura	0	2	195	60
Lithosiinae				
Lithosia lurideola (Common Footman)	0	0	84	51
Noctuinae				
Agrotis exclamationis	1	144	255	38
Nuctua pronuba	0	13	32	87
Ochropleura plecta	25	41	24	3
Number of species	5	8	10	8
Number of individuals	99	383	866	416
Margalef's index $[S-1]/\log_{10}^n N$	2.0	2.7	3.1	2.7
Simpson's index $(1 - \sum p_i^2)$	0.67	0.77	0.81	0.82
Alternative form of Simpson's index $(1/\sum p_i^2)$	3.0	4.3	5.3	5.6
Shannon–Wiener index $(-\sum p_i \log_e p_i)$	1.2	1.6	1.8	1.8

p_i is the proportion of the ith species in the sample. For example, 47 *Meristis trigrammica* were caught in May. p_i in this case is $47/99 = 0.4747$; $p_i^2 = 0.225$. See text for more details on these indices.

There are also various simple species richness indices based on the total number of species and the total number of individuals in the sample or habitat. For example Menhinick's index (1964) and Margalef's index (1951) are simple measures of species richness and are expressed, respectively, as:

$$D = S/\sqrt{N}$$

$$D = S - 1/\log N$$

where D is the index, S is the number of species and N is the total number of individuals. This is also a simple measure of mean population size and using the form $(S - 1)$, rather than S, gives a value of zero if there is only one species. Despite its simplicity, this and similar indices are affected by sample size and sampling effort. Furthermore, these indices could be misleading because they fail to take into account abundance patterns. For example, two samples (A, B) could have the same species richness (5) and same number of individuals (100) but the samples may differ in the proportional abundance of the species. This problem is considered in the following sections.

6.5   Biological diversity

Diversity or variety in nature can mean many things to many people. There is much confusion and misunderstanding with respect to the terms biological diversity and ecological diversity. The 1992 Convention on Biological Diversity gave rise to the popular use of the term or concept of 'biological diversity'. The Convention defined biological diversity as 'the variability among living organisms from all sources including, inter alia, terrestrial, marine and other aquatic ecosystems and the ecological complexes of which they are part; this includes diversity within species, between species and of ecosystems'. Unfortunately, biological diversity is often equated with diversity or variety of species. Diversity at the species level of biological organization is just one small component of biological diversity. Also unfortunate is the unnecessary use of the term biological diversity when other terms would be more relevant. For example, it makes nonsense to say 'that wetland has lots of biological diversity'. That is the same as saying 'that wetland has lots of nature'.

Biological diversity occurs at all levels of biological organization, from molecular to ecosystems. There are attempts to quantify biological diversity at various levels and from ecological taxonomic and other perspectives.

With reference to the use of the term biological, I suggest the following three rules:

1.   Define what you mean by biological diversity
2.   Ask yourself if you need to use the term biological diversity and whether or not there is a more appropriate term such as biota and nature
3.   If you have to use the term biological diversity, say at which level of biological organization you are referring to.

6.6   Ecological diversity

Long before biological diversity came into common use, ecologists had used the term ecological diversity. In ecological research, diversity may refer to species diversity, habitat diversity or the diversity of resources in a niche. In other words, ecological diversity embraces different kinds of variety.

One approach is to think in terms of alpha, beta and gamma diversity. Alpha diversity is the diversity of species within a particular habitat or community. Beta diversity is a measure of the rate and extent of change in species along a gradient from one habitat to another (or an expression of between-habitat diversity). Gamma diversity is dependent on both alpha and beta diversity and is the diversity of species within a geographical area. Here we are concerned mainly with alpha diversity and the different ways of measuring and expressing that diversity.

6.7   Species diversity indices

An index of species diversity can be based either on the number of species present and species composition without any measure of abundance (species richness index) or can be based on the species and abundance of species in a habitat or community (diversity index). In plant ecology studies, percentage cover of plant species in quadrats has had common use, including in long-term vegetation monitoring. For example, the late Joy Belsky (1985) in her vegetation monitoring in the Serengeti National Park, Tanzania, found that of several variables used, percentage cover of species in permanently marked plots was the most useful for long-term studies. The data in Table 6.1 from the study of trampling have also been expressed as percentage cover, but we can also see that the sample from the low-trampling site has greater species richness than the high-trampling site. So, how can we quantify this observation? Diversity can be based on percentage cover as a measure of abundance (Hill, 1973):

$$D = \frac{(\Sigma x)^2}{\Sigma x^2}$$

where D is an index of diversity and x is a measure of abundance of the species, in this case percentage cover. This measure of diversity was one of many suggested by Hill (1973) and was used in the 1970s for studies of the effects of trampling on vegetation (Liddle, 1975). One example is shown in Table 6.1.

The sequential comparison index was devised as a simplified method for non-biologists to estimate relative differences in biological diversity and is based on the chance of any one individual being the same species as the previous individual in a sample. In the field, individual organisms are recorded at intervals along a transect. Random samples taken back to the laboratory are scored in turn as being the same species or different species to the preceding individual. Samples with many species each represented by few individuals will result in many changes and a high index value. For example, a hypothetical sample has three species (A, B, C) and when arranged in rows the arrangement is as follows:

C	C	A	A	B	A	A	C	B	C	A	A	C
1	2		3				4	5	6	7		8

After scoring one for the first individual, the number of changes of one species to another is eight and the total number of individual organisms is 13. A simple measure of the variety or diversity in this sample may be expressed as x/N, where x is the number of changes and N is the number of individual specimens. In the above hypothetical example, D would be $8/13 = 0.6$.

Williams & Dussart (1976) suggested that this index be based on a random collection of 200 organisms and that random numbers are needed in the laboratory once the individual has been classified to species. The reliability of this and other measures of diversity could be quantified by determining the diversity of the sample several times. In other words, what are the chances of the same measure of diversity being obtained if the sample were to be mixed then arranged in another set of rows? A few simple tests can usefully be undertaken on a random set of letters to test the reliability of Cairn's measure of diversity (Cairns, 1979). In other words, it is useful to ask the question, how much variation could be expected to occur if the sequential comparison index was calculated several times using the same specimens but with a complete mixing of the specimens between each calculation?

In nature and based on relative abundance of different species, biological communities contain relatively few species that are represented by large numbers and many species represented by relatively few numbers. The theories of abundance and the ease with which it is possible to calculate the total number of species and total number of individuals provides a basis for calculating diversity. For example, the logarithmic series for species abundance in a sample is fixed by two variables, the number of species and number of individuals:

$$S = \alpha \log_e (1 + N/\alpha)$$

where S is the total number of species, N is the total number of individuals and α is the diversity index.

The alpha diversity index is most simply read off from a nomogram (Fig. 6.2).

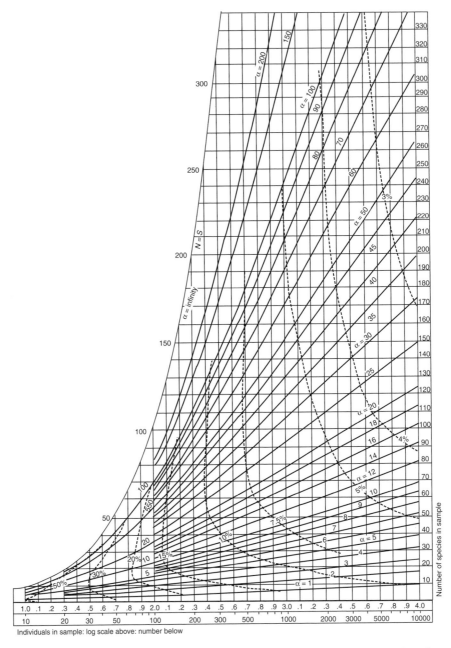

Fig. 6.2 Nomogram for determining the alpha diversity index for the number of species (S) and the number of individuals (N) in a random sample. (From Fisher *et al.* (1943) with permission of Cambridge University Press.)

6.8   Species diversity and equatability or evenness

We can see how the individuals are distributed amongst the moth species in Table 6.2 and how the percentage cover of the plants is distributed amongst the various grassland species in Table 6.1; however, is it possible to quantify the relative abundance and use this in the calculation of an index?

The way in which individuals are distributed among species is known as 'evenness' or 'equatability'. For example, consider an extreme hypothetical example of two samples, each of which has five species and a total of 100 individuals. In the first sample, each of the five species is represented by 20 individuals and in the second sample one species is represented by 96 individuals and the remaining four species by one individual each. The equatability is maximized in sample one and in the second sample there is an uneven distribution of individuals among the species.

Species diversity is greatest in the first sample and lowest in the second sample. This equatability or evenness can be expressed in the form of one of several indices (heterogeneity indices) where a more equitable distribution among species will give higher index values. In this context, species diversity is a function of species richness and the evenness with which individuals are distributed amongst the species.

If the numbers of all species can be recorded (not just samples taken) then a suitable index of diversity is Brillouin's index (1960):

$$H = \frac{1}{N}\log_{10}\frac{N!}{N_1!N_2!N_3!}$$

where H is the diversity index, N is the total number of individuals, N_i is the number of individuals of species i (here 1–3 species). This index has been derived for use where the total population can be measured (Pielou, 1966) and is not appropriate where samples of the population have to be used or where randomness of sampling cannot be guaranteed. It is an index that is not, therefore, subjected to limitations and error created by various sampling techniques. Brillouin's index would be calculated as follows. Assume the species A,B,C have the relative abundance of 4, 9, 5. Then

$$H = \frac{1}{18}\frac{(18 \times 17 \times 16 \cdots \times 2)}{(4 \times 3 \times 2)(9 \times 8 \times 7 \cdots \times 2)(5 \times 4 \times 3 \times 2)}$$

One example of how this index can be used was described in a paper by Ben-Eliahu & Safriel (1982) where they made comparisons of polychaetes from tropical and temperate intertidal habitats. Although the aim was not to monitor changes in diversity of polychaetes, their survey does provide a most useful

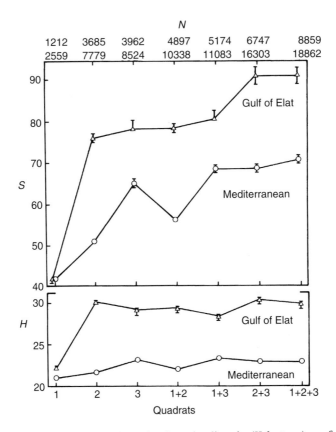

Fig. 6.3 Species richness (*S*, top) and species diversity (*H*, bottom) as a function of sample size where there are *N* individuals in samples (upper row, Gulf of Elat; lower row, Mediterranean). Circles (Mediterranean) and triangles (Gulf of Elat) stand for values obtained when all juveniles were allocated to their putative species. Vertical bars represent the ranges between minimal diversity obtained when unidentified juveniles were pooled with dominant congeners, and between maximal diversity by omitting the juveniles. (Redrawn from Ben-Eliahu & Safriel, 1982.)

example of the kind of data used as a baseline for monitoring. They considered species richness and diversity *H* as a function of sample size (Fig. 6.3) and found that the number of species and the diversity increased asymptotically with increased number of individuals.

In circumstances where it is not practical or not possible to determine the total number of individuals, then a random sample has to be used. There are many species diversity indices that are based on both the number of species and the proportional abundance of species. Two commonly used species diversity indices are Simpson's index (Yules) and the Shannon–Wiener index. Simpson's index *D* is calculated as follows:

$$D = \sum_{i=1}^{s} P_i{}^2$$

and

$$(P_i{}^2) \sim \left(\frac{N_i}{N_T}\right)^2$$

where D is the index of diversity (a measure of the probability that two randomly sampled individuals belong to different species) and P_i is the relative abundance of the ith species (N_i is the number of the ith species, N_T is the total number of individuals). This index is sometimes used in either of the following forms:

$$D = \frac{1}{\sum P_i{}^2} \quad D = 1 - \sum P_i{}^2$$

For example, and with reference to Table 6.2, the diversity of the sample of moths collected during May would be calculated as follows:

$$P_i{}^2 = \left(\frac{47}{99}\right)^2 + \left(\frac{4}{99}\right)^2 + \left(\frac{22}{99}\right)^2 \ldots = 0.33$$

This gives $1/0.33$, or 3.0, in one form and $1 - 0.33$, or 0.67, in the other.

Predicting correctly the next species of the next individual to be collected provides a basis for another index. The Shannon–Wiener index is calculated from the equation:

$$D = -\sum_{i=1}^{s} P_i \log_e P_i$$

where D is the species diversity index and P_i is the proportional abundance of the ith species in the same. The use of natural logarithms is usual because this gives information in binary digits ($\log_{10}$ of 100 is 2, $\log_e$ of 100 is 4.6, log 10 of 1000 is 3, $\log_e$ of 1000 is 6.9).

With reference to data in Table 6.2, the Shannon–Wiener diversity index for the sample of moths collected during May would be calculated as follows:

$$D = -\sum \left(\frac{47}{99} \times \frac{47}{99}\log_e\right) \times \left(\frac{4}{99} \times \frac{4}{99}\log_e\right) \ldots$$
$$= -\sum (0.47 \times -0.74) + (0.04 \times -3.2)$$
$$= -\sum (-0.35) + (-0.13)$$
$$= 1.2$$

There are some interesting values amongst the diversity indices for the samples of moths (Table 6.2). Eight species were caught in both June and August and so the species richness remained the same. The diversity index, if expressed as a simple relationship between the number of species and total number of individuals, was the same for those two months. However, when we come to look at Simpson's and Shannon–Wiener indices of diversity, we see that the August sample has a greater value, indicating greatest evenness in the August sample compared with samples taken during other months. The lower diversity index values in June can be attributed to less evenness or less equatability of the sample, particularly with regard to the large number of individuals of one species *Agrotis exclamationis*.

6.9　The Shannon–Wiener or the Shannon–Weaver index of species diversity?

In the literature, the 'Shannon index' is sometimes referred to as the 'Shannon–Weaver' index and sometimes as the 'Shannon–Wiener' index. When references from current research reports on species diversity indices are tracked back via the cited references, it is not unusual for Shannon to be listed in the bibliography (Spellerberg & Fedor, 2003).

A common reference for 'Shannon's' function H for species diversity is Shannon & Weaver (1962). This small book contained two sections or two reports that were republished versions of previous reports from about 14 years before. *The Mathematical Theory of Communication* (Shannon paper in the *Bell System Technical Journal* in 1948) was repulished with some corrections and additional references. The second, *'Recent Contributions to the Mathematical Theory of Communication'* (Weaver, the Rockefeller Foundation) had not previously been published in this form (a condensed version had appeared in *Scientific American* in 1949).

Thus, Shannon first published an account of the entropy H in 1948 and Weaver built on this in the second part of Shannon & Weaver (1962). Shannon acknowledged that 'communication theory is heavily indebted to N. Wiener for much of its basic philosophy and theory' and cited Wiener (1939, 1948, 1949).

The confusion in the name for the species diversity index H arises because of the joint authorship of Shannon and Weaver's 1962 book. Based on this, the index is sometimes incorrectly called the Shannon and Weaver index whereas, because Shannon built on the work of Wiener, it would be correct to refer to Shannon's index or the Shannon–Wiener index.

6.10　Variation in species diversity

In simple terms, maximum diversity (equatability) exists if each individual is a different species. Minimum diversity exists if all individuals belong to one species. It does not necessarily follow that a higher species diversity is 'better' than a lower species diversity, nor does it follow that high diversity indices can be interpreted as being a reflection of high 'quality' of habitat or state of the environment.

Some communities have a naturally low species diversity and the meaning of 'quality' needs to be defined when using a diversity index to express some state of the habitat or environment. To complicate matters, not all species diversity indices are based on the idea of equatability and two different indices can give contradictory values. It is always useful to calculate the theoretical extremes of a diversity index with collection of simple, hypothetical data.

Temporal changes in species diversity must be expected, whether it be on a circadian, tidal, seasonal or annual basis. Dills & Rogers (1974) gave one example of seasonal variations in diversity (Shannon–Wiener) from their research on macro-invertebrate community structure affected by acid mine pollution in small streams in Alabama (Fig. 6.4). Changes in species diversity will also occur over time as a result of changes in temperature levels. This was investigated by Dennis et al. (1979) and one example was that reported by Goldman in 1974 (Fig. 6.5) for species diversity of phytoplankton in Lake Tahoe, California. Goldman (1974) had shown that Lake Tahoe was subject to rapid and accelerated eutrophication. The analysis of Goldman's data by Dennis et al. (1979) showed positive relationships between species richness and the diversity of phytoplankton and water temperature.

6.11   Application of diversity indices

The nature and application of diversity indices and the advantages of one index over another have been discussed many times. It is useful to consider the advantages and disadvantages of using diversity indices, particularly in monitoring programmes or in relation to assessment of impacts on the natural environment.

Ecological monitoring has to start with some kind of 'benchmark' or baseline study. In such studies, measures of species diversity may be used alongside other ecological variables. For example, Debinski et al. (2000) working in the Gallatin National Forest and the northwest corner of the Yellowstone National Park, Montana (USA) used Shannon's species diversity index for plants, birds and butterflies to establish baseline studies in a range of six habitats (hydric to mesic xeric). They found that mesic meadows supported the highest plant species diversity. They found no significant difference in bird or butterfly species diversity amongst the meadow types. They concluded that it may be easier to detect significant differences in more species-rich taxa

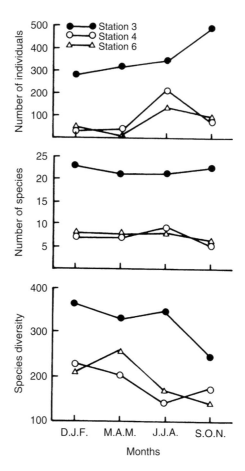

Fig. 6.4 Seasonal variations in the number of individuals, number of species and species diversity (Shannon–Wiener) at stations 3, 4 and 6 in Cane Creek Basin (Walker County, Alabama, USA) from February 1970 to January 1971. (Redrawn from Dills & Rogers, 1974.)

(e.g. plants) than those taxa represented by fewer species (in this case butter-flies and birds).

Diversity indices are simple mathematical expressions and from that point of view there is nothing special let alone mystical about them. One advantage of an index is that a lot of data can be summarized in one figure or set of figures.

At public inquiries or in environmental courts, an expert witness may seem to be giving very impressive evidence in the form of indices. However, indices need to be put into context and they also need to be qualified. That is, diversity indices tell us nothing about the type of distribution, stage or succession of species composition of a biological community.

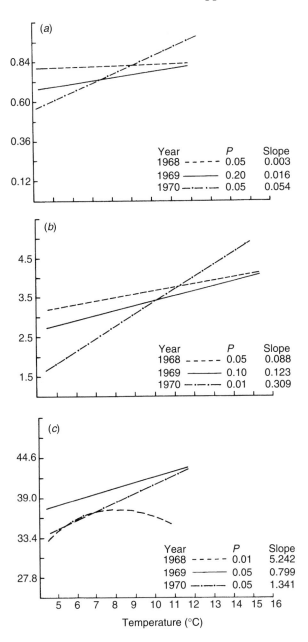

Fig. 6.5 Species richness and diversity of phytoplankton as related to temperature for Lake Tahoe, California. (a) Diversity calculated with Simpson's index; (b) diversity calculated with the Shannon–Wiener index; (c) species richness or number of species. (These graphs were prepared and modified from regression analysis by Dennis *et al.* (1979) using data from Goldman (1974).)

An index score is meaningless unless it can be put into context and, there-fore, we should ask what are the minimum and the maximum scores that might be obtained. How consistent are the values? What degree of statistical variance is there when the diversity is measured on several occasions? Variance of the diversity indices should be made known, or at least some measure of variation should be recorded.

Although an index has the advantage of being able to summarize a large amount of data in a simple form, it is perhaps useful to consider qualifying an index by using other data, especially when monitoring changes in biotic com-munities. The number of species in different group, numbers of individuals of particular species, total biomass and levels of biomass at various trophic levels are equally valuable attributes that can be used in monitoring.

It has long been known that, for various taxonomic groups, the number of species increases with increasing sample size or effort: that is, there is a species–area relationship. This relationship has important implications when interpreting species richness and species diversity. The biological explanations for the species–area relationship and the mathematical descriptions of the relationship have been well researched and reviewed (Spellerberg, 1991). For example MacArthur & Wilson (1967) developed models to explain species rich-ness on islands of various sizes as a function of immigration and extinction rates, and Williams (1964) developed the habitat diversity hypothesis, which assumes that as area increases, so more habitats are included and thus more species are recorded.

The species–area relationship can pose a serious problem in the use of diversity indices: we need to know if the sample size has any effect on the index. This can easily be investigated by pooling successive standard samples (i.e. from quadrats and not sweeps) and then calculating the index for each set of data. One example of a small effect of increased sample size when using the Shannon–Wiener index is clearly shown in Fig. 6.6. As a guide, a minimum of ten samples would be taken.

While diversity indices have had popular use in ecological monitoring, there are many conceptual, semantic and technical problems. For example, the use of a species diversity index assumes that all species have been sampled. This can rarely be the case and, therefore, the index will always be biased towards an underestimate of the diversity.

The few diversity indices mentioned here may be summarized as follows.

1.   Hill's diversity index:

$$D = \frac{(\sum x)^2}{\sum x^2}$$

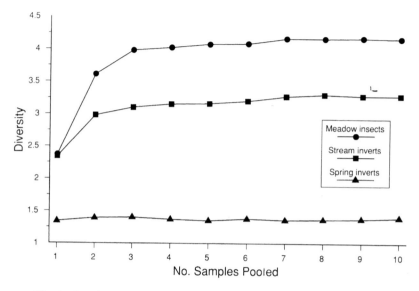

Fig. 6.6 Species diversity (Shannon–Wiener index) values obtained by pooling successive samples from 1 to 10. (Data from Wilhm & Dorris, 1968.)

This is one example of a series that may be difficult to interpret and are not widely used.

2.   Sequential comparison index:

$$D = \frac{x}{N}$$

This is very simple but there is a high sensitivity to sample size. The material must be distributed randomly.

3.   Williams' alpha diversity index:

$$S = \alpha \log_e(1 + N/\alpha)$$

This has low sensitivity to sample size.

4.   Brillouin's index:

$$H = \frac{1}{N}\log_{10}\frac{N!}{N_1!N_2!N_3!}$$

This is based on complete sampling.

5.   Simpson's index (Yules):

$$D = \sum_{i=1}^{s} p_i^{\,2}$$

Tends to give more weight to common species. The effect of sample size is low.

6.   Shannon–Wiener index.

$$D = -\sum_{i=1}^{s} p_i \log_e p_i$$

This is strongly influenced by species richness. The effect of sample size is low.

6.12   Similarity indices

Environmental managers, including ecologists, have been looking at ways of quantifying the extent of similarity and dissimilarity between communities or biotic samples for many decades. Consequently, a number of coefficients of similarity, dissimilarity and association have been devised. Similarity indices have had popular use in field ecology, particularly as a basis for quantifying similarities between communities and in the investigation of discontinuities between different sampling methods employed at any one particular site. These indices can also be used in comparing communities that have been subjected to pollution but the assumption has to be made that the communities being compared will have received the same types of pollutant. Comparison of samples from the same community taken at various intervals of time is one basis for the application of similarity indices in monitoring. Also, analysis of community structure, using similarity indices and cluster analysis, provides a basis for monitoring changes in a community.

6.13   Community similarity indices

Similarity indices have also been used as a basis for cluster analysis, showing affinities between samples and thus providing a baseline for monitoring programmes. Thus, similarity indices can be used for the analysis of data from various communities but can also be used to quantify differences between successive samples from one community. For convenience, the term 'sample' is used but this could also mean 'community'.

Simple but effective, binary community similarity indices, and those which are based only on the number of species whether common or rare, include Sorensen's index (Sorensen, 1948) and Jaccard's index (Jaccard, 1902). The latter was originally known as the coeffieient of floral community and is probably the earliest index of similarity. These indices are calculated, respectively, as follows:

$$C = \frac{2w}{A + B} \times 100$$

$$J = \frac{w}{A + B - w} \times 100$$

where C or J is the index of similarity, w is the number of species common to both samples (or community) and A is the number of species in sample one and B is the number of species in sample two. Here the scores have been multiplied by 100 to give a percentage scale. A value of 0% indicates complete dissimilarity whereas a value of 100% indicates maximum similarity between samples (or communities). Both these indices increase with increasing sample size. As they both give equal weight to all species, they place too much significance on the rare species in the samples.

The abundance of the different species can be considered using a modification of the Sorensen index:

$$C_n = \frac{2jN}{(aN + bN)}$$

where aN is the total number of individuals in sample A, bN is the total number of individuals in sample B and jN is the sum of the lower of the two abundances recorded for species found in both samples. That is, if 10 individuals of one species were in community A, but in sample B there were 20 individuals of the same species, then the value of 10 would be included in the sum and so on.

An index of community similarity based on percentages is the percentage community similarity index (Brock, 1977), expressed by:

$$P_{SC} = 100 - 0.5 \sum (a - b)$$

where P_{SC} is the index of community similarity, a and b are for given samples as percentages of the total sample. The absolute value of the differences is summed overall. For example looking at the data in Table 6.3, the P_{SC} for samples A and C would be calculated as shown in Table 6.4. The P_{SC} value from Table 6.4 would be $100 - 0.5(55.5)$, which is 72.25.

During the late 1970s, an extensive programme of research on Water Boatmen (Corixidae) led Savage (1982) to devise an index of similarity based on percentages (proportions) because he intended to reflect relative numbers of individuals of different species in different habitats and not actual numbers. In connection with that research on water boatmen, Savage looked at the classification of lakes and used a form of a similarity index as follows (Box 6.1):

$$I = \frac{[(a_1 \times b_1) + (a_2 \times b_2) \ \ldots \ (a_n \times b_n)]^2}{(a_1{}^2 + a_2{}^2 \ \ldots \ a_n{}^2) \cdot (b_1{}^2 + b_2{}^2 \ \cdots \ b_n{}^2)}$$

Table 6.3 *Benthic macro-invertebrate samples taken at monthly intervals from one stream*

Species	Samples and abundance of species[a]					
	A	A_1	B	C	D	E
Oligochaeta						
Haplotaxius sp.	4	4	3	2	2	0
Lumbriculus variegatus	60	64	54	57	51	0
Stylodrilus sp.	1	2	0	0	6	0
Tubifex sp.	65	66	81	83	196	108
Chaetogaster sp.	25	24	40	44	31	0
Crustacea						
Ascellus aquaticus	3	3	2	3	1	0
Gammarus pulex	19	20	2	4	3	1
Insecta, Diptera						
Chironomous sp.	5	3	20	11	30	18
Spaniotoma sp.	21	18	4	5	4	0
Tanypus sp.	46	45	5	10	10	0
Insecta, Ephemeroptera						
Ephemera sp.	3	3	1	0	0	0
Species richness	11	11	10	9	10	3
Number of individuals	252	252	212	219	334	127
Community similarity indices						
Sorensen's index	1.0	0.95	0.9	0.95	0.43	
Percentage community similarity index[b]		97.2	69.5	72.3	59.4	60.9
Pinkham and Pearson index		0.8	0.41	0.44	0.33	0.08
Morista index		1.0	0.8	0.9	0.7	0.48

[a] Samples A_1 and A were taken at about the same time to provide data for a baseline similarity index. For calculation of similarity indices, A_1, B, C, D and E are compared with A. Small quantities of industrial effluents entered the stream after samples A and A_1 had been taken and a large pollution incident occurred in the interval between samples D and E.
[b] Calculation of data is illustrated in Table 6.4.

where I is the index of similarity and a_1 is the percentage of individuals of a given species at a particular site and b_1 the percentage of individuals of the same species at the other site ('sites' could be replaced by 'samples' in monitoring or surveillance); a_2 and b_2 are the respective percentages for the second species and so on up to any number of species. Table 6.5 shows data for two lakes (Hanmer

Table 6.4 *Illustration of the calculation for the percentage community similarity index using data in Table 6.3*

T$_1$	A (%)	C (%)	Difference between % of a and b
Haplotaxius sp.	4 (1.6)	2 (0.9)	0.7
L. variegatus	60 (23.8)	57 (26.0)	2.2
Stylodrilus sp.	1 (0.4)	0 (-)	0.4
Tubifex sp.	65 (25.8)	83 (37.8)	12.1
Chaetogaster sp.	25 (9.9)	44 (20.1)	10.2
Total			55.5

Table 6.5 *Data for six species of corixid* Sigara

Lake	Percentage of sites					
	scotti	*distincta*	*concinna*	*praeusta*	*dorsalis*	*falleni*
Hanmer Mere	0 (a_1)	26.7 (a_2)	1.3 (a_3)	17.3 (a_4)	18.7 (a_5)	36.0 (a_6)
Oak Mere	1.5 (b_1)	2.2 (b_2)	0 (b_3)	9.7 (b_4)	85.1 (b_5)	1.5 (b_6)

See text for definition of a, b, 1–6 and use in calculating Savage's index.

Mere, Oak Mere) and six species of the corixid *Sigara*. From these data, the index would be calculated as follows (Savage 1982):

$$I = \frac{[(0 \times 1.5) + (26.7 \times 2.2) \; \dots \;]^2}{(0^2 + 26.7^2 \; \dots \;) \cdot (1.5^2 + 2.2^2 \; \dots \;)} = 0.1$$

If the same species are present in identical percentages at two sites then the index of similarity is 1.0; if there are no species in common then the index would be 0.0. Further comments about the application of Savage's work are made in Ch. 9.

This index of similarity can be used with numbers of individuals rather than percentages. For example, Morris (1969) used the above index in the form:

$$I = \frac{[\sum(m_i \times n_i)]^2}{\sum m_i^2 \times \sum n_i^2}$$

where m_i is the number of specimens of species i recorded from the first site or sample and n_i is a similar term for species from the second site or second sample.

Box 6.1   Use of corixids in the classification of a lake

The following is an example of the use of similarity indices, especially Savage's (1982) similarity index.

In Savage's account of the use of corixids in the classification of lakes, he used his indices of similarity to construct a dendrogram, based on a method used by Mountford (1962).

The following example shows how the four categories of sites on the right side were derived: Esthwaite (E); Llyn Coron and Llyn Hendref (W); North West Midlands meres, group A (A); and North West Midlands meres, group B (B) (Savage, 1982). The dendrogram for the last two groups published was by Savage & Pratt (1976). The indices of the four categories (sites) are set out in tabular form:

	E	W	A	B
E		0.524	0.600	0.322
W			0.928	0.847
A				0.750
B				

The highest index is selected from the table and the pair of sites (A and W) corresponding with this index are combined to form a single group (AW). The indices between this new group and each remaining site or group are then calculated:

$$I(E{:}AW) = \frac{I(E{:}A) + I(E{:}W)}{2} = \frac{0.600 + 0.524}{2} = 0.562$$

and

$$I(B{:}AW) = \frac{I(B{:}A) + I(B{:}W)}{2} = \frac{0.750 + 0.847}{2} = 0.798$$

A reduced table is made:

	E	AW	B
E		0.562	0.322
AW			0.798
B			

The highest index is selected and the procedure repeated:

$$I(E{:}AWB) = \frac{I(E{:}A) + I(E{:}W) + I(E{:}B)}{3} = \frac{0.600 + 0.524 + 0.322}{3} = 0.482$$

The final table is thus:

E	AWB
E	0.482
AWB	

The dendrogram is drawn as shown below. The procedure may be further illustrated algebraically for five sites; A_1, A_2, A_3, B_1 and B_2 comprising two groups with indices $I(A_1A_2A_3)$ and $I(B_1B_2)$. The index of similarity between these two groups is

$$I(A_1A_2A_3{:}_1B_2) = \frac{I(A_1{:}B_1) + I(A_1{:}B_2) + I(A_2{:}B_1) + I(A_2{:}B_2) + I(A_3{:}B_1) + I(A_3{:}B_2)}{3 \times 2}$$

The general equation for the index of two groups of sites $A_1 \ldots A_m$ and $B_1 \ldots B_n$ is:

$$I(A_1 \ldots A_m{:}B_1 \ldots B_n) = \frac{I(A_1{:}B_1) + \ldots I(A_1{:}B_n) + \ldots I(A_m{:}B_1) + \ldots I(A_m{:}B_n)}{mn}$$

This is only one of a number of possible methods of calculation.

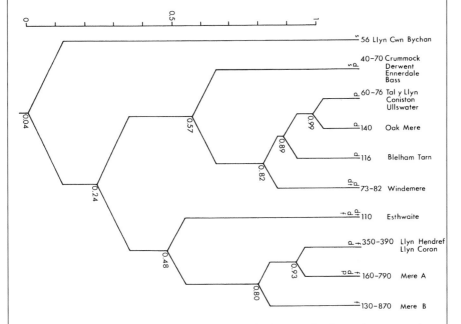

Construction of dendrograms for use with similarity indices. Species that make up more than 20 per cent of the corixid fauna in each lake or group of lakes are indicated: s, *scotti*; d, *dorsalis*; dt, *distincta*; p, *praeusta*; f, *falleni*. Conductives are indicated by the lake names (μS/cm at 25 °C).

The Pinkham and Pearson index of community similarity (Pinkham & Pearson, 1976) is based on actual and not relative abundance of the species and is expressed by the following:

$$P = \frac{1}{k} \sum \left[\frac{\min(X_{ia}X_{ib})}{\max(X_{ia}X_{ib})} \right]$$

where P is the index of community similarity, X_{ia} and X_{ib} are the abundance of species i for the respective samples (the smaller number being divided by the larger for each species), k is the total number of comparisons or different taxa in the two samples.

The Morista index of community similarity (Morista, 1959), although more complex in its calculation, has advantages with respect to sample size and is expressed as follows:

$$C = \frac{2 \sum n_{1i} n_{2i}}{(\lambda_1 + \lambda_2) N_1 N_2}$$

where

$$\lambda_1 = \frac{\sum n_{ji}(n_{ji} - 1)}{N_j(N_j - 1)}$$

and C is the index of community similarity, N is the number of individuals in sample j, n_{ji} is the number of individuals of species i in sample j.

The 'subsample' from data on stream macro-invertebrates (Table 6.3) provides us with some data with which to demonstrate the use of Sorensen's, Pinkham and Pearson and Morista community similarity indices. Note that five samples were collected from the same site at intervals of one month and an additional one sample (A_1) was taken within a few days of the first sample (A) to act as a baseline in the interpretation of the results. After the first two samples were taken, small quantities of industrial effluent entered the stream and then in the interval between samples D and E being collected, larger amounts of effluent entered the stream. The number of species, the number of individuals and community similarity indices are shown in Table 6.3.

Sorensen's index of community similarity is easily calculated from the number of species only. Looking at the information in Table 6.3, it is perhaps not surprising that samples B, C and D are very similar to A but that E is clearly dissimilar. The same trend is obtained for the percentage community similarity index.

The Pinkham and Pearson index is calculated as follows for samples A and B:

$$P = \frac{1}{11} \sum \frac{3}{4} + \frac{54}{60} + \frac{0}{1} + \frac{65}{81} \cdots = 0.41$$

The Morista index for samples A and B (Table 6.3) is calculated as follows:

$$\lambda_i = \frac{(4 \times 3) + (60 \times 59) + (1 \times 0) \ldots}{252 \times 251}$$

$$C = \frac{2 \sum (4 \times 3) + (60 \times 54) + (1 \times 0)}{(0.429) \times 252 \times 212} = 0.8$$

6.14   Application of community similarity indices

When interpreting indices of community similarity, it is important to have a baseline index and information on the sample size. That is, in addition to calculating the indices for samples taken at various intervals in time, a value indicating maximum similarity should be calculated from two samples taken at one station at about the same time. This provides a baseline index with which to compare indices calculated at later intervals. For example, and with reference to the benthic macro-invertebrates (Table 6.3), A and A_1 are two such samples that were taken as a basis for calculating a baseline similarity index.

Sample size does affect similarity indices. The Morista index seems least affected by sample size and species richness. In one detailed analysis, Smith (1986) favoured the use of Sorensen's index as a good binary index of community similarity and various forms of the Morista index as a good quantitative index.

The Pinkham and Pearson index is particularly sensitive to rare species and has been criticized as not being sensitive enough to variation in the dominant species. By way of comparison, the percentage community similarity index has a greater response to variation in dominant forms and the relationship between dominant and semi-dominant forms. As with the species diversity indices, it would seem not only useful to make a careful selection of the index but also to employ more than one index in case any particular one method gives misleading results.

6.15   Ranking, classification and ordination

The ranks of the proportional abundance of different taxa in two samples can provide a basis for comparison, using non-parametric tests such as Spearman's or Kendall's rank correlation coefficients (see Huhta (1979) for application of Kendall's rank correlation test). Although the former test is described here for completeness, comprehensive statistical books should be consulted.

Absolute measurements of species abundance are not required, since it is the relative importance of each species that is used in the calculation of the rank correlation. This is an advantage in monitoring programmes when large samples or a large sampling effort is required.

Spearman's rank correlation is calculated as follows:

$$r_s = 1 - \frac{6 \sum d^2}{N^3 - N}$$

where r_s (-1 to $+1$) is the correlation coefficient, d is the difference in the magnitude of the rank of a species or group of taxa in the two samples and N is the total number of species or groups of taxa in the paired comparison. Ties of rank are given mean values and where a species occurs in the first group but not in the second, it is put last in the rank of the second list. Testing whether or not there is a statistical difference between samples is possible but depends on the number of species involved.

Using data in Table 6.6 Spearman's rank correlation would be calculated as

$$r_s = 1 - 6 \times 587.5/6840 = 1 - 0.515 = 0.48$$

Classification of data by way of cluster analysis provides a visual or graphic method for examining and interpreting levels of affinities between groups. In other words, on a collection of samples we may wish to determine which are most closely related on the basis of similarity indices. This could be useful for establishing baseline data on affinities between communities and, later, similar analysis may indicate the extent to which perturbations have affected the affinities. There are many kinds of cluster analysis and one good example of the use of this method of analysis is the work of Savage (1982), who followed a method of analysis previously used by Mountford (1962). Indices of similarity were used to construct a dendrogram to show the affinities between communities of water boatmen (Corixidae) in a series of lakes. The details of the calculation are given in Box 6.1.

Cluster analysis using the data from Jaccard's index of community similarity has had popular use in research on pollution of freshwater communities. For example, in an examination of the effects of pollutants on the macro-invertebrate fauna in a small river, Scullion and Edwards (1980) assessed various methods as a basis for biological surveillance. Jaccard's index was used to quantify the extent of similarity between sampling stations. Arithmetic means of monthly indices were then calculated for each pair of sampling stations and these were clustered using the average linkage method. The results of this analysis can then be expressed in the form of a dendrogram, as shown in

Table 6.6 *Comparison of the relative abundance of plant species in an unmown grass sward and a sward mown once a year in August*

Species	Mean percentage cover		Rank		*d*
	Unmown	August mown	Unmown	August mown	
Holcus lanatus	32.4	30.8	1	1	0
Dactylis glomerata	15.0	13.7	2	3	1
Festuca rubra	9.1	13.0	3	4	1
Agrostis sp.	8.3	7.6	4	5	1
Plantago lanceolata	8.2	1.5	5	9	4
Arrhenatherum elatius	5.9	0.3	6	12	6
Centaurea nigra	4.6	15.0	7	2	5
Malva moschata	2.8	0.0	8	18	10
Heracleum sphondylium	0.6	1.2	9.5	10	0.5
Rumex acetosa	0.6	0.0	9.5	18	8.5
Hypochaeris radicata	0.5	0.3	11	12	1
Phleum pratense	0.4	0.2	13.5	14	0.5
Anthoxanthum odoratum	0.4	3.2	13.5	7	6.5
Ranunculus acris	0.4	0.0	13.5	18	4.5
Trifolium pratense	0.4	0.3	13.5	12	1.5
Vicia sativa	0.3	0.1	16	15.5	0.5
Ranunculus bulbosus	0.1	0.1	17	15.5	1.5
Lotus corniculatus		5.0	18.5	6	12.5
Achillea millefolium		1.8	18.5	8	10.5
Litter	9.6	5.3			
Bare ground	0.3	0.6			

d, difference in the magnitude of the rank of a species in the two samples.

Fig. 6.7. The conventional dendrogram is in the form of a rectangular grid with the level of affinity on the ordinate and the groups on the abscissa. In Fig. 6.7, it can be seen that the stations C, D and H form one level of affinity. These were the unpolluted sampling stations, while sampling stations E and F were influenced by coal siltation and G and B (which the analysis has not grouped together) by acid pollution.

Despite the wide use of cluster analysis as a method of examining similarity measures, there has been some criticism of the apparent and not real objectivity of cluster analysis. In an assessment of 11 similarity measures used to assess the impacts of clam digging on the fauna of an intertidal mudflat, Hruby (1987) concluded that too many subjective choices are made in selecting algorithms and methods of interpretation. To reduce the number of subjective choices

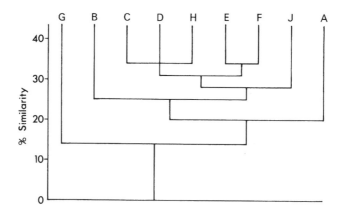

Fig. 6.7 A classification (dendrogram) showing station similarity based on average monthly Jaccard's similarity indices. (Redrawn from Scullion & Edwards, 1980.)

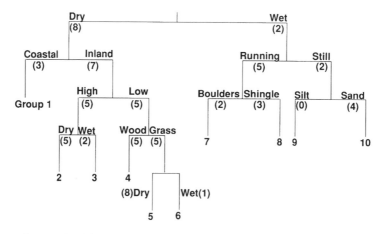

Fig. 6.8 Classification (by TWINSPAN) of 248 carabid localities in northeastern England. The numbers in parentheses refer to the number of indicator species at each division. (Redrawn from Luff *et al.*, 1989.)

made, Hruby used analysis of varience of the similarity matrix as an alternative to clustering.

Classification of communities has also been undertaken using two-way indicator species analysis (TWINSPAN). One interesting example using this method is the classification of carabid beetles from 248 localities in northeastern England undertaken by Luff *et al.* (1989). TWINSPAN is a FORTRAN program for arranging multivariate data in ordered two-way tables and it produces a hierarchical classification by repeatedly dividing data into groups. This classification of carabid beetles produced a hierarchical classification by dividing the data into two groups, followed by repeated division of each group. This was continued until each further division produced two end groups that were not ecologically distinct. The analysis resulted in the identification of ten ecologically meaningful habitat groups (Fig. 6.8), which could usefully serve as a baseline dataset for a monitoring programme.

Ordination is a multivariate technique of analysis particularly suitable for large datasets and is now commonly used in ecological research and analysis but less so in monitoring studies. Nevertheless, the technique can usefully be noted here, along with ranking and classification. The multidimensional arrangement of both sites and species in the same multidimensional space can be used as a basis for describing site characteristics of importance to community structure or it can be used to describe preferred environments of those species present. For example, Luff et al. (1989) used the FORTRAN program known as DECORANA (detrended correspondence analysis) to ordinate the species of carabids from the 248 localities. Using this analysis, they plotted the centroids of each habitat group against three DECORANA axes and then the position of each centroid and the relative distance from one another were calculated. The calculate distances between end groups is a measure of similarity between each habitat group.

7

Planning and designing ecological monitoring

7.1 Introduction

The aim of this chapter is to introduce the main requirements for ecological monitoring. There are some generic aspects of ecological monitoring that should always be considered and these are discussed below. There may also be specific aspects that are relevant to certain taxa and these are found in publications dealing with standard ecological methods. For example, standard methods for monitoring have been published by the Smithsonian Institute Press (e.g. *Measuring and Monitoring Biological Diversity: Amphibians; Measuring and Monitoring Biological Diversity: Mammals*).

7.2 Standardized methods

Standard methods for monitoring birds have been developed by the BTO in the UK (see p. 74). There are standard methods for community-based ecological monitoring of aquatic communities (see Ch. 8). For example, a monitoring manual for stream periphyton has been prepared for use in New Zealand (Biggs & Kilroy, 2000).

In many countries, information on standard methods for surveys and for monitoring may be obtained from government environmental or conservation agencies (e.g. Sykes & Lane, 1996) and also from NGOs conservation.

The use of standard methods for monitoring is a prerequisite for the success of monitoring programmes that are within and between countries. For example, the Heritage Network, which extends throughout North America and into South America, includes many participating organizations and groups. There is collaboration and commitment but of particular importance is the use of standard inventory monitoring methodology.

7.3   Communication and technical language

The success of any ecological monitoring will be determined in part by the quality of the communication between the various interested parties. In particular, it is the responsibility of the ecologist to ensure that there is good interpretation of technical terms. Terms such as indicators, biological diversity and environmental health will mean different things to different people. A good investment would be the time taken to ensure that all interested parties are involved in the discussions. One example of communication in ecological monitoring is by Schiller et al. (2001). They assessed the communication of ecological indicators and in doing so noted that such communication is an interdisciplinary problem. Furthermore, they found that communication is more than just about translating words. It is about developing a language that 'fits' with people from different disciplinary backgrounds and interests.

7.4   Data and unplanned monitoring

The data for monitoring and surveillance can be obtained either retrospectively or from a planned programme. In the case of some monitoring programmes, the choice of variables is limited because the data come from work that previously may not have been designed as a monitoring programme. Although a planned monitoring programme should be more sophisticated in terms of the choice of variables and in terms of the quality of the data, it does not necessarily follow that all retrospective monitoring is less successful than planned monitoring.

The example of the long-term monitoring of impacts of grazing Wildebeest in the Serengeti National Park (Belsky, 1985), mentioned in Ch. 6, shows how good use can be made of information that was not necessarily collected for monitoring. That example is as relevant today as it was 20 years ago. Had the increases in population levels of the Wildebeest been predicted, then the monitoring would have been designed in a different manner. As it was, and in addition to field data collected by the late Joy Belsky in 1982, the monitoring had to make use of reports of vegetation collected on previous occasions. Use was also made of aerial photographs of permanent plots established midway through the period of increased grazing as part of the Serengeti Ecological Monitoring Programme. In her research, however, Belsky (1985) successfully assessed vegetation changes and on that basis was able to suggest the best methods for a continuation of that monitoring programme.

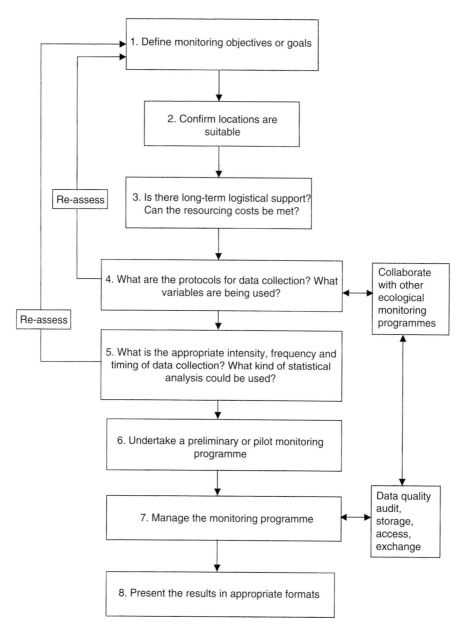

Fig. 7.1 Conceptual plan for a monitoring programme. Arrows indicate direction of steps that could be taken.

There is no doubt at all that planning and design are an important part of any monitoring programme. That being the case, good use could be made of a conceptual plan or framework for the monitoring programme, whether it be monitoring environmental quality via organisms, or monitoring populations, habitats or ecosystems. One conceptual plan was described in Ch. 2 (see Fig. 2.1, p. 34).

Indeed a conceptual plan or framework is so important for both monitoring and environmental assessments that there would appear to be a need for more research on conceptual plans. Although it would be difficult to establish a conceptual framework for all kinds of biological and ecological monitoring, the model suggested in Fig. 7.1 is of a very general nature and is meant to suggest good, basic practices.

7.5   The format and structure of a biological monitoring programme

Defining the objectives or goals

What seems to be an obvious first step, that of defining the objectives of an ecological monitoring programme, needs to be considered very carefully because all other components of the monitoring plan will be dependent on the objectives. It is the objectives that should discipline the sampling programmes that are undertaken.

Different types of monitoring have been designed by some organizations and in that case the objectives will vary according to the type of monitoring. For example, the Department of Conservation in New Zealand carries out ecological monitoring to enable it to described the state of indigenous nature and how it is changing. Monitoring is seen as an important management tool that is used to assess how the Department is progressing towards its national priority outcomes. Three types of monitoring are undertaken (Box 7.1): result monitoring, outcome monitoring and surveillance monitoring.

Some authors distinguish between scientific objectives and management objectives (Yoccoz *et al.*, 2001): scientific objectives focus on developing an understanding of the behaviour and dynamics of the ecology whereas management objectives are useful in making decisions about changes in policy. The decision about monitoring for science or for management will have implications about what to monitor.

A broad statement of objectives might initially be useful so long as there was subsequent further qualification of the objectives following a consideration of other aspects of the monitoring plan, such as facilities for collection, storage and analysis of data. Indeed, realization of the full objectives may not be possible until baseline survey work has been undertaken: without some biological and ecological information, defining the objectives becomes limited in its scope. However, a broad statement of the objectives might take the form of one

Box 7.1   Different types of monitoring for nature conservation: the three types of monitoring used by the Department of Conservation, New Zealand

Terrestrial habitat monitoring is a key area for New Zealand's Department of Conservation. The monitoring enables the Department to detect changes in 'health' and the functioning of ecosystems as well as enabling regular reporting on their condition. Monitoring is, therefore, seen as an important tool to assess how the Department is progressing towards its National Priority Outcomes. The vision is given by the following statement. 'Through ecological monitoring, the Department of Conservation expands its understanding of the ecology of the ecosystems we manage and how key species interact with their environment in space and time. By "managing to learn" our monitoring enables us to intervene effectively and efficiently to maintain, restore, and enhance biodiversity.'

The Department of Conservation, New Zealand conducts three main types of monitoring to help to determine when and how to intervene and protect biodiversity:

1.   *Result monitoring.* This is a measure of change in the pressure or disturbance affecting key conservation assets as an outcome of the Department's management. Result monitoring conducted at the same sites over time can be useful in measuring pest population dynamics and is often used instead of outcome monitoring in driving management decisions. The effectiveness of this approach is contingent on identifying damage thresholds at sites where pest density reaches a critical point or threshold that triggers severe damage to the conservation asset.

2.   *Outcome monitoring.* This is the measurement of change in the indigenous conservation asset of interest, rather than dynamics of the exotic agent causing disturbance to the asset. It is usually a repeated measure tracking 'health', productivity and/or regeneration.

3.   *Surveillance monitoring.* This is monitoring carried out in the absence of specific management actions. Surveillance can provide information on the condition of conservation assets.

An example of monitoring: the Wellington Conservancy

The Wellington Conservancy is a region in the lower North Island of New Zealand. It also includes (for management reasons) the Chatham Islands, which lie to the east of the South Island.

Result monitoring

There are nine result-monitoring projects of weed control. Residual trap-catch monitoring of possum (*Trichosurus vulpecula*) numbers is also undertaken. The brush-tailed possum was introduced from Australia and has had severe impacts on threatened plants and consequently has had significant impacts on indigenous fauna.

Outcome monitoring

A focus has been on the damage caused by possums. The aim is to monitor the condition of selected native plant species after possum control. This monitoring focuses on foliar condition and phenology.

Surveillance monitoring

A baseline for monitoring ecological change was established in the form of permanent vegetation plots in 1958–1959. The data have provided a record of the plant community dynamics in the continue presence of exotic browsing animals.

Summary

Overall, this strategy has the aim of integrating different types of monitoring within key sites so as to allow the best possible ecological interpretation of results from different management treatments, including no management. Within the strategy, the importance of appropriate physical and electronic systems is recognized for managing historic, current and future data.

(a)

(b)

(c)

(d)

The effect of an exclosure built near Wellington in 1951 (*a*) and then assessed in 1971 (*b*). The results demonstrated that such 'control' could have very clear benefits and would have implications for ecological succession (*c*, *d*). These show the recovery over 18 years following sustained ungulate control.

(Data from Urlich & Brady (2003). Photographs from the NZFS/DoC
Ecological Monitoring Collection, courtesy of Steve Urlich.)

of the following: to determine the abundance of bird populations in relation to agricultural practices; to provide baseline data on which to establish polices for management; to assess effects of forest ride management on dispersion and dispersal of reptiles; to quantify characteristics of an aquatic ecosystem in relation to management of water quality; or to determine the effects of the development of an oil terminal on the physical and biological environment.

The Canadian Parks Service established goals for monitoring ecosystems that were quite specific (Woodley, 1993):

- to measure and detect changes in the 'ecological integrity' or state or 'health' of ecosystems that a national park has been established to protect
- to measure the effects of specific threats to the ecosystem
- to provide data on the state of the national parks to other national and international agencies
- to provide a database that can be used for the preparation of reports on the state of national parks.

In doing so, Woodley (1993) recognized that there were deficiencies in knowledge about some of the ecosystems, that there were occasional unexpected disturbances in the ecosystems and that stresses on ecosystems vary in space and in time. He also suggested that a range of related social factors should also be monitored.

One example of planning objectives well in advance was the monitoring undertaken by British Petroleum and the Field Studies Council at the Sullom Voe oil terminal, Shetland Islands (Westwood *et al.*, 1989). By the time oil exploration in the North Sea was well advanced, a decision was made to bring oil from the East Shetland Basin by pipeline to Sullom Voe.

A Sullom Voe Advisory Group was established and planning for the monitoring of the effects of the development began before construction of the terminal commenced. A wide range of baseline studies was initiated and the monitoring programme was audited by a specially established committee, which, in turn, reported to the Shetland Oil Terminal Environmental Advisory Group.

Monitoring included work on the physical and biological components of the water, sediments and shores around Sullom Voe. The rocky-shore fieldwork was undertaken by staff of the Field Studies Council. The seabird populations were considered to be a special component of the environment and these have been monitored by a special working party. Such a large development inevitably brought the developers and the local community together, and the resultant monitoring programme could not have been better planned. What is important is that planning and baseline studies took place, a lesson to be learnt for many other large and small developments undertaken by the oil industry.

There is perhaps one final component of an ecological monitoring programme that needs to be included in the planning and subsequent stages, which is that there should be a measure of the 'effectiveness' of the programme. The effectiveness should be measured against the objectives of the programme and indeed such measures of effectiveness need to be undertaken at intervals during the programme to provide feedback. The requirement for deliberate and systematic adjustments of management and research questions is something that has been advocated by Hoenicke *et al.* (2003) in their monitoring of San Francisco Bay. In Rainer Hoenicke's experience (personal communication), the management questions are the foundation for building conceptual models that can bring scientists and decision makers together regarding the current understanding of how an ecosystem works. He goes on to say:

> ... obtaining general agreement about 'desirable states' is really crucial, since it is the starting point for any adaptive management process. For example, the Montreal Process and the indicator developments for temperate and boreal forests (Santiago Declaration)

used increasingly specific definitions of what is considered 'sustainable' management or use of resources. Goals and desired states can be determined and continually adjusted on trend lines that either point towards or away from a sustainability benchmark.

Finally, Rainer Hoenicke suggested the following components for adaptive management (personal communication):

- identify desired states and establish conceptual models of how things might work
- act to adjust the system and move it towards a more desirable state
- monitor to see if it worked
- accompany the monitoring with research to test specific hypotheses and adjust your conceptual model.

Monitoring localities

Having defined the objectives, it is necessary to consider where the monitoring will take place. The locality for monitoring needs to be secure in the sense that there needs to be certainty that the locality can continue to be used as a suitable site for monitoring. A secure location is equally as important as a location that has been the subject of previous recording and research. Some protected areas such as nature reserves or national parks fulfil these require- ments and may, therefore, be suitable, secure locations for long-term monitor- ing programmes.

In earlier ecological monitoring studies, there were occasional difficulties in locating monitoring site after a passage of time. It was perhaps not surprising that little research had been undertaken on effective, permanent marking of transects, quadrats and other data-collection sites. Today, monitoring sites locations can, in many terrains, be located with considerable precision using the global positioning system (GPS; e.g. Hulbert & French, 2001). However the use of GPS for locating sites in mountainous terrains still has limitations.

When the NSF was considering the establishment of long-term ecological research sites (NSF, 1977, 1978), it gave priority to sites with the following characteristics:

- where productive and useful short-term research has already been conducted
- representative of major ecosystems and populations
- considered important to major ecological issues
- where active scientists have shown interest
- protected and accessible for research purposes

- that are pristine ecosystems where crucial variables can be monitored
- that are not pristine but are unaffected by undesirable impacts and meet the other criteria.

In choosing sites and whether that locality would provide the most 'useful' sources of data, the question has to be asked, 'Are the sites representative of the ecological communities or ecological systems that are to be monitored?'

The spatial scale has to be considered as also should the necessity to distinguish between changes in managed and unmanaged systems. For example, and in relation to conservation objectives for woodlands in southern England, Kirby & Thomas (2000) concluded that changes in ground flora as well as woody layers in both managed and unmanaged stands should be monitored.

Reference or 'benchmark' sites are needed but may be difficult to identify because of the normal variability that takes place from year to year. Control sites are technically impossible because there is no control possible over background variability. The consequences are that only large-scale changes set against reference data are detectable in the short term and trends of change may take many years to identify. The situation may be further complicated by the limited amount of knowledge about the biology or life cycles of some species, even common species. One reference site and one other site would probably not be sufficient. There is generally a need for several sites for ecological monitoring or for long-term ecological research.

There is much to be said for using existing long-term monitoring sites but the following are a few questions that may need to be considered:

- are the localities suitable for monitoring, particularly in terms of the objectives?
- will the locality be secure for the duration of the monitoring?
- are the localities representative?

Some biological monitoring depends on data collected from precisely the same sites, and permanent quadrats may have been established for this purpose. That being the case, not only should the locality be secure and representative but it is important that the sampling sites of the permanent quadrats can be located on each occasion data are collected.

It is important to ensure that data will continue to be collected in a uniform manner and that methods of data storage and retrieval will be adequate. Continuity is of an essence, and the longer the monitoring programme the greater the need to ensure that there is the support for that continuity.

A good strategy to adopt in any research is to think about the methods of analysis before the data are obtained, because there is the possibility that data can be accumulated in a form that is not suitable for analysis. Such a strategy may, therefore, require a preliminary investigation. At this stage, it is important to consider the intensity and frequency of the sampling that would be required to detect change for a given level of precision.

Long-term logistical support and resourcing

From my experience, I would estimate that approximately 80–90% of monitoring programmes fail or are abandoned because of lack of resources. There has to be a long-term logistical, financial and resourcing commitment.

Monitoring may be undertaken for long periods of time, during which the people initially involved in establishing the programme may change. Obviously, there needs to be coordination of the programme so that records are not only deposited safely but are accessible and understandable to successive people involved in data collection. Standardization of methods as well as clear and precise descriptions of the methods used are of paramount importance for the continued success of any monitoring programme.

The advances in information technology should be exploited as fully as possible. The internet will increasingly provide data entry and checking routines and ways of storing and accessing data. There are now many innovative ways of data storage and data exchange.

While some biological monitoring programmes may require only one person or a few people to make regular data collections, other monitoring may require much larger datasets or data collected simultaneously from different regions of the country. It is for this reason that many monitoring programmes have drawn on the help of volunteers for the collection of data and the following are but a few examples (see Ch. 8).

The Field Studies Council in the UK has drawn on the help of schoolchildren for data gathering. Some universities and colleges have made use of students in ensuring regular data collection. The UK BRC, even from its earliest beginnings, relied very much on amateur help for collection of data. The British Society Lichen Mapping Scheme, the BTO common birds census and many other monitoring programmes rely on the work of amateur naturalists.

Two programmes concerned with the state of coastal areas have made use of support in the form of many, many volunteers. These two programmes contribute to a European-wide series of coastwatch projects in which ten countries play an active role. The Coastwatch programme collects information on the distribution and extent of the UK's coastal habitats and intensity of human activities. These basic data will provide a baseline for future monitoring of coastal habitats. The other coastal programme is the Norwich Union Coastwatch UK. This was established in 1987 and is part of the wider Coastwatch Europe Initiative. Each year since 1989, there has been a national survey to collect information on the state of the coasts. Funding for this particular programme become an issue in 1996 when Norwich Union withdrew its funding (a lesson in the need for an assurance that there will be long-term logistical and resource support).

In the conceptual plan or framework (Fig. 7.1), there is reference to the need to ensure that there will be logistical support for the monitoring programme. One of the greatest problems facing biological monitoring, especially long-term monitoring, has been the lack of financial and administrative support. Much ecological research has been based around short time periods of three or four years, and opportunities to extend that research are rare. Although the advantages, benefits and need for more long-term monitoring projects have become obvious during the last decades of the twentieth century as a result of many pressing environmental issues, there is still a need to undertake structuring of the monitoring programme to make it cost effective.

Long-term ecological monitoring (more than three years) requires a commitment not only in terms of human resourcing but also in terms of funding. Few agencies are prepared to commit to long-term funding and this has been a major barrier for ecological monitoring programmes, and indeed long-term ecological research. The few examples of environmental monitoring that have withstood tests of time and changes in government and government restructuring are those programmes that affect everyone and are perceived as being relevant to everyday activities. For example, monitoring climate change and daily reporting of weather has continued in the face of many logistical and financial changes. It is a pity that the relevance of some ecological monitoring to everyday life has not been made apparent.

Data collection protocols and ecological variables

Perhaps the most basic aspect for a monitoring protocol is agreement on baseline information or data. Ecological monitoring is to do with trends, and as such, trends have to be compared with something. It is essential, therefore, that a first step is to obtain baseline data or a benchmark.

In nature, the unexpected is bound to occur. Stochastic processes are part of nature and have to be expected as part of ecological monitoring and long-term ecological research. That which was not predicted may well occur, and consequently there may have to be a reassessment of the protocols and ecological variables. Similarly, long-term environmental changes have implications for the use of previously selected variables in ecological monitoring. For example, climate change is beginning to have confounding effects in ecological monitoring programmes.

What to monitor and over what spatial scale? There is a wide range of variables and processes, ranging from indicator species to productivity, that can be used for

Table 7.1 *Ecological variables, indicators and indices used for forest biological diversity: the following are part of a set derived by the Forest Liaison Group in the process for implementation of the 1992 Convention on Biological Diversity*

Group	Aspects covered
International	Forest cover
	Forest condition
	Protected areas
Core set	Area of natural forest
	Area of natural forest as a proportion of the total
	Change in natural forest over 10 years
	Protected areas (International Union for the Conservation of Nature classes)
	Protected area by eco-region
	Number of forest-dependent species
	Proportion of forest-dependent species at risk
	Area of forest managed to prioritize conservation
	Air pollution levels exceeding forest critical loads
	Legislation to protect biological diversity
Detailed national information	Mapped details of forest type
	Mapped details of protected areas
	Percentage and extent in area of forest types
	Level of fragmentation and connectiveness
	Number of forest-dependent species
	Population levels of indictor species
	Area of forest containing endemic species that is cleared annually
	Percentage of annual natural regeneration
	Human disturbance index
	Main threats to forest biological diversity

biological monitoring and the use of many of these has been described in previous chapters. As part of the process used to identify the variables or the indicators or indices to be used in the ecological monitoring programme, it is helpful to assess existing initiatives. Indeed, there may be a case for establishing the new programme as part of an existing monitoring programme.

International initiatives that have led to identification of ecological variables for biological diversity include the Convention on Biological Diversity and the Montreal Process. A convention liaison group on forest biological diversity has, for example, established a set of variables, indicators and indices (Table 7.1). The Montreal Process, although more concerned with variables that meet the needs of reporting on carbon stocks (see also p. 6), has a range of variables that relate to diversity at different biological levels of organization (Table 4.4, p. 132).

The information collected must be scientifically defensible and to achieve this the methods of data collection need to be selected carefully as should the methods of analysis. The frequency of data collection has implications for analysis and this, in turn, has implications for the defensibility of the results. However, there is a difficulty here, which is that environmental variables will show normal variability from year to year. The multifactorial sources of variability make it difficult to identify long-term trends, which could be affected by pollution or other perturbations.

For example, some butterfly numbers can fluctuate considerable from year to year and, therefore, it is of interest to know the extent to which a monitoring programme is able to detect any trends. When there are large fluctuations from year to year (caused by weather, availability of food plants, and predators), it is difficult to detect such trends. Van Strien et al. (1997) have assessed the statistical power of butterfly monitoring programmes (British butterfly monitoring scheme, Netherlands monitoring scheme) to detect trends and they also attempted to estimate the variance components in the data. Several different approaches for the assessment of statistical power are discussed in their paper. For the British programme, they reported that, for 37 out of 51 species studied, a decrease of 50% or less is detectable with a power of 80% within a 20-year period. For the Netherlands Programme, they reported that a decrease is detectable for 29 out of 47 species.

Ideally, the choice of variables and processes should have a wholly ecological basis, but logistic limitations (finance, time and effort) may override such considerations. A good example of balancing sampling effort against scientific defensible data is described by Andersen et al. (2002) with their work on ants as indicators of pollution (Box 7.2). Indeed they have found that simplification of data can lead to superior indicator performance (Andersen, personal communication).

Many would support the suggestion that ecological monitoring programmes should be as simple as possible. The simpler the data, the easier it is to collect in a cost-effective manner and the more likely it is that the monitoring programme will continue for a long period of time. Related to this is the suggestion that variables that are too specific and are focused on today's concerns may become too limiting in the future (Allen et al., 2003a,b). That is, if the monitoring is relevant to the long term, then it is important not to focus the data on today's explicit concerns.

Rob Allen (personal communication) has stressed that monitoring should not be specific to today's needs and, therefore, it is best simply not to be too specific. However, the issue of how specific or how general relates to different kinds or scales of monitoring. For example, if the goals of a large-scale monitoring programme are wide ranging, then it is probably important not to be too specific and to allow for an evolving programme. If the programme is related to specific activities such as mining, then it is probably best to be specific and simple.

Box 7.2 Ants as biological indicators of SO_2 pollution: assessing trade-offs between simple monitoring procedures and reliability

In practice, the costs of ecological monitoring have to be considered (indeed it is essential) when designing an ecological monitoring programme. The use of biological indicators (and indices) need to be assessed with care in respect to both the costs and the scientific credibility.

Compared with vascular plants and vertebrates, relatively few terrestrial invertebrate taxa have been used successfully as biological indicators. However, several species of ants and ant communities have long been used as biological indicators (Andersen, 1990). In Australia, ant species richness and species composition have been used as indicators of restoration success and also for off-site mining impacts and forest management.

In Australia, Alan Andersen and colleagues have undertaken a comprehensive assessment of the response of ant communities to the SO_2 emissions from the Mt Isa copper and lead smelting activities. Intensive surveys resulted in 174 species and revealed several ecological criteria that were affected by levels of SO_2.

Building on the previous intensive surveys, Andersen and colleagues then went on to assess the extent to which a much simplified approach using ants as biological indicators could produce the same results. The simplified approach gave much the same results. In particular the results revealed the following:

- different habitats support distinct ant faunas
- ant abundance declined with increasing SO_2
- species richness declined with increasing SO_2
- species composition varied with levels of SO_2
- several common species showed clear abundance patterns in relation to SO_2
- species responses varied according to their biogeographical affinity.

Alan Andersen and his team have concluded that comprehensive sampling is not required when using ant communities as biological indicators of pollution. They add that the main issue is not about being comprehensive but about being reliable.

The copper and lead smelter at Mt Isa in the Australian semi-arid tropics. (Photograph by Tony Griffiths.)

The northern meat ant *Iridomyrmex sanguineus*, a dominant species throughout the Australian semi-arid tropics. (Photograph by Greg Miles.)

*(Material adapted with permission from Andersen et al., 2002. Journal of Applied Ecology, **39**, 8–17.)*

The collection of data should be non-destructive. This is not always possible and in such cases steps should be taken in the sampling procedure to ensure that you do not end up measuring and monitoring the effects of the monitoring programme.

Because data collection may be time consuming and expensive, methods for collection of data from the field or assemblage of data from other sources should be considered along with the choice of parameters.

Simple variables may well be more cost effective as well as meaningful in monitoring programmes and these aspects were included in recommendations for long-term ecological monitoring made to the NSF (1977, 1978) that measurement techniques should have the following characteristics:

- simplicity and reliability: so that studies made at different sites or times or by different investigators may be compared with confidence
- stability: sites unlikely to change drastically over a period of decades or when subject to rigorous intercomparison when techniques change.

To these we must add the need for standardized methods. Standardized methods have been developed for long-term research on soils (Robertson *et al.*, 1999) but perhaps less so for long-term ecological research in general. However, some protocols for data collection and indeed selection of ecological variables have been developed. For example, an early set of protocols or 'rules' with special reference to woodland ecosystems is shown in Table 7.2. These rules were suggested by Peterken & Backmeroff (1988) as a result of their ecological monitoring work on woodlands. Much later, protocols have been established for LTER and the ILTER network (Ch. 2). The UK ECN has established protocols for standard measurements at terrestrial sites (Sykes & Lane, 1996).

There is a need for a permanent record of the protocols or precise methodology that can be made available to any changes in staff involved. If there are any changes in the protocols then such changes need to be clearly recorded and clearly linked to the monitoring being undertaken at that time. An example of the use of straightforward and simple variables is described in the Countryside Commission exercise on monitoring and management of wildlife habitats on farms (Matthews, 1987). One of the many activities supported by the Countryside Commission as part of the Demonstrations Farm Project was the creation of a pond in a wet area of 'waste land'. The results of monitoring have been described by Usher (in Matthews, 1987) and the monitoring was based on the use of plant species richness, abundance and composition (Table 7.3). On the basis of simple comparisons between 1979 and 1984, it was possible to give a fairly

Table 7.2 *A set of rules for ecosystem monitoring, with special reference to woodland ecosystems*

1.	Any variable or process which can be readily measured and dated may be valuable in detecting changes in ecosystems
2.	Long-term monitoring must be supported by administrative continuity otherwise the programme may simply be overlooked or forgotten
3.	Facilities are required to ensure (i) survival of records and duplicate copies or records, (ii) markers locating the transect or quadrat and (iii) that the programme is known to exist
4.	Repetitive recording is obviously necessary and although it may not be necessary at regular intervals, further records should be taken after or prior to any formative events
5.	The monitoring locality should be inspected regularly (annually for a woodland) even if information is not collected
6.	Although objectives of the monitoring need to be defined, recording aims should be open-ended. The basic systematic record should be supplemented with casual adjuncts, which have a habit of being valuable at a later date; this is because we do not know how the data could be applied in the future
7.	Simple variables and processes well recorded are more valuable than poorly recorded complex variables and processes: it is better to record something rather than nothing
8.	Representative records and replicates should be established if possible but even an unrepresentative sample may be valuable in the future analysis
9.	Regular analysis and preparation of reports, even at early stages in the monitoring, help to improve the methods for data collection and help to refine the objective; these reports also serve as a reminder of the programme

From Peterken & Backmeroff, 1988.

detailed but informative account of the changes that had taken place. On reflection, Michael Usher made the following observations (personal communication).

- If you want both amateurs and volunteers to help with monitoring, then it needs to be kept simple.
- If more than one person is doing the monitoring, then it is absolutely essential that the methods are defined precisely. In the example given, r (rare) was 1–10 individual plants or one clump in a feature (hedge, pond, field etc. (i.e. a feature was an entity that had been mapped); o (occasional) was 11–30 plants, or two or three clumps; f (frequent) was 31–100 individuals or 4–10 clumps, and so on. In other words, define the rules precisely and carefully so that they can be understood by anyone doing the monitoring.

Table 7.3 *A simple and effective basis for monitoring the plant species of a pond during establishment and management of wildlife habitats on demonstration farms; here, results are taken from a study of a new pond established from wet wasteland at Hopewell House, UK: the plant species composition has changed and species richness has increased from 28 to 34*

Species	1979 abundance	1984 abundance
Agropyron repens (Common Couch)	●	f
Agrostis stolonifera (Creeping Bent)	●	a
Alopecurus geniculatus (Marsh Foxtail)	a	●
Arrhenatherum elatius (False Oat-grass)	●	a
Artemisia vulgaris (Mugwort)	●	r
Chamaenerion angustifolium (Rosebay Willowherb)	a	●
Cirsium arvense (Creeping Thistle)	f	a
Dactylis glomerata (Cock's-foot)	●	a
Deschampsia cespitosa (Tufted Hair-grass)	f	a
Epilobium hirsutum (Great Willowherb)	o	a
E. palustra (Marsh Willowherb)	f	●
E. montanum (Broadleaved Willowherb)	●	o
Equisetum arvense (Field Horsetail)	o	●
Galeopsis tetrahit (Common Hemp-nettle)	a	o
Gallium palustre (Common Marsh-bedstraw)	a	●
Glyceria declinata (Small Sweet-grass)	a	●
Heracleum sphondylium (Hogweed)	o	r
Holcus lanatus (Yorkshire Fog)	●	o
Iris pseudacorus (Yellow Iris)	●	o
Juncus acutiflorus (Sharp-flowered Rush)	a	●
J. effusus (Soft Rush)	o	a
Lolium perenne (Perennial Rye-grass)	●	r
Lotus uliginosus (Great Bird's-foot Trefoil)	o	●
Odontites verna (Red Bartsia)	●	r
Phleum pratense (Timothy)	o	o
Phalaris arundinacea (Reed Canary-grass)	a	●
Plantago major (Greater Plantain)	r	●
Poa trivialis (Rough Meadow-grass)	r	●
Polygonum amphibium (Amphibious Persicaria)	a	f
Polygonum persicaria (Common Persicaria)	●	o
Potentilla anserina (Silverweed)	a	r
Ranunculus repens (Creeping Buttercup)	f	a
Ranunculus ficaria (Lesser Celandine)	o	r
Rosa canina agg. (Wild Rose)	●	o
Rubus fruticosus agg. (Bramble)	●	o
Rumex crispus (Curled Dock)	o	●
Rumex obtusifolius (Broadleaved Dock)	r	r

Table 7.3 (cont.)

Species	1979 abundance	1984 abundance
Salix cinerea (Grey Willow)	●	o
Salix fragilis (Crack Willow)	●	r
Scrophularia nodosa (Common Figwort)	●	o
Senecio jacobaea (Common Ragwort)	●	r
Solanum dulcamara (Bittersweet)	o	r
Sparganium sp. (Bur-reed)	●	?
Stachys sylvatica (Hedge Woundwort)	o	o
Taraxacum officinale agg. (Dandelion)	●	r
Trifolium pratense (Red Clover)	●	o
Urtica dioica (Common Nettle)	f	f
Veronica beccabunga (Brooklime)	f	●

●, absent; r, rare; o, occasional; f, frequent; a, abundant.
From Matthews (1987) with kind permission of the Countryside Commission.

- As this was a demonstration farm, they needed qualitative information that could easily be understood by visiting farmers, students, etc. The monitoring was therefore designed accordingly; if they had wanted a more scientific dataset, then a more intensive and quantitative scheme would need to have been devised.

Most if not all ecological sampling methods have a bias and have limitations. The bias of any sampling method must be determined and the limitations of the sampling methods should be qualified. Sources or error during data collection must be identified and acknowledged. For example, when measuring diversity, the sampling effort or the size of the survey will have implications for the results. Yoccoz *et al.* (2001) have drawn attention to two potential sources of error that should be considered when estimating biological diversity. One is a detection error because few survey methods permit the detection of all individual organisms or even species. The second source of error involves the inability to survey large areas and the need to take samples that draw inferences about the large area on which the sample is based.

There may also be sources of error arising from species identification. People trained with taxonomic skills are essential but it is also useful to consider building in checks such as confirmation of species identifications.

The choice of time-series analytical techniques should be made at the same time as the selection of variables and processes. The disadvantage of not following this procedure is that many data can be collected that are not easily analysed or are not in a form that can easily be examined and presented in reports.

Apart from the common conventions for time-series data analysis described in any appropriate statistical book (e.g. Diggle, 1990) or monitoring books with substantive statistical information (Downes *et al.*, 2002), there is no right or wrong way for presenting data. Several alternatives may be investigated before a suitable form is adopted, but it may be useful to consider who will be reading the reports and who will be making recommendations on the basis of the reports.

Collaboration, communication and partnerships

Ecological monitoring is not just about ecology. Ecological monitoring in practice involves many disciplines. Furthermore, there are many agencies, organizations and groups of people who are interested in or who will be affected by environmental change. There are many groupings involved in environmental change. Therefore, throughout any ecological monitoring programme there has to be partnerships and good communication.

The success of any ecological monitoring programme depends partly on good planning, partly on the logistical support to continue the monitoring programme over the appropriate period of time and partly on coordination with other related programmes. In the 1980s, a US Federal Marine Pollution Plan concluded that there was a need in marine monitoring for improved coordination among the existing monitoring programmes as well as for more effective compatibility and communication. At that time, steps necessary for good levels of coordination and management include the following (Wolfe & O'Connor, 1986).

1. Develop an active inventory of regional and local monitoring activities to provide programme details for all users.
2. Establish uniform formats for a suite of selected monitoring parameters to facilitate accession and analysis of information.
3. Establish quality assurance systems to assure comparability of data programmes and regions.
4. Establish a national network of database management systems for storage, dissemination and analysis of data.

More recently, the importance of partnerships was discussed at an International Conference on Ecosystem Health at the Quetico Centre in Canada (see also p. 54). There, in answer to the question 'what does it take to develop a partnership?', Sonia Ortega suggested the following:

- set collective goals
- have a shared purpose
- standardize measurements and also ideas
- continue communication.

Frequency and timing of data collection

Together with careful selection of variables and processes there may also be a need for judicious selection of timing and frequency of data recording. Where organisms are involved, phenological aspects must be taken into consideration as well as the spatial and temporal patterns of distribution.

The frequency of data collection will depend, in part, on the objectives of the ecological monitoring programme and the questions being asked. Perhaps more importantly, the frequency of data collection and the amount of data should be determined by the statistical analysis. That is, for different types of analysis there will be a need for a certain level of change or number of cycles of data. Power analysis on the pilot studies will be important in helping to decide on the intensity and frequency of sampling.

Feasibility or pilot study

One key reason for a pilot study is to re-assess the field methods being used and also to review the methods in relation to the goals of the monitoring programme. An example of the value of a pilot study was an assessment of the methodology developed by the UK ECN for long-term monitoring of macrophytes in rivers and streams. A pilot study of the ECN methods took place at five sites in 1997 (Scott *et al.*, 2002). The results included some practical difficulties in the methods:

- problems with the definition of the area to be surveyed
- the definition of 5 m sections for watercourses that curved substantially
- estimates of vegetation percentage cover.

In addition, the pilot study provided an opportunity to re-assess the lengths of the watercourses to be surveyed. A primary reason given for smaller survey lengths was the overall aim of the ECN monitoring. This is a good example of the value of re-assessing the methods in relation to the objectives.

One other key reason for a pilot study is to obtain an unbiased estimate of the variance in the ecological data. Therefore, the feasibility study must include a preliminary data-gathering exercise, which might also provide the baseline data. Planning biological monitoring cannot take place without biological information and, therefore, data may also need to be assembled from published sources or from preliminary field surveys.

A useful purpose of the pilot study is that it allows an opportunity to gather general information about the biology of the area. In its most simple form, this information could be in the form of species lists for selected taxa such as flowering plants, birds and spiders or it could be in the form of distribution of sensitive habitats and protected species. More detailed information might extend to the distribution and the population size and structure of the selected taxa.

To be able to quantify changes that may be the result of management, it is necessary to have a basis for comparison. As already emphasized, good baseline data, assembled as part of the monitoring plan, are an important prerequisite for successful biological monitoring. We can think in terms of baseline data as that information collected in the same place and on the same basis as subsequent data collection whereas reference data may have been collected previously. For example, data from monitoring of a population in relation to management could usefully be compared with populations trends observed elsewhere. This has been done with some data from the NCC Butterfly Monitoring Scheme (Pollard et al., 1986), one of several countrywide butterfly ecological monitoring programmes (Pollard et al., 1995). In order to assess the results of butterfly habitat management, an index of abundance of each species of butterfly on a nature reserve can usefully be compared with regional trends (Fig. 7.2). In addition to these comparisons with regional trends, Pollard (1982) also made comparisons within a transect on the nature reserve to try and identify the cause of any departure from regional or national trends.

The feasibility or pilot study is a good stage at which to re-assess the goals of the monitoring and to ask whether the information collected can be applied and used to answer the questions posed. Perhaps the wrong information is being collected or perhaps the goals are too ambitious.

A baseline survey will also help in an assessment, for example, of which species will be most useful in the monitoring exercise. Many of the species may be identified easily by those with a basic biological knowledge, but there will be many species that can be identified only with the help and expertise of specialists and taxonomists. Some monitoring programmes have been modified only because it was soon realized that identifying species and the skill required for identification are all too scarce. The need for more people skilled in taxonomy and systematics has become clear as a result of more and more environmental assessment and monitoring work.

The pilot study may also be a good opportunity to assess the skills and knowledge of the people who will undertake the data collections and species identifications. Perhaps some training may be required depending on the skills and expertise needed.

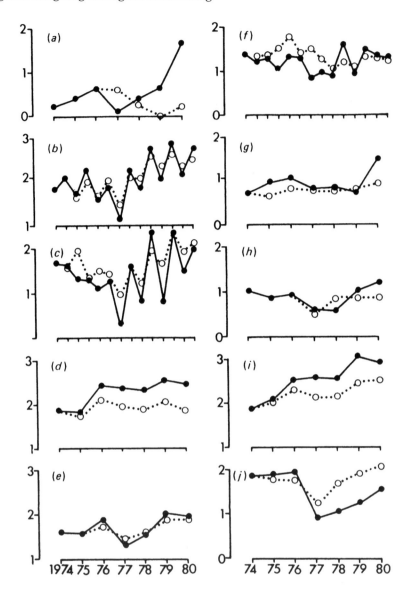

Fig. 7.2 Butterfly monitoring in the UK during the 1970s. Index values (log values) at Monks Wood (—) were compared with regional or national trends (. . . .). The starting point for regional trends is the Monks Wood index value for 1974 and for national trends the Monks Wood index value for 1976. Monks Wood index values are excluded from the regional data. (a) Grizzled Skipper (*Pyrgus malvae*); (b) Green-veined White (*Pieris napi*); (c) Peacock (*Inachis io*); (d) Gatekeeper or Hedge Brown (*Pyronia tithonus*); (e) Small Heath (*Coenonympha pamphilus*); (f) Brimstone (*Gonefteryx rhamni*): (g) Orange Tip (*Anthocharis cardamines*; (h) Speckled Wood (*Pararge aegeria*); (i) Meadow Brown (*Maniola jurtina*); (j) Ringlet (*Aphantopus hyperantus*). (Redrawn from Pollard, 1982.)

Managing the ecological monitoring programme

It may come as a great relief to many people when the ecological monitoring programme has commenced. Such relief is probably only a reflection that there has been much planning, assessment and re-assessment of the goals and the methods. Once established, and in addition to regular data collection, the programme offers the opportunity to continue to confirm such things as:

- good communication between the interested parties
- that protocols for data collection are adhered to
- that the infrastructure is in place to support the programme.

Added to these is the continued need to address adequate data storage, data searching and data exchange and communication. In particular, it is essential to make best use of opportunities provided by the internet.

Analysis and presentation of results

Imagine that data on a species population has been collected with the objective being to assess the effects of habitat management. This could have been a rare species. Imagine that the numbers have overall increased. It would be tempting to believe that the management has been the direct cause of the increase in population. Demonstrating cause and effect is very difficult and very often can only be achieved with carefully managed experiments.

An example from Allen *et al.* (2003b) demonstrated some limitations with regard to analysis and interpretation. In New Zealand, mortality of some tree species has previously been used as an indicator of Brushtail Possum (*Trichosurus vulpecula*) foraging impacts. However, in one analysis of the impacts of possums, it became difficult to identify the effects of possums because possum density was related to plant composition, reflecting soil fertility. Dead trees can relate to soil fertility gradients and it is difficult, therefore, to identify the effects of the possum gradients.

There is a growing number of long-term ecological studies that have provided data for detailed analysis. The kind of analysis undertaken in the case of some of these long-term ecological studies may act as a guide for analysis of data from ecological monitoring. For example, English Nature has published a pilot analysis of long-term ecological change (1971–2000) in some British woodlands (Smart *et al.*, 2001). This pilot analysis was on data from 14 broadleaved woodlands and two native Scots Pine (*Pinus Sylvestris*) woodlands. The goal was to test analytical techniques against the variation present in 16 pilot sites. The analysis included the following:

- change in age–class distribution of trees and shrubs (based on classes defined by trunk diameter at breast height (1.4 m))
- change in basal area of trees and shrubs
- change in derived vegetation and environmental variables between and within sites (using Ellenberg fertility scores, species richness, competitor and ruderal scores)
- soil pH
- soil organic matter
- Frequency of ancient woodland indicator species.

One aspect of ecological monitoring easily forgotten until much later in the monitoring programme is the presentation of data. Simple summaries of data can often be useful, particularly when several variables have been considered. Tabulation of data may be necessary simply to make available all the data for assessment at a later date. Graphing the data can be a simple but effective first way of presentation but there is more than one way to show relationships in a graph.

One old but good example of alternative ways of presenting data comes from a study by Haefner (1970) in which mortality of Sand Shrimps (*Crangon septemspinosa*) was monitored in relation to the combined effects of temperature, salinity and oxygen. A summary of the data is shown in Table 7.4 and although not a large set of data, it is not immediately obvious which combination of variables causes least mortality. Haefner then put the data into simple graph form, two examples of which are shown in Fig. 7.3, and the combined effects of temperature, salinity and oxygen become immediately obvious.

The range of statistical and graphical computing packages now available provides an excellent basis for data presentation. Those not familiar with the range of packages available are encouraged to attend appropriate computing courses and read around the subject (e.g. Downes *et al.*, 2002).

7.6   Ben Stout's 'good, bad and ugly'

A useful summary of some main points can be found in an overview and assessment about defining monitoring questions and objectives that was once described as 'the good, the bad and the ugly' (Stout, 1993). Under the heading of 'the ugly', Benjamin Stout drew attention to the fact that enthusiasm may decline and programmes may become abandoned. Under the heading of 'the bad', he referred to the maxim 'the road to hell is paved with good intentions'. He then goes on to refer to the common factor in monitoring programmes and that is changes in protocols. Under the heading of 'the good', he suggested that a good monitoring programme would:

Table 7.4 *A summary of four-day mortality data recorded for the Sand Shrimp* (Crangon septemspinosa) *during exposure to low dissolved oxygen concentrations at 12 different temperature–salinity combinations*

Temperature (°C)	Salinity (%)	Percentage mortality		
		Male	Non-ovigerous female	Ovigerous female
5.0	4.95	100	100	100
5.0	14.78	42.1	41.7	50.0
5.0	25.48	6.7	11.1	23.1
5.0	45.44	22.2	41.7	7.7
15.0	4.92	100	100	100
15.0	14.69	40.6	27.3	50
15.0	24.88	19.4	21.4	30.8
15.0	46.79	44.1	30.0	50.0
23.5	4.97	100	100	100
23.5	14.42	70.4	33.3	90.0
23.5	24.22	30.0	52.9	81.8
23.5	45.07	92.6	90.9	100

From Haefner, 1970.

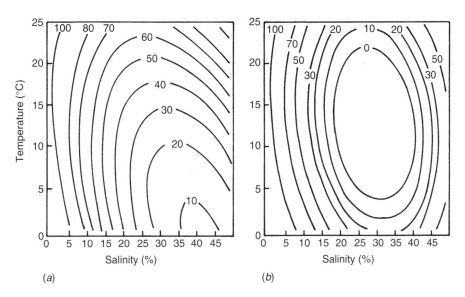

(a)   (b)

Fig 7.3 Estimation of percentage mortality of ovigerous Sand Shrimps (*Crangon septemspinosa*) based on data from 12 combinations of salinity and temperature in (a) low concentrations of dissolved oxygen and (b) aerated water. (Redrawn from Haefner, 1970.)

- be structured and developed so that it will be supported for its duration by science and politics
- have clear objectives
- have rigorous plans for analysis
- have a sampling protocol that takes into account variation at spatial scales
- have protocols that produce chronologically collected data which will allow for rigorous testing.

Ben Stout (personal communication) has seen little evidence that things have changed very much. He gives as an example the Pacific northwest forestry industry and makes the observation that the original sampling of the Northern Spotted Owl and Marbled Murrelet (on which decisions were made) were flawed. Thus 'poor monitoring can lead to economic chaos'.

8

Community-based ecological monitoring

8.1   Introduction

Whether they are called public environmental monitoring, community environmental monitoring, citizen science monitoring, lay environmental monitoring or volunteer environmental monitoring, they all mean the same thing. They are environmental monitoring programmes that promote or encourage groups of people (local urban communities, rural communities, schools) to take part in environmental monitoring programmes. The term 'community-based monitoring' is used here in a generic sense to refer to all of the above groups of people.

The term community-based environmental monitoring has been used to refer to the collecting, analysing and interpreting of environmental data by lay people (Lambie, 1997). Indeed, in circumstances where there is a lack of 'specialists', there may be good reasons for a community role in monitoring and management. Danielsen *et al.* (2000) described such circumstances where there is a need for simple monitoring programmes in countries that lack specialist staff.

The word community in this chapter is used to refer to human communities (a collective group of humans). This chapter is about the role of various groups of people such as schools, colleges, village groups, local communities and Agenda 21 groups, and their role in monitoring ecological change.

Perhaps some of the earliest examples of communities of people recording changes in nature can be seen in cave drawings. In some caves, a series of drawings indicate changes in abundance of various species of animal. Recording changes in nature is, therefore, not a new idea and such practices are embedded in many cultures throughout the world.

Ecological monitoring programmes are, generally speaking, designed and managed by scientists in conjunction with government departments or research organizations. However, ecological monitoring, recording and surveillance of nature is undertaken by many individuals, groups of people and communities.

Included in this chapter are some examples of ecological monitoring projects that have been or are being undertaken by individuals and local community groups.

8.2   Why community-based ecological monitoring?

In today's world, why would any community take part in an ecological monitoring programme? What role, if any, do community-based ecological monitoring programmes have to play in environmental management? What are the advantages and disadvantages of community-based ecological monitoring?

It is likely that the Rio Earth Summit, by way of its products, gave support for the growth in community-based environmental studies and possibly environmental monitoring. In the Rio Declaration, for example, there is reference to community involvement with regard to environmental issues in Principle 10.

> Environmental issues are best handled with the participation of all concerned citizens, at the relevant level. At the national level, each individual shall have appropriate access to information concerning the environment that is held by public authorities, including information on hazardous materials and activities in their communities, and the opportunity to participate in decision-making processes. States shall facilitate and encourage public awareness and participation by making information widely available.

Agenda 21 was and in some countries continues to be a blueprint to encourage sustainability into the twenty-first century. Section 111 is about strengthening the role of 'major groups' including women, children and youth, indigenous peoples, NGOs, workers and their trade unions, business and industry, the scientific and technological community, and the role of farmers. At the risk of taking quotes out of context, there are several sections in Ch. 36 of Agenda 21 that appear to prompt community involvement. The following could be included.

> Promoting education, public awareness and training. A. Re-orientating education towards sustainable development. Basis for action. Activities 36.5e. Relevant authorities should ensure that every school is assisted in designing environmental work plans, with the participation of students

and staff. Schools should involve school children in local and regional studies on environmental health, including safe drinking water, sanitation and food and ecosystems and in relevant activities, linking these studies with services and research in national parks, wildlife reserves, ecological heritage sites, etc.

Promoting education, public awareness and training. B. Increasing public awareness. Basis for action. Activities 36.10c. Countries and regional organizations should be encouraged, as appropriate, to provide public environmental and developmental information services for raising the awareness of all groups, the private sector and particularly decision makers.

Promoting education, public awareness and training. C. Promoting training. Activities 36.17. Countries should encourage all sectors of society, such as industry, universities, government officials and employees, NGOs and community organizations, to include an environmental management component in all relevant training activities, with emphasis on meeting immediate skill requirements through short-term formal and in-plant vocational and management training.

These three sections all seem to point towards the need to raise environmental awareness amongst all groups and more widely throughout the community. It is tempting to suggest that these prompts would have given some support for the growth in community-based environmental monitoring programmes.

8.3   The growth of community-based ecological monitoring

Community-based monitoring of the changes in the environment is not a new idea invented by Western science. First peoples have been undertaking various forms of environmental monitoring for many hundred of years. The monitoring may be different to the Western scientific approach but the aim is usually the same and that is environmental sustainability. Many first peoples work and live off the land and notice subtle changes in the environment from day to day, month to month and year to year. Time, while important to meet the expectations of the larger society is still followed in the natural cycle of living, sometimes if only for a few days at a time. For example, Glenn Nolan (Chief, Missanabie Cree First Nation) described what environmental monitoring means to him and his friends and colleagues (Box 8.1). Four main points emerged from his discussions but in conclusion their ability to 'see' changes comes from a willingness to listen to nature.

Box 8.1   First peoples and ecological monitoring

1.   Our understanding of the land comes from the ability to 'see' changes as they occur. This involves constant awareness of our physical and spiritual surroundings. We notice when animals including insects, birds and fish behave out of the normal context of their environment. For example snakes coming out of wintering sites too soon, insects flying around a month after they would normally be out and about.

2.   We notice changes in animal temperament and we try to understand the reason behind those changes. Sometimes those changes are due to the belief that humans have done something to disturb the natural rhythm, and the spirits of those beings (whatever they are, mammal, bird, fish, insect) are repaying us for our neglect or our treachery.

3.   Weather patterns are changing at an increasing rate. Our knowledge of the weather allows us to feel the changes, even if we are not from that part of the world. As an example, my family was on vacation in Greece in March and my son (19 years old) commented a number of times on the strange feel of the wind and the weather in general. After some discussions, we were told that it had been a long time since weather like that was remembered on the southern island of Santorini.

4.   Some of our knowledge comes from watching our fathers and mothers follow routine and listening to their stories of what has happened in the past (stories are passed down generation to generation over countless years). Many of our teachings about what is normal or usual or likewise comes from our intuition, our feelings of how the world should be. I'm not sure where this comes from but it is a very strong and persistent part of our lives.

In conclusion our ability to 'see' changes comes from our willingness to listen to nature and to relate it to our experiences from the past, including the past of our ancestors.

The best way for you to understand what I am talking about is to travel with me for a time and then you will see for yourself the changes that I see on Mother Earth.

(Kindly provided by Glenn Nolan (Chief, Missanabie Cree First Nation)

after consultation with his friends and colleagues.)

Keeping a personal diary of natural history events has been undertaken by many people. Such diaries are the basis of natural history and ecological monitoring. One example from the eighteenth century is *Gilbert White's Diary* (Box 8.2).

Over the last few decades there appears to have been a growth in community-based ecological monitoring. Examples range from water-quality monitoring to fish monitoring and from butterflies to birds. Contributing to this growth could be an increasing awareness of the state of the environment and a desire

---

Box 8.2 Extract from Gilbert White's Diary

This is an entry from Gilbert White's *Diary* or *Calendar of Flora* dated 1766. The diary is a wonderful example of natural history recording with its detailed observations of events (phenology) and it is also an important piece of biogeographical information. Gilbert White was the Curate of Selbourne in the south of England.

(This entry is from A Nature Calendar *by Gilbert White edited by Wilfred Mark Webb, FLS, and published in 1911 by the Selbourne Society, London. Published with kind permission of The Selbourne Society.)*

Fig. 8.1 A diver conducting a REEF fish survey in the Caribbean. (Photograph by Paul Humann. Reproduced with permission. Copyright REEF 2003.)

by local communities to take actions themselves or at least contribute to the data collection. There are community-based monitoring programmes of beaches, rivers and urban wildlife. Community-based water-quality monitoring is very common. For example, Au *et al.* (2000) described a methodology for 'public monitoring' of toxicity in waterways in Canada. The methods they developed were based on several variables including total coliform per 100 ml water sample and toxicity. These methods have been used by high school students.

Some of the community-based ecological monitoring programmes can be supported by people around the world. One example of this is the Reef Check Programme (Fig. 8.1, see p. 260), which was founded in 1997 by Gregor Hodson. Reef Check is a simple and rapid monitoring programme supported by volunteers around the world. The main goals are to raise public awareness about the value and state of coral reefs, to obtain scientific data on the state of reefs and to provide managers with the resources to manage reefs at a community level.

Some communities may wish to collect their own data as a basis for ecological monitoring because they or their actions have come under criticism. For example, some farming practices have come under criticism as being the cause of local pollution. The question often arises as to who is to blame for the deteriorating conditions of rivers and streams in rural environments. Land owners may feel uncomfortable about criticism of their farming practices and rather than have government authorities collect monitoring data, the farmers may wish to collect the data and have some ownership of that data. This may then satisfy the farmers as to the 'integrity' of the data.

Statutory requirements for ecological monitoring coupled with the growing costs of ecological monitoring programmes may result in communities and government bodies working together. Indeed, there seems to be a willingness on behalf of some local government agencies to encourage communities to take ownership of local ecological monitoring.

Another contributing factors for the growth in community-based ecological monitoring could be the realization that ecological monitoring can contribute to environmental education. Learning about the environment outside the classroom is not nothing new; nevertheless, some community-based ecological monitoring programmes provide education opportunities for local schools as well as providing data for local authorities.

While there may have been a growth in community-based ecological monitoring, there is a debate about whether ecological monitoring should be promoted through education or by enforcement measures. Should the needs of monitoring the environment be driven by regulation or be driven by educational programmes? In some countries, there seems to be the attitude that regulation is not the best approach and that it is better to provide encouragement and guidelines. This topic is a subject for a book in itself and indeed the underlying rational would be worthy of some research. What are the motives of a government who avoids environmental regulation in favour of other methods?

8.4   Community-based ecological monitoring in practice

The US EPA has a particular interest in monitoring and assessing water quality. They provide fact sheets about volunteer monitoring and EPA's support for volunteer monitoring. There is also a national survey of volunteer monitoring programmes. Volunteer monitoring resources available from the US EPA include:

- *National Directory of Citizen Volunteer Environmental Monitoring Programmes.* EPA 841–B–98–09, 1998
- *Volunteer Estuary Monitoring: A Methods Manual.* EPA 842–B–93–004, 1993
- *Volunteer Lake Monitoring: A Methods Manual.* EPA 440/4–91–002, 1991
- *Volunteer Stream Monitoring: A Methods Manual.* EPA 841–B–97–003, 1997
- *Volunteer Water Monitoring: A Guide for State Managers.* EPA 440/4–90–010, 1990.

The publication *The Volunteer Monitor* (Box 8.3) is the national newsletter of volunteer water monitoring and is distributed by River Network.

Box 8.3 *The Volunteer Monitor*

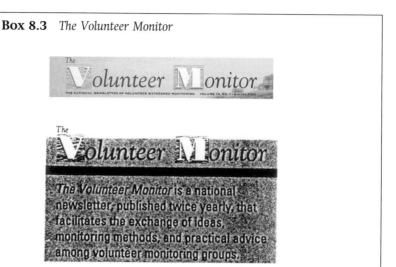

The Volunteer Monitor is a valuable source of information, guides and assessments of community-based environmental monitoring programmes. A different volunteer monitoring programme serves as a coeditor for each issue. The articles have included topics related to monitoring streams, lakes, wetlands, estuaries, beaches, reefs and working with schools. Other articles have been directed at monitoring flora and fauna, monitoring restoration projects, working with youth, using the Clean Water Act and community outreach. Each issues of the newsletter lists specific resources to the articles in that particular issue. For example, the issue on monitoring fauna contained a number of suggested resources for volunteers monitoring macro-invertebrates. The issue of monitoring restoration suggested resources for that activity and so on.

Rhode Island Watershed Watch

The Department of Natural Resource Sciences at the University of Rhode Island, USA, has coordinated a volunteer lake monitoring programme since 1988 (Gold *et al.*, 1994). The objectives included:

- to determine if volunteers operating under the Rhode Island Watershed Watch Protocol collected data that was statistically similar to that collected by professionals using the US EPA approved protocol
- to develop a quality assurance/quality control field visit protocol
- to develop and implement an advanced water-quality monitor training programme.

Gold *et al.* (1994) reported that there was no statistical differences between samples collected by the different groups. However, the amount of variability between samples collected by the different groups differed for each parameter. Chlorophyll *a* had some of the largest differences while pH had some of the smallest differences. In general, therefore, the data collected by volunteers were found to be as representative as data collected by professional water-quality staff.

Taiga Net

Taiga Net is a cooperative environmental and community network based in Yukon Canada and managed by the Arctic Borderlands Ecological Knowledge Society (see http://www.taiga.net/). Amongst the wide range of projects are those aimed at identifying environmental trends, for example:

- indicators of change in the Arctic using human roles in reindeer/caribou systems
- population status of bird species and some whale species
- abiotic indicators.

The Korup Project in southwest Cameroon

The diurnal primates of the Korup area of southwest Cameroon have been the focus of a community-based monitoring programme (Box 8.4). This is a multilateral project supported by WWF, the Government of Cameroon and the European Union. Of particular interest is the need for an ongoing environmental education programme to complement the ecological monitoring of the wildlife.

Field Studies Council (UK)

The Field Studies Council (a company limited by guarantee) has the main aim of increasing environmental understanding for all. That aim is achieved by a rich range of courses at their field centres, by publications and by research. A national network of Field Studies Council sites has provided a useful network for various ecological monitoring programmes. The Field Studies Council has collaborated with other environmental and educational organizations such as the WATCH Trust for Environmental Education in order to promote countrywide projects. For example, the WATCH Acid Drops Project (a project devised to allow children throughout the country to monitor rainfall and acidification) was organized by the WATCH Trust in association with the Field Studies Council.

Community Streamwatch (Australia)

The Streamwatch programme for 'living streams' based in Melbourne was one of the many waterwatch, streamwatch or riverwatch community-based

Box 8.4   Community-based wildlife monitoring in southwest Cameroon (the Korup Project)

The Korup Project is a non-profit-making non-governmental organization that assists the government of Cameroon to implement its policy on conservation of biological diversity. The Korup National Park and its surrounding support zone make up the Korup Project Area.

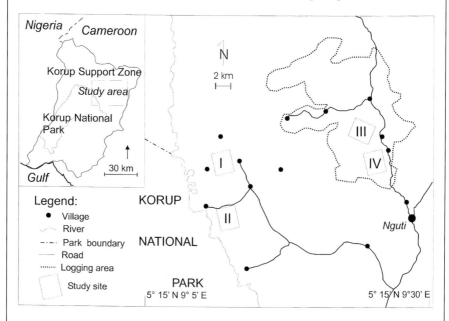

The target species include diurnal primates, large birds and some insectivorous birds. The aim of this project is to combine conservation efforts both within and around Korup National Park. A community-based programme commenced in 1999 with the aims to:

- obtain data on the status and distribution of the species
- facilitate long-term documentation of population trends
- investigate the impact of logging on the wildlife.

Surveys have been carried out by four teams, each composed of three field workers experienced in animal tracking and hunting and who were selected from villages in the local vicinity. Most team members were formerly hunters. Today they are setting up measures to regulate wildlife hunting including a complete ban on endangered species. They are backed by the traditional council.

Data are collected using distance sampling and in each of the study sites there are permanent transects. Data are used from both visual and acoustic encounters.

It is suggested that for effective management of primates and other wildlife in this region, the conservation of the Korup National Park needs to be linked to the management of wildlife in the support zone. Wildlife management in the Korup Support Zone needs to be combined with a long-term environmental education programme.

The Cameroon Biomonitoring Network (CBN) is a relatively new network of existing Cameroonian conservation projects and aims to develop an official forum for conservation in that country.

A survey team carrying out wildlife surveys. All team members shown here (former hunters) are from the village Mgbegati at the northeastern Park border.

Two species in need of urgent conservation action are the Drill, *Mandrillus leucophaeus* (left), and the Crowned Monkey, *Cercopithecus pogonias* (right).

(Information and photographs kindly provided by Dr Mattias Waltert.)

programmes established around the world during the 1990s. There is a National Waterwatch Advisory Committee in Australia and the Streamwatch programmes are undertaken in partnership with a number of organizations. The objective of Streamwatch is to 'increase community awareness, understanding and ownership of stream environmental issues and to assist communities and management agencies to take actions to address identified problems'.

The REEF volunteer fish monitoring programmes

Originating in about 1993, the REEF volunteer fish monitoring programme was developed by the Reef Environmental Education Foundation. In an account of the programme, Pattengill-Semmens & Semmens (2003) described volunteers collecting data on fish abundance and distribution when diving and snorkelling (see Fig. 8.1). Standardized methods were employed for the monitoring and the data were stored in publicly accessible databases on the website (http://www.reef.org). Applications of the data include:

- analysis of fish–habitat interactions in the Florida Keys national marine sanctuary
- development of multi-species trend analysis methods to identify sites of management concern
- evaluation of no-take zones in the Florida Keys.

8.5   The watches

Throughout the world there are many examples of 'the watches' (Box 8.5). Typically these are community- or school-based monitoring programmes that have been established to 'watch', observe and record changes in environmental and biological conditions. The variety amongst 'the watches' is impressive, with many established around aquatic systems.

Many of the watch programmes have been established by regional government authorities and include within their programme some aspects of community-based monitoring. For example, In Australia, the Streamwatch programme was established to provide comprehensive monitoring of the state of Melbourne's waterways. It was established in 1993 and is now coordinated by Melbourne Parks and Waterways with support of the Melbourne Water Corporation.

An important part of Streamwatch is the development of indicators environmental health that can be understood by the wider community. Water-quality indicators (Table 8.1) include:

- *Escherichia coli* (pathogens)
- heavy metals

Box 8.5   Examples of community-based environmental and
nature surveillance and monitoring: the 'watches'

EnviroWatch.com^{tm}: monitoring of environmental data

EnviroWatch.com^{tm}: monitoring of environmental data
NatureWatch (Canada)
Coast watches:
   Coastwatch Europe
   Coastwatch Atlantic
Saltwatch Australia
Water, river and stream watches:
   Waterwatch Australia
   Waterwatch (Isaac Centre for Nature Conservation, New Zealand)
   WaterWatch (England and Wales)
   Alberta Riverwatch
   Molalla Riverwatch (Oregon)
   Illinois Riverwatch (part of Illinois Ecowatch Network)
   Hoosien Riverwatch (Indiana)
   National Riverwatch (UK)
   Streamwatch (NSW)
   Streamwatch (UK)
   North Carolina Streamwatch
Icewatch (part of the NatureWatch series, Canada)
Soilwatch
PlantWatch (part of Canada's NatureWatch)
Treewatch (Quebec)
Butterfly watch
Monarch watch
WormWatch (Canada)
FrogWatches
   FrogWatch USA
   FrogWatch Canada
   FrogWatch Australia
Turtlewatch
Birdwatches
   Birdwatch Australia
   Birdwatch Canada
   Birdwatch Denmark

Table 8.1 *Indicators of water quality as reported in the 1993–1994 water-quality monitoring programme organized by Streamwatch in Melbourne, Australia*

Indicator	Explanation
Escherichia coli	The concentration of this bacterium is measured to indicate levels of faecal contamination in waterways; although most strains of *E. coli* are not considered dangerous, other organisms of faecal origin can cause illness
Toxicants (heavy metals)	Concentrations of heavy metals in waterways are monitored because they can harm aquatic life forms; in addition to direct monitoring of toxicants, deformities in non-biting midges are monitored as bio-indicators of contamination
Nutrients	Compounds of nitrogen (N) and phosphorus (P) are released when organic matter such as leaves, faeces and food scraps decomposes and nutrients are also derived from fertilizers, detergents and leachates from landfill sites; high nutrient levels are associated with excessive plant growth and algal blooms (including toxic blue–green algae)
Suspended solids	Particles of soil or other substances in high concentrations result in muddiness, reduced light penetration and lower photosynthetic activity, which retards plant growth and can result in lowered levels of dissolved oxygen; suspended solids are also associated with the transport of nutrients, toxicants and pathogens
Turbidity	Turbidity is a measure of how easily light passes through a sample of water; high turbidity values are associated with waters that are coloured or muddy
Biochemical oxygen demand	Oxygen is absorbed when organic matter decomposes and the biochemical oxygen demand is a measure of organic matter in a stream
Dissolved oxygen	Dissolved oxygen is monitored because waters with very low concentrations of dissolved oxygen cannot support life

- nutrients
- suspended solids
- turbidity
- dissolved oxygen and biochemical oxygen demand
- pH.

The indicators of stream conditions and management performance include:

- aquatic invertebrates
- sediment contamination
- stream environment condition
- stream amenity
- litter
- community perceptions (Table 8.2).

Table 8.2 *Attributes of waterways used to calculate the 'nearby waterways satisfaction index'*

1.	Free from rubbish and visible litter
2.	Reasonably healthy, pollution-free water
3.	A pleasant, natural-looking environment
4.	Suitable provision of bicycle and walking tracks
5.	Adequate access to the waterway and near-by public land
6.	A feeling that you would be personally safe walking by yourself in the vicinity of the waterway during the day
7.	Adequate provision of public toilets in areas near the waterway
8.	Dogs are adequately controlled so they are not a nuisance or threat to you or your family
9.	Sufficient signs and directions
10.	Sufficient barbeque and picnic areas near the waterway
11.	Sufficient drinking fountains in waterway area
12.	Adequate provision of rubbish bins in the areas near the waterway
13.	Adequate mowing of grassed areas to control snakes and rats
14.	Presence of bird life and other wildlife

From Streamworth, Melbourne, Australia.

The survey of community perceptions has been undertaken to establish a benchmark of public attitudes towards waterways against which improvements can be made and to help to quantify community satisfaction with the way waterways are managed. Two indices were developed to monitor public perceptions, one of which was the 'nearby waterway satisfaction index'. This was based on 14 criteria shown in Table 8.2.

The interest in community-based environmental monitoring led to the establishment of Streamwatch (see above).

A very good example of a watch programme is the Canadian NatureWatch. NatureWatch is a collaborative project between the Canadian Nature Federation and EMAN. (The programmes can be accessed through www.naturewatch.ca/.) NatureWatch is a suite of programmes that encourages schools, community groups and individuals to become 'citizen scientists and engage in the monitoring of soil, air and water quality'. NatureWatch includes a diverse group of programmes including PlantWatch, IceWatch, FrogWatch and WormWatch.

8.6   What value have the community-based environmental monitoring programmes?

Perhaps the most general question that could be asked of the 'watches' and all the other community-based environmental monitoring programmes is why.

Is there any value in involving the public? Are these many and varied naturewatch programmes all in vain? Can these activities help to improve or restore the state of the environment? Does the information collected have any sound scientific basis? If not, then perhaps there is a danger in using data from these programmes.

One of the acknowledged benefits of community-based environmental monitoring programmes is environmental education. Taking part in the programmes almost certainly helps to promote greater awareness of the state of the environment and the changes that are taking place. The Canadian NatureWatch series includes the following in their promotion:

> As the Canadian landscape changes due to global warming, ozone depletion and human activity, no one is better positioned to monitor the effects of these changes than the Canadians who live, work and recreate in local communities. If we are to respond to these changes and conserve our environment, as a nation, we need to listen to what they have to say. Encouraging the public to take an active role in ecological monitoring will not only make them better environmental stewards and advocates, it will increase their awareness and support of scientific research.

There are several powerful statements in this promotion including the perception that communities can make a contribution to monitoring the state of the environment.

8.7   Conditions for success

What makes a community-based environmental monitoring programme succeed or fail? What are the criteria for judging success or failure? The fact that these programmes are reliant on volunteers means that there has to be an agreed protocol and there is a need for standardized methods. For example, Frogwatch USA (a partnership between the National Wildlife Federation and the US Geological Survey Patuxent Wildlife Research Centre) relies on volunteers. It was established in 1998 in response to the evidence showing decline in amphibian populations. The volunteers can register on-line a wetland of their choice and once they have become familiar with the frog and toad vocalizations they can monitor their site using the protocol available on the website (www.frogwatch.org).

Danielsen *et al.* (2000) have proposed a simple system for monitoring biological diversity in protected countries where there is a lack of 'specialist staff' and in their recommendations they include the following:

- standardized recording of routine observations
- fixed point photography

- line transect surveys
- focus group discussions.

They have found that this approach is useful in countries that are embarking on shared management of natural areas with local communities.

The question about what makes a programme successful was addressed by James Lambie (1997) as part of his research on community-based environmental monitoring. His broad conclusions are summarized in Table 8.3. He identified several conditions considered to be generic for success. These include participant safety, clear objectives, training, specialist input, communication and feedback, long-term resourcing, a coordinated structure, partnerships or collaborative programmes, testing of the methodology, maintaining novelty, data management, reliability of the data and credibility of the programme.

Other criteria for success were more related to specific cases. These criteria included promoting the programme to the wider community, individuals who are committed to the monitoring programme, participant responsibility for the programme, participant ownership of the programme, the possibility of tangible rewards and conflict management.

Overall, these criteria seem to be common sense, but nevertheless it is almost certain that environmental monitoring programmes have been abandoned because criteria such as those mentioned in Table 8.3 were not considered and acted on in a rigorous and practical manner.

8.8   A future for community-based ecological monitoring?

Since the early 1990s, and throughout the world, there has been more and more emphasis on monitoring the state of the environment. These activities have included the continuation of existing programmes and the establishment of new programmes for ecological monitoring. Monitoring requires resourcing and one way to help to achieve monitoring objectives with limited budgets is to establish community-based environmental and ecological monitoring programmes.

It is likely that there will be a growth in community-based monitoring programmes and that these programmes will generate data that will be used as a basis for decision making. If this is true, there are likely to be education and training implications.

Community-based ecological monitoring may be an integral part of wider community-based monitoring and evaluation. Indeed some of the frameworks for community-based monitoring and evaluation are appropriate for the remit of ecological monitoring.

Table 8.3 *Conditions for success of community-based monitoring programmes.*

Conditions	Features
A. Generic conditions	
Ensuring participant safety	This is a generic condition for the success of all community-based environmental monitoring programmes; in some countries there may be a legal requirement to ensure safety in the field
Clear objectives	All monitoring programmes must have objectives; it is, therefore, considered essential that the objectives of the environmental monitoring programme are agreed and clearly defined
Training	It is essential that training programmes are incorporated into the programme to ensure that individuals are able to complete their assigned tasks in a successful manner
Specialist input	Despite good training, there will be a need for some specialist input at various stages of the monitoring programme; the availability of such specialist advice, training and analysis must be assured before the programme commences
Communication and feedback	This is considered to be essential for maintaining commitment from individuals and thus continuity of the programme
Resourcing the programme	It is considered essential to secure a long-term financial basis for the programme; this may require multiple resourcing in order to ensure that there is continuity of at least some levels of support
Coordination and structure	The need for coordination for the programme is essential as is some kind of structured management
Partnerships	This refers to the programme being undertaken by a community in partnership with an organization such as a university or government research institution; such partnerships provide opportunities for specialist input and can contribute to the credibility of the programme
Testing the methods	It is recommended that the programme be tested; a pilot programme can be used to identify processes, methods and protocol that could endanger the success of the programme at a later stage
Maintaining novelty and variety	There is a risk that participants will become bored with the programme, resulting in participants not completing their allocated tasks, being careless in their tasks or abandoning the programme
Data management	A formal process for data collection, analysis, storage and communication is essential for maintaining continuity

Table 8.3 (cont.)

Conditions	Features
Being able to reach 'realistic conclusions'	This relates to the reliability of the data and instituting some method of data quality controls; if the data are unreliable (that is despite the same conditions of data collections, different data are collected) then there will be difficulties in reaching realistic conclusions
Programme review and evaluation	This is part of the organizational process and is necessary to ensure that continuity is maintained, the programme is meeting expectations and that the programme remains within the financial constraints
Credibility of the programme	This relates to data quality assurance. That is the methods are standardized and clearly described, and the data are collected and analysed in a standardized and clearly described manner. To achieve this, it is considered that there be peer review of the programme, the methods and analysis of data
B. Case-specific conditions	
Community 'outreach'	This refers to the promotion of the monitoring programme and its outcomes to the wider community; this is case specific and therefore not generic to success
Committed participation	Although not generic for success, it is important to identify individuals who have some commitment to the programme
Participant involvement in programme determination	Although not considered generic to success of environmental monitoring programmes, it is considered that it is essential if the participants take responsibility for the programme
Participant ownership of the programme and data	Ownership needs to be built on involvement of individuals in the design and direction of the programme; although not generic to the success of all programmes, a sense of ownership could be important in specific cases
Tangible rewards	The need to provide some tangible rewards or benefits was thought to be essential in some cases but not generic to success
Should participants be paid?	This may be appropriate in some circumstances to help to ensure the continuity or commitment to the programme
Conflict management	Arguments and conflicts may arise and these could endanger several aspects of a monitoring programme; it is suggested that processes for informal conflict management be identified and made available when necessary

From Lambie (1997) with permission of James Lambie.

For example, at the *International Conference on Ecosystem Health* held in 2002 at the Quetico Centre in Canada, there were some discussions about community-based monitoring and about information design so that data are linked to policy making (Ringius, 2002). One of the contributions to that conference was from Elizabeth Kilvert and was entitled 'Citizen science: community based monitoring'. With reference to engaging citizens she referred to the following (Box 8.6):

- the means to improve access to and awareness of environmental information
- the means to foster communities of practice on specific environmental topics
- the tools to foster environmental involvement at the local level.

Another example comes from the World Bank (2002) and is described as a report aimed at practitioners and entitled *Sleeping on our Own Mats: An Introductory Guide to Community-based Monitoring and Evaluation*. The title was prompted by the Moré proverb 'The one who sleeps on a borrowed mat should realise that he is sleeping on cold, cold ground' (Burkina Faso).

During 2002 and 2002, World Bank staff and collaborators took part in participatory research in 18 villages in Niger, Benin and Cameroon to develop a locally appropriate monitoring and evaluation system. The purpose was to increase the ability of communities to assess and track local developments while at the same time making adjustments to their local development plans. Some people would call this an adaptive management approach. Community-based monitoring and evaluation is defined in this guide as being 'monitoring and evaluation of community development by an interested community, so that the community can make independent choices about its own development'.

Five stages were suggested for the implementation of community-based monitoring and evaluation:

- preparation
- introduction of the monitoring and evaluation concept to communities
- development of the community's monitoring and evaluation work programme
- monitoring of the development activities
- evaluation of and re-appraisal of local development.

The system is shown in Fig. 8.2 as a flow chart and this relates to the cycle of events, which includes planning, implementation and monitoring, evaluation and re-appraisal.

Box 8.6 Citizen science: community based monitoring: a contribution to the *International Conference on Ecosystem Health 2002* by Elizabeth Kilvert

Design of monitoring should include:

- Visioning process including community demographics, concerns and needs
- Consultation and outreach (cast net as wide as resources will allow)
- Inclusion of all identified partners
- Standardized methods, good statistical design and sound research from the start.

Mapping is:

- An interpretation communication tool
- Resource dependent
- Useful for communication, analysis, reward and miscommunication
- A more integrated systematic approach, as opposed to many wasted resources and efforts.

Data management requires:

- Database design – this is product dependent
- Species analyst
- Data warehousing verification and accessibility (complex in a multi user–contributor situation)
- Historic records – rigour of data
- Supporting resources for the above needs, for academic institutions, museums, naturalists.

Engaging citizens:

- Means to improve access to, and awareness of environmental information
- Means to foster communities of practice on specific environmental topics
- Tools to foster environmental involvement at the local level.

Holding governments accountable:

- Data to support community identified outcomes
- Data to support national environmental indicators
- Comprehensive, credible and continuous reporting on the state of the environment and the state of environmental management in Canada.

Strengthening basis for public policies:

- Means to set priorities for information development, through dialogue between users and producers
- Tools to integrate environmental information from various sources
- Tools to integrate environmental information with social, economic and health information
- Tools to explain and predict the connections among environmental change, human action, and human well-being.

Monitoring and assessment in support of decision making

- Enable timely access to and effective application of relevant, credible, integrated environmental data and information in support of decision making by all Canadians, through a coordinated, cooperative network of government agencies, the private sector, academia, non-government organizations, aboriginals, and others.

Quetico Centre can contribute:

- Programs that go outside of institutional walls in order to provide service, resource and function to the community
- Research by an ethical and trusted institution that can provide a balance of information and retain objectivity around difficult topics and questions
- Generation of data, information or assessments that can be used to communicate to decision makers (institutionalization and lobbying).

(With permission from Elizabeth Kilvert, Program Coordinator, EMAN,
Environment Canada, Burlington, Ontario, Canada.)

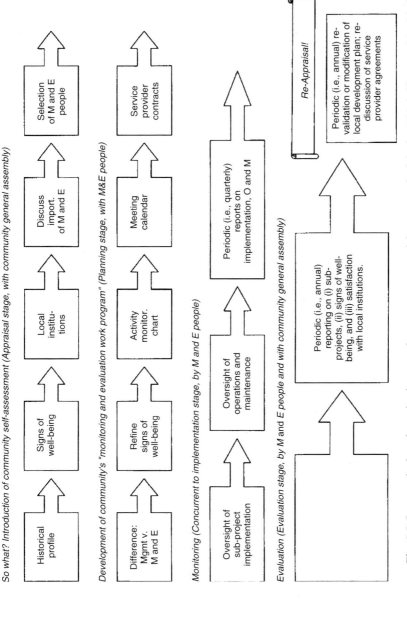

So what? Introduction of community self-assessment (Appraisal stage, with community general assembly)

Historical profile → Signs of well-being → Local institutions → Discuss import. of M and E → Selection of M and E people

Development of community's "monitoring and evaluation work program" (Planning stage, with M&E people)

Difference: Mgmt v. M and E → Refine signs of well-being → Activity monitor. chart → Meeting calendar → Service provider contracts

Monitoring (Concurrent to implementation stage, by M and E people)

Oversight of sub-project implementation → Oversight of operations and maintenance → Periodic (i.e., quarterly) reports on implementation, O and M

Evaluation (Evaluation stage, by M and E people and with community general assembly)

Periodic (i.e., annual) reporting on (i) sub-projects, (ii) signs of well-being, and (iii) satisfaction with local institutions. → Periodic (i.e., annual) re-validation or modification of local development plan; re-discussion of service provider agreements

Re-Appraisal!

Fig. 8.2 Community-based monitoring complements strong, participatory methodologies. This is illustrated in this flow chart and shows where each stage occurs. M, monitoring; E, evaluation; Mgmt, management; O, oversight. (With permission from the World Bank publication (2002) *Sleeping on our own Mats: An Introductory Guide to Community-based Monitoring and Evaluation*.)

271

Table 8.4 *World Bank project indicators of well being*

Time \ Sign of well-being	Past — Proportion of the community which has access	Past — Reference base	Present — Proportion of the community which has access	Present — Reference base	Future — Proportion of the community which has access	Future — Reference base
Children at school	☺ ☺ ☺	10	☺ ☺ ☺ ☺ ☺	10	☺ ☺ ☺ ☺ ☺ ☺	10
Other indicators	x	y	x	y	x	y

From World Bank, 2002 with permission.

Indicators have also been included such as indicators of 'well-being'. A good example is the number of children at school compared with the number who could be at school. The results are presented in a way that best communicates the information (Table 8.4).

Much of what is contained in this World Bank guide to community-based monitoring and evaluation could well be applied to community-based ecological monitoring projects. However, there are missing elements, including the setting of objectives or targets and a process for change should the objectives or targets not be met. It is no use monitoring for the sake of monitoring and it is no use recording progress unless there is some objective and some way of changing the conditions if the objectives are not met.

9

Ecological monitoring of species and biological communities

9.1   Introduction

The aim of this chapter is to provide some examples of applications of ecological surveillance and ecological monitoring. The examples range from projects directed at conservation to monitoring the condition of streams to the effects of pollution on coastal communities. Some examples of long-term population studies are described in order to set the scene and to emphasize the relevance of long-term ecological studies in ecological monitoring.

9.2   Long-term population studies

Two examples of long-term animal population studies have already been mentioned: the population studies of herons initiated by Lack in 1928 and that is continued today by the BTO. The other example is the Rothamsted Insect Survey, which continues to undertake long-term studies on many species of insects (see p. 71). Another example of classic long-term insect population studies is the study of forest insects established in Germany in the early 1800s (Klimetzek & Yue, 1997), one example of which is shown in Fig. 9.1.

One of the earliest long-term plant ecological experiments commenced in Nigeria in 1929 (Hopkins, 1962) and was still continuing in 1968 and is possibly continuing in some form today (Brian Hopkins, personal communication). There are several long-term ecological monitoring studies of grasslands. For example, Dunnett & Willis (2000) described the monitoring of roadside verges over a period of 39 years.

In the USA, a database of all long-term vegetation studies utilizing permanent plots in North America has been compiled at the Institute of Ecosystem

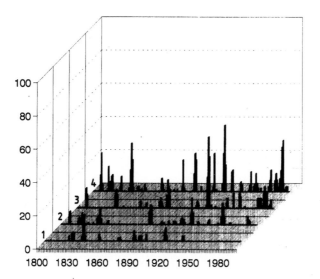

Fig. 9.1 Outbreaks of the Pine Beauty Moth (*Panolis flammea*) in Germany from 1801 to 1990. Regions are Pfalz (1), Oberfranken (2), Oberpfalz (3) and Mittelfranken (4). (From Klimetzek & Yue (1997) with permission.)

Studies, New York Botanical Garden. There are now many long-term plant population studies that have been undertaken in North America. For example, Harcombe *et al.* (2002) used data collected over 18 years from a permanent plot in southern hardwood forests in Texas to investigate forest stand dynamics. In West Virginia, Kery & Gregg (2003) have studied orchids and used long-term study plots. One intensively studied site is the Hubbard Brook site, which was initiated when the Forest Service established research programmes to look at hydrology in forests under different management and harvest methods. In the 1960s, that research was further extended to include chemical budgets and then later ecosystem processes.

Long-term plant studies have typically been undertaken at ground level but there are many examples that have been based on changes detected from aerial photographs and later GIS. In Japan, for example, Fujita *et al.* (2003) have used aerial photographs spanning 32 years to analyse changes in the canopy of old broad-leaf forest in the Tatera Forest Reserve of the southwestern region of the island of Tushima (Fig. 9.2). An analysis of the forest gaps and the size and age of the gaps led Takashi Fujita and collegaues to conclude that long-term and large gaps are essential for the establishment of deciduous broad-leaved and shade-intolerant pioneer species.

Examples of long-term plant population studies in Europe include the experiments at Rothamsted (p. 71) and the studies on the Breckland Plots (East Anglia, England) established by Watt in 1936. In southern England, the ground flora in

Fig. 9.2 Temperate old-growth evergreen broad-leaved forest on the southwestern tip of the island of Tushima, Japan. An example of the aerial photographs used by Takashi Fujita and colleagues in Japan. (Photographs kindly provided by Takashi Fujita.)

Fig. 9.2 (Cont.)

Wytham Woods has been the subject of ecological surveillance over many years. Wytham Woods is predominantly deciduous and includes area of different ages and management types. Kirby & Thomas (2000) analysed the changes in the ground flora in Wytham Woods from 1974 to 1991. They found no changes in the total number of species and the mean richness per plot. Ancient woodland indicators as a group showed less change between years than species associated with open glades and grassland patches.

Most long-term ecological studies are probably of interest only to ecologists and environmental managers. Few people in the wider community are likely to be interested. However, a long-term study of daffodils is likely to attract interest across a wide section of the community. In 1977, a wild daffodil colony in West Dean Woods, England became the subject of a long-term ecology study. The aim was to assess the effects of shade and management options such as swiping (mowing with a rotary blade; Hopkins, 1999). Some of the results are shown in Box 9.1.

Box 9.1   Changes in density of mature wild daffodil (*Narcissus pseudonarcissus*) bulbs in West Dean Woods Nature Reserve, West Sussex, UK over 21 years (1979–2000)

Counting daffodils.

Brian Hopkins concluded the following from the 21 years of study. The project was designed to assess the effects of shade and management options such as swiping (Hopkins, 1999) on the population of wild daffodils. Daffodils continued to increase in density on three previously unswiped plots and appeared to be at equilibrium density on the two swiped plots. (Swiping is mowing with a rotary blade.) In wooded conditions, the current management of swiping seems ideal for conserving daffodils.

In open conditions, the grassland appears to have allowed for some increase. Two management alternatives are suggested. The first is to plant some standard oak trees. The trees would in time increase the shade; this, in turn, would help to suppress the grass. The second is to maintain present management.

Flowering varied greatly from year to year and appeared to be affected by weather and the previous spring conditions.

The five graphs below show the mature bulb density in plots varying in vegetation and management activity. In each, the point indicates the means and the dotted lines either side are the 95% confidence limits. The continuous line is the regression analysis, which is extrapolated as a dotted continuation.

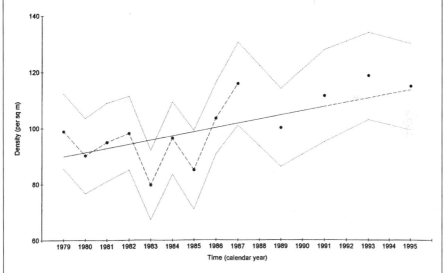

Plot 1: dense woody plants, unswiped.

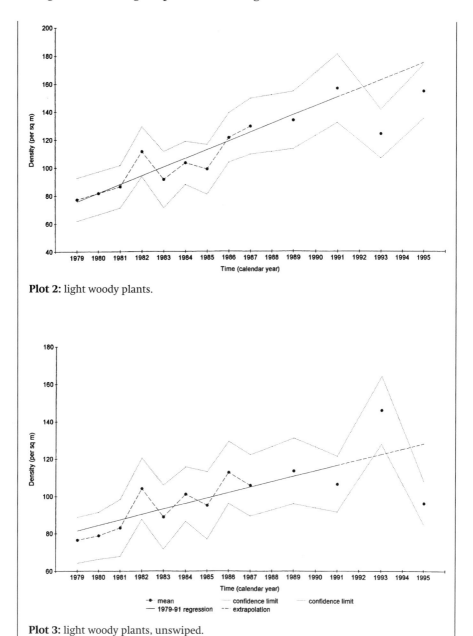

Plot 2: light woody plants.

Plot 3: light woody plants, unswiped.

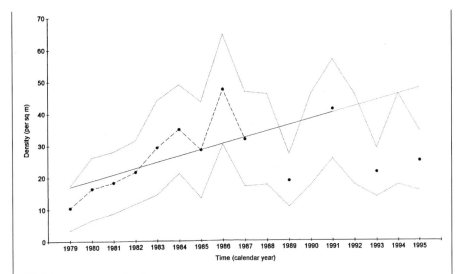

Plot 4: open area, swiped.

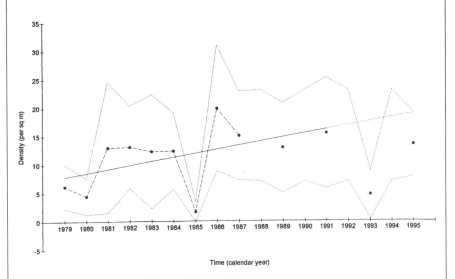

Plot 5: open area, unswiped from 1977 to 1993.

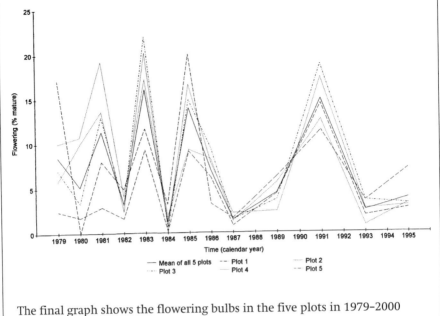

The final graph shows the flowering bulbs in the five plots in 1979–2000 (annual records only were available until 1987).

9.3   Monitoring birds

Birds are very often perceived as being more 'attractive' than other animal taxa and, in general, birds are more easily observed than some other animal groups. Possibly for these reasons, the abundance and population size of birds has long been of interest to many people as well as organizations and societies throughout the world. More importantly, as far as monitoring is concerned, birds are useful indicators of land-use change, as is shown in responses of bird communities to agricultural landscapes and agricultural practices. Agricultural chemicals are bound to be mentioned in any accounts of the biological effects of agriculture and in Ch. 1, I have already commented on Ratcliffe's (1980) classic study of the effects of DDT on eggshell thinning and I have also discussed the role of the Environmental Change Network in relation to land use (p. 63). Here we are concerned more with changes in avian populations, some of which with careful analysis can be attributed to changes in the agricultural landscape and agricultural practices.

Sampling, recording and interpretation

Changes in bird population levels and distribution have been recorded for many years and we are fortunate in having long-term data on some species.

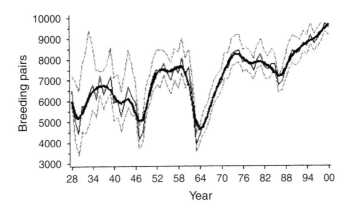

Fig. 9.3 Monitoring of Grey Heron (*Ardea cinerea*) population levels in England and Wales 1928–2000. (From the British Trust for Ornithology website with permission.)

A classic example is the population counts of Grey Herons (*Ardea cinerea*), which commenced in 1928 and is the longest census of any bird made in the UK. These data provide an excellent basis for analysis and interpretation (Fig. 9.3). There are of course regional variations in the data but on a national basis note that although the Heron population fluctuates, it does so within certain limits. Lack (1966) referred to this as 'restricted fluctuations' and also drew attention to the rapid recovery following severe winters. It is also interesting to note that rates of change in population levels do vary and tend to be greatest when levels of population are well below the mean. Could this mean that levels of reproduction in these Herons can be affected by population levels and/or population density?

The Heron numbers are higher than ever before and perhaps benefit from reduced pollution, persecution, warmer winters and possibly better levels of freshwater fish (Gibbons *et al.*, 1993; Marchant *et al.*, 2004). The occasional dramatic declines in Heron population levels could, in general terms, be attributed to cold winters, but such an explanation is an oversimplification. In the absence of data from mortality studies, we would need to be careful when we attempt to interpret these data because apparent changes in population levels could be attributed to a combination of ecological and observer factors. Causes of changes in population levels are as difficult to analyse as is the identification of patterns and trends. These Heron data are so variable as to demonstrate that long-term studies are a prerequisite for the detection of trends.

As with counts of other animals, there will always be sampling errors and sources of inaccuracy that must be identified and taken into consideration in the final figures. Location of all birds on a count and counting accuracy are two basic sources of error. In addition to these basic considerations, monitoring bird population levels requires decisions about sampling, time, duration and size of sampling.

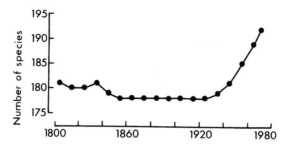

Fig. 9.4 Number of bird species (other than those introduced or reintroduced since 1800) breeding regularly in Britain and Ireland in each decade from 1800 to 1972. (From Sharrock, 1974.) Later totals would be approximately 202 for 1968–1972 and 204 for 1988–1991. (Based on Gibbons *et al.*, 1993.)

Rothery *et al.* (1988) analysed counts of guillemots (*Uria* spp.) and their work is a good example of how to make these decisions when planning and analysing monitoring schemes. They counted samples of Guillemots in demarcated cliff plots in 27 colonies (ranging in size from 108 to 64 000) around Britain and Ireland and were especially interested in assessing validity of sample size or plot size as a reliable guide to total population size. Several underlying sources of variation in the counts, including simple day-to-day fluctuations and seasonal changes in populations, contributed to the coefficient of variation of approximately 10% for plots of 200 birds or more. Analysis of detailed data from the Isle of May suggested that the recommended model Guillemot monitoring scheme should consist of five plots of 200–300 birds counted on 10 days. The timing of the counts should take into consideration the biology of the species at each colony. Since male birds depart at about the time when the young leave, Rothery *et al.* (1988) suggested that counts need to be completed before the young leave, sometime in June. Clearly such monitoring programmes require a good knowledge of the ecology and biology of the species, but Rothery *et al.* (1988) also noted that interpretation of changes observed in monitoring requires additional information from detailed demographic studies of the biology and ecology of the species.

Interpretation of a national change in avian species richness can be equally as interesting as interpreting data for a single species. For example, the data in Fig. 9.4 show changes in avian fauna from 1800 to 1972 in Britain and Ireland. More recent counts (based on Gibbons *et al.*, 1993) would be approximately a total number of species of 202 for 1968–1972 and of 204 for 1988–1991. Here, we could usefully ask not only what are the reasons for this change but also how the data were collected and what are the limitations of these data. More recently climate change has surely had an effect. Does the apparent change in species richness reflect a change in the number of species and if so what

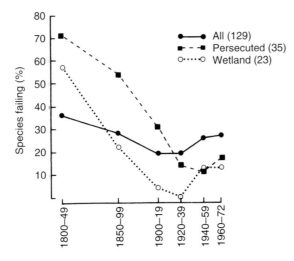

Fig. 9.5 Percentage of all bird species failing (decreasing in number or range contracting) in each of six periods during 1800–1972. (From Sharrock, 1974.)

could be the explanation? The apparent rapid increase (about five species per decade) could be the result of two activities, increased number of designated areas of protected habitats and increased interest in birds (that is increased observer effort), both activities that have long been promoted by the BTO, the RSPB and other organizations.

A general analysis of Britain's avian fauna was undertaken by examining changes in those bird species deemed to be 'failing' (species decreasing in numbers or whose ranges are contracting) or being 'successful' (increased numbers or expanded range). For this analysis, Sharrock (1974) looked at the status changes (increasing in numbers or expanding range; decreasing in numbers or diminished range; no change) for 129 species of birds in Britain over six unequal periods (Fig. 9.5).

The 'persecuted' species include Great Crested Grebes (*Podiceps cristatus*; feathers used on hats), Gannets (*Sula bassana*) and gulls (eggs collected for food), Woodpigeons (*Columba palumbus*; pests of crops), Goldfinches (*Carduelis carduelis*; trapped for cage birds) and Red-backed Shrikes (*Lanius collurio*; killed in the name of game preservation). The percentage of species 'failing' seems to decline up until 1939, after which it is followed by some increases, especially in the wetland species. Reduction and damage to habitats has probably been the major contributing factor resulting in more wetland species 'failing'.

A more recent example of cross-species analysis is the indicator graphs that both BTO and RSPB provide to the UK government. These can be found on www.sustainable-development.gov.uk/indicators/headline/h13.htm.

Bird census and surveillance programmes

Bird census activities, both amateur and professional, range from local and occasional recording to the large and well-organized national programmes. Not surprisingly the different census programmes have different objectives and different methods, a subject that has been well debated in periodic meetings on bird census and studies. There is an intrinsic interest in bird census programmes, but in addition some bird species may be used as indicators of the state of the environment, for example in monitoring marine ecosystems. More recently, data on bird numbers have become part of the UK government's sustainable development indictors programme.

Well-organized long-term census programmes are a prerequisite for monitoring and the following are examples of census programmes that have provided valuable information for monitoring the distribution and abundance of birds. In assessing these examples, I find it useful to ask:

- what is the objective?
- how is the information collected?
- How was the information and analysed?

These questions provide a necessary basis for the subsequent interpretation of the data. Information on other census programmes can be found in some of the BTO publications.

In North America, there are bird-nest recording and census programmes, ringing programmes and various counts and records of birds throughout the continent, organized by a wide range of groups. Government organizations in North America such as the US Fish and Wildlife Service and the Canadian Wildlife Service sponsor the Breeding Bird Survey and also bird-ringing programmes. The national Audubon Society sponsors bird census programmes including a breeding bird census established in 1937, a winter bird population study which commenced in 1948 and a Christmas Bird Count, which was established in 1900 and has since proved very popular amongst its participants. Migratory bird populations are the subject of three classes of survey to assess population size as a basis for monitoring and management (Martin *et al.*, 1979). In the first class, statistically valid attempts are made to count all individuals within a sampling unit and this applies to waterfowl breeding counts in the USA. The second class is based on surveys of birds calling and/or observed within a sampling unit; one species that is the subject of this class of survey is the Mourning Dove (*Zenaida macroura*). The third class described by Martin *et al.* (1979) is represented by total counts, within designated sampling areas, of species such as wintering waterfowl.

There is a wide range of bird distribution and population recording, census and surveillance activities in the UK. For example, the BTO is the sole organization responsible for bird ringing and is the major contributor to bird census and surveillance work in the UK via the Common Birds Census, the Waterways Bird Survey, the Nest Record Scheme and by an integrated population monitoring programme that embraces several programmes.

The BTO Atlas Project has recorded the distribution of both breeding and wintering birds for many years (Sharrock, 1976; Lack, 1986; Gibbons *et al.*, 1993) and the distribution maps of each bird species (based on presence or absence in each 10 km² of the National Ordnance Survey Grid) can be used for baseline information in monitoring programmes. The Game Conservancy administers the National Game Census, the aim of which is to monitor the number of game shot and thus derive population trends for species such as Pheasant (*Phasianus colchicus*), Red Grouse (*Lagopus lagopus*), Grey Partridge (*Perdix perdix*), Brown Hare (*Lepus capensis*) and Wood Pigeon. The National Game Census has continuous records that were established in 1961 but also has records going back to 1880 (e.g. Tapper, 1992). The Wildfowl and Wetlands Trust participates in censuses of wildfowl, both at an international and national level and the RSPB has monitoring programmes including those directed at threatened species.

Data collection, analysis and interpretation

There are many methods of data collection for bird census studies (Bibby *et al.*, 2000). Recording (by sight and sound) species along a transect or within a defined habitat are commonly used techniques. There are many variables that may affect the results, such as amount of time spent recording, rate at which the transect or habitat is walked, time of day or year, weather conditions and the experience of the recorder.

Point count techniques such as the variable circular plot technique (Reynolds *et al.*, 1980) are popular with North American observers and provide a standardized technique for estimating bird numbers. Basically, this technique consists of making series of timed stops during which the distance at which each bird is first detected is estimated. For each species, an effective detection distance is determined, beyond which detections drop off sharply. Numbers and densities are calculated from records within the effective detection distance.

If one of the aims of the census is to prepare distribution maps, which may later be used for monitoring changes in distribution and relative abundance, then many participants in many locations is a necessary prerequisite. Bird counts such as the Christmas Bird Count in the USA are very popular. With the use of specific locations, each of which covers a 24 km radius, and a standard eight-hour recording time at each site, the Christmas Bird Count generates sufficient information for preparation of a bird atlas giving both distribution and relative abundance data (Fig. 9.6).

Data on more than 600 species from 1282 count sites taken over a period of 10 years were analysed in detail by Root (1988). She advocated the use of average values to reduce any spurious effects that could be attributable to observer's expertise, time and effort spent recording and direct effects of weather. Indirect effects of weather could include diminished participant motivation and increased effect of predators on woodland birds when conditions are favourable for soaring.

Average density values for the various species at each of the 1282 sites were calculated as follows:

$$\bar{x} = \left[\sum_{i=1}^{y} (I/Hr)_i \right] / y$$

where $\bar{x}$ is the average density, I is the number of individuals seen at a given site, Hr is the total number of hours spent counting by the groups of observers in separate parties at a given site, and y is the number of years the count took place. Values for various years are summed.

For ease of computer plotting, the density values were made to range between 0 and 1 for each species by dividing the average values at each site for a given species by the average value at the site with the maximum abundance or the value which was greater than 99% of all the values for a given species (maximum abundance value).

An example from Root (1988) clarifies this calculation. The highest value for the Common Raven (*Corvus corak*) was 6.67 individuals seen per party-hour, while the next three highest abundances were 18.24, 15.76 and 13.02 individuals per party-hour. Instead of using the absolute maximum, Root used the average value which was 99% greater than all other values for a given species. This value for the Common Raven is 13.02 and the values higher than this were given a value of 1.0 for the computer mapping. An example of computer mapping for the Black-billed Magpie (*Pica pica*) is shown in Fig. 9.6. This kind of mapping has become much more sophisticated with advances in GIS.

Bird census studies can provide information on patterns of distribution and relative abundance but usually not population size because, with the exception of some species with small populations, it is impossible to count all individuals. The BTO's Common Birds Census has addressed this problem in an interesting manner. The original objective of the census, which commenced in 1961 (the year before publication of Rachel Carson's book *Silent Spring*), was to

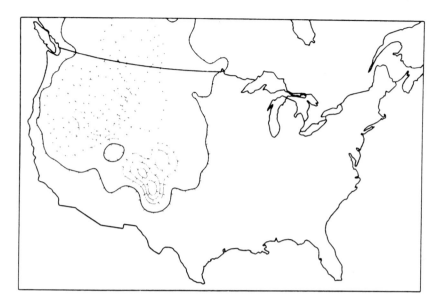

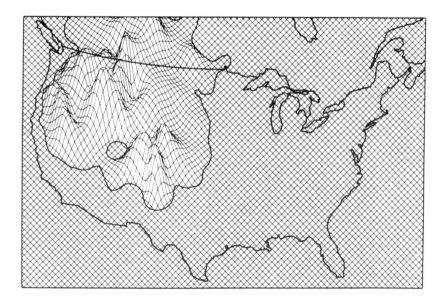

Fig. 9.6 Contour and three-dimensional maps of the winter distribution and abundance patterns of the Black-billed Magpie (*Pica pica*). The four contour intervals are 20%, 40%, 60% and 80% of the maximum abundance value used in the normalization process. For the Black-billed Magpie, this value is 18.30 individuals seen per party-hour. (Reproduced from Root (1988) with kind permission of the author and the University of Chicago Press.)

monitor any adverse effects on bird populations that may have been caused by agricultural chemicals, especially pesticides. The census was designed to detect and measure population changes of various bird species (not only common species) in woodland and on farmland. This was undertaken by the Common Birds Census until 2000, when the monitoring function was taken over by the Breeding Bird Survey, which had previously commenced in 1994.

The Common Birds Census was based on data collected mainly from woodland and farmland plots that were visited on several occasions during March to July. The farmland plots were of many types but were at least 40 ha, and preferably 60 ha, in area. The woodland plots included many kinds of semi-natural vegetation, each at least 10 ha in area. Records of all birds heard or seen were plotted onto large-scale maps provided by the BTO.

The field data from the census was collated, recorded and analysed in a number of different standardized ways and the index of population change for a period of up to 30 years proved popular when analysing trends in populations. The index of population change was relative to an arbitrarily chosen datum year where the index was set at 100. The datum year was previously 1966 but was then set at 1980 for most species (Fig. 9.7).

It is important to note that data collection and analysis are highly standardized, so that any temporal trends observed are not the result of subtle changes arising out of differences in data collection or analytical methods that might occur over time.

The Common Birds Census index (and the Waterways Bird Survey index) were originally calculated by the chain method, that is data on the number of territories from each plot was paired with those from the same plots in the previous season. Such a method helped to eliminate variation that might have been caused by census accuracy and observer turnover. The counts were then summed across all pairs to produce an overall estimate of percentage change. This estimate was then applied to the previous year's index value.

This chain method is no longer used but in its time it provided a useful basis for assessing and monitoring population change at a national level. Later, the so-called Mountford model was used, which was based on a six-year moving window (e.g. Peach et al., 1994). This was used from about 1990 to 1999. Population changes are now modelled using a generalized additive model, which is a log-linear regression model that incorporates a smoothing function to remove short-term changes cause by severe weather or error in measurements.

With regard to interpretation of the index, it was interesting to consider the overall changes (which exclude regional differences) with respect to various species. The explanations for changes in bird populations levels are many and include the effects of weather, especially severe winter weather, changes in

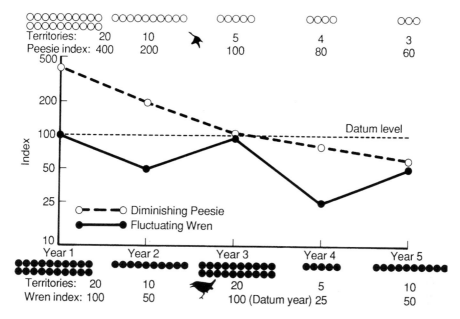

Fig. 9.7 The basis of the old Common Birds Census index as illustrated by population indices for two hypothetical species: the Diminishing Peesie (*Vanellus disparator*; broken line) and the Fluctuating Wren (*Troglodytes updownii*; continuous line) over a period of five years. The index was set to an arbitrary value of 100 (in this case year 3) and was a measure of a species change in abundance relative to the datum year. In some years, these two hypothetical species had the same number of territories but the index value would differ for each species. The Peesie had a higher value throughout despite the lower number of territories in years 3 to 5. Indices are no longer calculated in this way.

rainfall patterns and habitats at the overwintering locations of migrants, habitat changes in the UK and the effects of pollutants.

The following (Fig. 9.8) are examples reproduced here with permission from the BTO website.

The Whitethroat (*Sylvia communis*), a palaearctic migrant, is a species associated with heathlands and commons with good scrubland vegetation. The dramatic decline during 1968 and 1969 has been attributed not to loss of habitat but to mortality when wintering in the Sahel zone of western Africa, where there had been severe droughts. The population of this species now appear to be stable.

The index for the Wren (*Troglodytes troglodytes*) is outstanding, with a ten-fold increase between 1963 and 1974, suggesting that this species has a unique ability to respond to favourable conditions. Winter weather is a major factor in the mortality of this species.

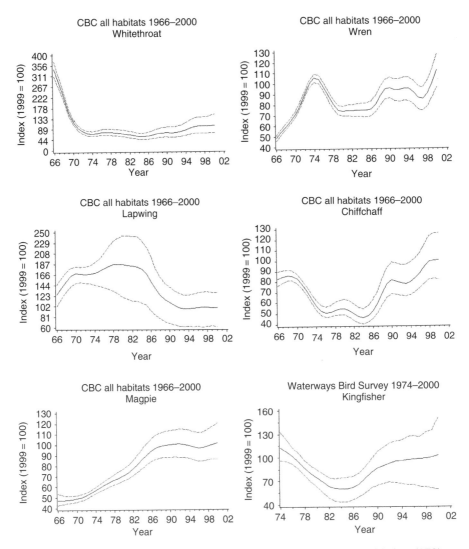

Fig. 9.8 Examples of graphs produced by the British Trust for Ornithology (BTO) Common Birds Census and Waterways Bird Survey index. Note that although these can be used for surveillance of species, there cannot be numerical comparisons between species. (From the BTO website with kind permission.)

Lapwings (*Vanellus vanellus*) were commonly found over much of lowland UK but suffered a decline after the 1962–1963 winter. In general, there seems to have been a decline in southern areas (resulting from changes in farming practices and drainage of damp meadows) but an increase in the north, possibly owing to an ameliorating climate. Conservation of this species could usefully be aimed at establishing traditional wintering sites such as large, old, fertile pastures where organic farming is practised.

The index for the Chiffchaff (*Phylloscopus collybita*) shows a dramatic decline, then a fluctuation leading to an increase. The decline in the early 1970s is possibly a reflection of loss of habitat, which is typically old, mixed deciduous woodland with some tall trees but with a good undergrowth or shrub and scrubby woodland edges. Winter conditions may also be an important factor for this species.

Following persecution for protection of game earlier in the twentieth century, the Black-billed Magpie has an increasing population in both farmland and woodland habitats and is now stable. It appears that this species has been able to adapt to wooded landscapes in urban areas. There is some evidence to show that higher densities occur in urban areas where there is a greatest number and varied species composition of tree species.

The Kingfisher (*Alcedo atthis*) is especially vulnerable to severe winters and was the worst affected of all British birds in the winter of 1962–1963. Other factors such as pollution of waterways and increased recreation along waterways have contributed to the longer-term decline of this species. There appears to been some recovery.

In general, changes in agricultural practices would seem to be of major importance for a number of bird species. Habitat structure and diversity (of woodlands, hedges, scrub, ponds), the introduction of autumn-sown cereals (1974–1977) and the increasing use of chemicals have all contributed directly or indirectly to change in populations of many farmland birds. Increased pressures of recreation have probably contributed to a more recent decline of some woodland species.

Underlying these brief interpretations is, of course, a wealth of data made possible first by the Common Birds Census programme and more recently by the Breeding Bird Survey. These programmes are, in many ways, an excellent example of the kind of infra-structure necessary for long-term studies and ecological monitoring.

The integrated population monitoring programme of the BTO (Baillie, 1990) aimed to identify changes in population variables that require conservation action based on an understanding of normal patterns of variability. Yet how confident can we be about lack of spurious, random fluctuations in, for example, the old Common Birds Census index? Early on, Moss (1985) addressed that question by way of 20 simulations for data collected over a 20-year period and found that random fluctuations gave rise to at most 25% deviation over 20 years. This was much smaller than the variations in population levels that had been estimated using the Common Birds Census data. Moss also looked at the relationship between the index values (published in the journal *Bird Study*) and population density values for plots where the area of the plot was known. In some cases, there was a high correlation between density of breeding pairs and the Common Birds Census index, suggesting that the index does reflect actual density in an accurate manner. However, in one of the examples chosen, the Spotted Flycatcher (*Muscicapa striata*), there was no relationship between the index and the population density over a 20-year period as a whole. Moss (1985) attributed this to the fact that new plots coming into the census on average held more Spotted Flycatchers than those plots dropping out and so the index drifted downwards.

These analyses by Moss (1985) related specifically to the chain index method-ology which is no longer used. Marchant (personal communication) has raised the point that census the representativeness of the plots may be a more import-ant issue. Fuller *et al.* (1985) have shown that, despite the selection of sites being left to observers, the farmland sites in the Common Birds Census were repre-sentative of farming types at least across southern and eastern UK.

Variables, such as weather conditions during population counts and the level of observer expertise, need to be considered when assessing data from census programmes. The Common Birds Census provides a rich source of data for analysis of these variables. Variations in weather conditions have been found to be too small to bias the results in the Common Birds Census (O'Connor & Hicks, 1980) and whereas expertise was found to be an important variable in individual censuses, observers of varying levels of skill tended to make similar estimates of bird population change between years (O'Connor, 1981).

Census and monitoring of birds in the UK have now been incorporated into the Countryside Survey 2000 and this will allow further analysis of changes in populations and distribution. There is a report available on the Countryside Survey 2000 website.

Some bird census programmes are directed at particular species. The National Game Census, established in 1961, is one example. It aims to monitor the status of several game species, including birds, and the value of this census has been steadily increasing with each succeeding year. This census is based on numbers of game shot across the UK. For many years, landowners and shoot managers have recorded details of the bags, and although the number of birds shot is not linearly related to actual bird density, it does provide a useful index to monitor year-to-year changes and trends. It is assumed that the reliability of the data is improved with increased sample size and the Game Conservancy looks for 10 to 20 records for each county. Over 500 farms and estates return records and of those over half return records for the Grey Partridge and over 80% return records for the Woodcock (*Scolopax rusticola*). Census forms ask for exact totals for all game taken on the shoot in the preceding 12 months as well as numbers for any game released.

The two examples shown here from the summary data in the *Annual Reviews of the Game Conservancy* show that there are apparent and sometimes dramatic changes in the census returns (Fig. 9.9), which in this case are expressed as numbers of birds per area of habitat. Results from Game Conservancy monitoring studies on the Grey Partridge have shown dramatic declines in returns from shoots. The decline in numbers shot mirrors declines recorded by spring counts of breeding pairs from long-term monitoring by both the Game Conservancy and the Common Birds Census. These declines seem to be attributable to declines in food for the chicks, which require insect-rich diets. Insect species richness and diversity in agricultural ecosystems is greatly affected by the use of insecticides, especially if the insecticides are not selective. It would appear that intensive use of insecticides has been a major factor contributing to the decline in the Grey Partridge. Monitoring programmes such as this provide information not only about game birds but also about the impacts of chemicals on agricultural ecosystems. However, interpretation of data needs to be undertaken with care. For example, the apparent increase in Woodcock may not reflect a true increase in numbers. The increase in the returns of birds shot is partly a reflection of the increase in Pheasant shooting. Most Woodcock are shot during Pheasant drives and the larger numbers of Woodcock shot may be biased by this increase in hunting effort.

9.4   Monitoring for conservation

Any of the above examples of long-term studies could provide data as a basis for designing methods and policies for conservation. The same data could possibly be used to help to identify the conservation status of a species. The conservation status has previously been qualified and quantified in what have been called *Red Data Books*.

The Red Data Books

In 1966, the IUCN Survival Service Commission was established to undertake a programme of action to prevent the extinction of plant and animal species and to ensure that viable populations were maintained in natural habitats. In order to achieve these ambitious aims, information was required for many species but that information was sometimes scattered throughout the world in many reports, scientific journals and books. The concept of a new system for bringing together information on various species came in the form of *Red Data Books*. The synoptic information in these books, culled from many sources, represented a basis not only for devising conservation measures but also for monitoring the status of the species.

(a)

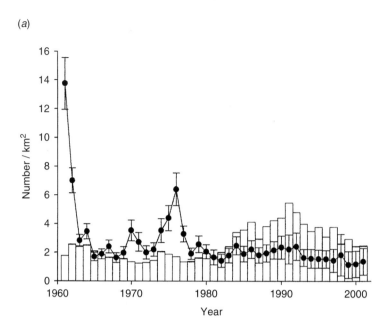

(b)

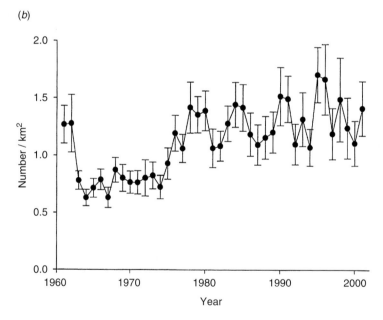

Fig. 9.9 Game Conservancy data for (a) the Grey Partridge (*Perdix perdix*) and (b) the Woodcock (*Scolopax rusticola*). (Graphs kindly provided by Dr N. J. Aebischer.)

National Red Data Books

Many years ago, several countries, including the UK (Perring & Farrell, 1983), Russia (Borodin, 1978) and New Zealand (Williams & Given, 1981) produced their own *Red Data Books* as a basis for monitoring and assessing the conservation needs of various species. In addition to these books, there have been many lists of threatened and endangered plants produced for several regions or countries, including Africa (Hall *et al.*, 1980), Australia (Leigh & Boden, 1979), Europe (Council of Europe, 1977), North America (Fairbrothers & Hough, 1975; Kershaw & Morton, 1976) and Russia (Malyshev & Soboleuska, 1980).

The *British Red Data Book* on vascular plants (Perring & Farrell, 1983) included numerical information on plant species abundance as part of the assessment of threatened status (there was a similar volume for Ireland: Curtis & McGough 1988). The information in these *Red Data Books* was based on the change in the presence of a species in $10\,\mathrm{km}^2$ over a certain period of time. The rate of change in abundance can be quantified and used along with other information, such as number of locations in protected areas, as a basis for calculating the threatened status of that particular species.

Rare plant monitoring and conservation in the USA

The number of plant-monitoring projects to be initiated in the USA has grown steadily. In 1987, an assessment of the efficiency and accuracy of 172 rare plant monitoring projects was undertaken by Palmer. That assessment (which basically asked the question, what kind of monitoring is being done, by whom and where?) was initiated to promote communication among people involved in rare plant monitoring and to present a summary of the 'state of the art' in current work. Perhaps not surprisingly, she found that different populations of the same species were being monitored in different locations using different ecological criteria. Information on the 172 monitoring projects came from 71 contacts and represented a good sample from the Natural Heritage programmes and the Nature Conservancy programmes. In addition to claims of no monitoring by 25% of all contacts, three basic types of monitoring programmes for rare plants were identified: inventory, survey and demographic.

The inventory method used was simply a population count repeated at intervals, usually each year. The survey method was based on transect or grid sampling in order to estimate population characteristics or obtain data on the entire population. The demographic studies involved marking or mapping individual plants in order to follow the status of those individuals.

The costs involved in these monitoring programmes may sometimes be seen as a trade-off in terms of the effectiveness of that monitoring method. Palmer (1987) noted, for example, that, although an inventory is simple and not time consuming, data on reproductive biology may be overlooked in inventories and this could lead to erroneous conclusions about the status of a species. Palmer's examination of 172 monitoring projects for rare plants in the USA led her to suggest 14 recommendations for rare-plant monitoring (Table 9.1): recommendations that realistically consider biological parameters as well as logistic parameters.

One particularly interesting application of species monitoring is that undertaken to ensure that management practices on nature reserves do not harm those qualities that originally led to designation as a nature reserve. For example, within the monitoring programmes promoted by the Nature Conservancy in the USA, there are some that are directed at fulfilling legal obligations (S. C. Buttrick, personal communication). Occasionally, the Nature Conservancy leases rights on a nature reserve such as for grazing or for hay. Baseline data and then monitoring programmes are established to ensure that the activities do not harm the protected species. In some instances, such as on the Katharine Ordway Sycan Marsh Reserve in Oregon, Black Terns (*Chlidonias nigra*) require surface water of varying characteristics throughout their reproductive cycle and the Sandhill Crane (*Grus canadensis*) requires a diversity of habitats ranging from open water for nesting to tufted hairgrass communities for foraging. Number of breeding pairs and reproductive success of these birds are just two of the variables used in the monitoring programmes.

The importance and relevance of rare-plant monitoring became a much discussed issue in North America in the mid 1980s and some of the techniques used for rare-plant monitoring in North America have been discussed in several papers published in the *Natural Areas Journal*, see for example Palmer (1986) and Bowles *et al.* (1986). A need for better communication was then noted by several members of the Ecological Society of America and as a result the *Natural Areas Journal* agreed to become a forum for a rare-plant monitoring network. The centre for that network is the Holcomb Research Institute, Butler University.

Monitoring the status of a species can be based on a subjective assessment of the information available in the database or alternatively precise ecological parameters can be used. However, the information about each species in the databases is not uniform and, therefore, the precise ecological parameters for monitoring a species will vary. For some species, monitoring may be based on numbers; for others it may be based on life tables and levels of recruitment.

In New Zealand, Given (1989) suggested one approach for the frequency of monitoring threatened plant species and that approach is as valid now as it was in 1989 (David Given, personal communication). The frequency of monitoring depends in part on the biology of the species being monitored (Table 9.2) and, therefore, any strategy for rare-plant monitoring needs to be structured and

Table 9.1 *Palmer's (1987) recommendations for rare-plant monitoring*

Stage	Recommendations
Planning	1. Choose species or communities whose degree of rarity or endangerment warrant monitoring (Palmer, 1986)
	2. Conduct preliminary sampling to determine an adequate number, size and shape of plots or transects needed to estimate parameters at the site (Green, 1979)
	3. Describe results of preliminary sampling, goals and methods in a proposal for external review by experts in designs for field research; propose the method of data analysis to ensure that the methods are appropriate for the plant
Use of monitoring types	4. Choose the inventory method for populations under no current threat in order to follow plant numbers over time and to make a preliminary assessment of need for further work
	5. Choose the survey method for species under no current threat to follow such indicators of performance as the number of reproductive plants in the entire population
	6. Choose the survey method to estimate numbers in very large populations or communities; follow numbers in stage classes (seedlings, juveniles, adults) if possible
	7. Choose the demographic method for sparse or threatened populations
	8. Record plant sizes and follow all stage classes, including seedlings, in demographic studies; use a stage–class transition matrix to identify critical stages and to predict changes in the population resulting from their management
	9. Choose the demographic method when little is known about the life history of the species, or when detailed information about individual location, size or reproduction is otherwise pertinent to the project goals
	10. Employ experimental manipulation in demographic or survey methods to test management treatments when there is a suspected threat to the site or to a life stage
Sampling and experimental design	11. Randomize sampling within each stratum in the habitat in survey and demographic studies
	12. Replicate experimental units in studies that compare or test management procedures
	13. Employ a design that acknowledges constraints in randomization when the sampling is not totally randomized
Publications and reports	14. Write proposals for review and annual progress reports; ultimately publish the results in a journal

Table 9.2 *Suggested frequency of monitoring data for rare-plant monitoring*
(a) For flora in New Zealand

Field	Frequency of monitoring	
	More often	Less often
Distribution	Few	Widespread
Population size	Small	Large
Habitat type	Seral	Climax
Life form, species type	r- (annual, etc.)	K- (long lived, perennial, etc.)
Population structure	Imbalanced	Balanced (good percentage of age classes)
Reproduction	Poor	Good
Breeding system	Self-incompatible, dioecious	Self-compatible, monoecious
Pollinating vector	Specialized	Unspecialized
Threats	Actual, greater	Potential, lesser
Protected areas	None or small	Numerous
Cultivation	None or little	Widespread
Management	No deliberate programme	A deliberate programme
Taxonomic distinctiveness	Greater (endemic)	Less (subspecies, variety)

(b) For flora in the Indiana Dunes National Lakeshore

Plant population type	Frequency of monitoring
Annual, biennial or rare	1–2 year intervals for reconnaissance monitoring
Perennial herbs, grasses	2–3 year intervals for reconnaissance monitoring
Dominant herbs, grasses	2–4 year intervals for reconnaissance monitoring
Trees or shrubs	4–5 year intervals for reconnaissance monitoring
Community dominants	5–10 year intervals for quantitative monitoring
Widespread plants not under threat	3–4 year intervals for quantitative monitoring
Rare plants or irregularly flowering plants	2–3 year intervals for quantitative monitoring
Successional	2–3 year intervals for quantitative monitoring

From Giren (1983) for New Zealand.
From Bowles et al. (1986) for Indiana, USA.

designed accordingly. A flow chart showing operations for such a possible monitoring scheme is illustrated in Figure 9.10.

9.5   Fisheries and ecological monitoring

There has long been a need to have sound ecological monitoring of fish populations. Between 1950 and 1969, the catch in the world's fisheries rose steadily at about 6–7% annually, from 20 million to 65 million tons (WECD, 1987). The fish stocks became more and more depleted, and after 1970 the average annual growth in catches fell to approximately 1%. It was estimated at that time that the world's catch had been depleted by as much as 24% more than it might have been with effective management of fish stocks.

The trend continues; for example, Pinnegar *et al.* (2002) in their research on long-term changes in the trophic level of the Celtic Sea fish community showed, significant declines in the mean trophic level of survey catches from 1982 to 2000. All of the largest fisheries are now almost certainly severely depleted. If ever there was a good example to justify the need for biological monitoring as a basis for calculating sustainable use of a natural resource, then overfishing of the world's oceans could not be a better example.

In broad and simple terms, overfishing is a world problem and although not entirely a biological problem (politics and economics are part of the problem), ecological monitoring is one of the key arguments for the urgent and unavoidable need for conservation of fish populations and sustainable use of those populations. A Marine Stewardship Council (MSC) scheme to allow traders to label products as having been sustainably taken is a small but important step: but an important step only as long as the label is audited fully and honestly (Pearce, 2003).

One of the methods used for monitoring fish stocks can be summarized as follows. Many fish stocks have been assessed using a simple analysis called the virtual population analysis (VPA), which is based on total mortality rate (natural mortality plus fishing mortality) and the proportion of fish surviving from one year to the next. The VPA is unreliable for the most recent years, partly because the natural mortality rate is usually guessed, but surveys may be used to estimate stock sizes in any current year. The survey methods include fishing surveys, acoustic surveys and egg and larvae surveys.

Fishing surveys can be used to obtain abundance indices of various fish species, particularly the juveniles that have not been taken by the fishing industry. Estimating juvenile abundance is vitally important for projections of stock size, especially where there is much dependence on population recruitment. In some cases, the survey area is divided into statistical rectangles

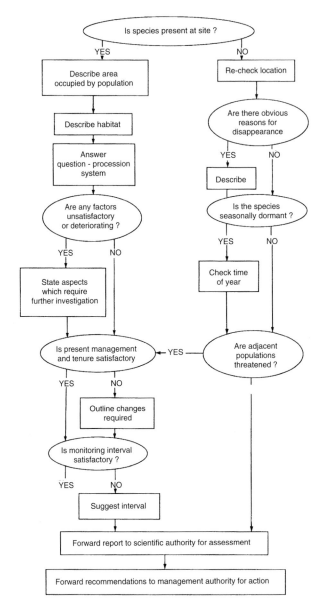

Fig. 9.10 Monitoring flow diagram for plant conservation in New Zealand. (With kind permission of David Given.)

and one haul is taken from each rectangle. International cooperation is one important key to success in these fishing surveys. An index of abundance for each major fish species is then calculated using the number of fish caught per hour of trawling. Such a survey method tends to give most reliable information for demersal species (living near or at the bottom), though useful information on pelagic stocks has also been obtained from bottom-trawl surveys. Variable weather conditions, the heterogeneous distribution of the fish and the trawling methods do not help to make fish surveys reliable, and there is some evidence that the catchability has changed on certain surveys (Armstrong & Cook, 1986).

Acoustic surveys of pelagic fish stocks have also been employed: the strength of the reflected signal is used to estimate the biomass of the shoal of fish. To do this, the acoustic target strength of the fish must be known and so experiments with caged and wild fish have to be undertaken. Problems that have been encountered with this method include mixtures of species and the difficulty of detecting fish near the surface or at the bottom.

Egg and larvae surveys have long been used as a basis for monitoring fish stocks, but not all species are suitable for this method. Not only must the spawning season be known but it must also be reasonably consistent from year to year. When the distribution of the spawning has been established, a decision then needs to be taken whether or not to estimate the absolute number to provide an index of the stock size. In waters around the UK, egg surveys have been used to assess Mackerel (*Scomber scomberus*) stocks and larvae surveys have been used to obtain indices of Herring (*Clupea harengus*) stocks.

9.6   Freshwater biological monitoring

Biological and chemical monitoring of freshwater can and should supplement each other, though either could provide reasonable indications of the effects of pollutants such as organic effluents. Hynes illustrated the chemical and biological effects of organic pollution very clearly in 1960 with a now much published figure (Fig. 9.11). The use of biological organisms for monitoring water pollution has sometimes been criticized because it is a lengthy process (and therefore expensive), there is a lack of standardization of techniques, interpretation of biological data is complex and there is a lot of variation in the organism's response. Most of these and other disadvantages have been dismissed as myths regarding barriers to biological monitoring techniques.

Chemical monitoring has become very sophisticated, largely in response to the high standards enforced by some agencies such as the US EPA. Advances in computing and associated technology have now made it possible to detect very small concentrations of some pollutants within a very short period of time.

However, chemical monitoring when used alone has its limitations and these limitations can be attributed to the following: difficulties associated with temporal and spatial sampling, potential synergistic effects and the possibility of

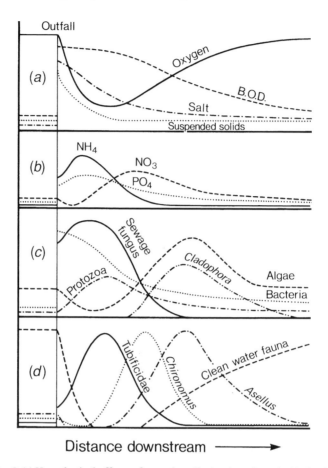

Fig. 9.11 Hypothetical effects of organic pollution in a river: (*a*, *b*) physical and chemical changes; (*c*) changes in microorganisms; (*d*) changes in larger organisms. B.O.D., biochemical oxygen demand. (Redrawn from Hynes, 1960.)

bioaccumulation in ecosystems. That is, without continuous chemical monitoring there is a possibility that unusually high levels of pollutants (unpredictable events) could remain undetected yet have serious effects on the aquatic ecosystem. In the absence of biological monitoring, the combined effects of pollutants cannot be assessed and chemical monitoring used alone may indicate deceptively low levels of pollutants that, if accumulated, could lower water quality below required standards and be harmful to aquatic life.

Surprisingly, little is known about the long-term effects of most chemical pollutants on aquatic ecosystems and associated biological processes. The impacts of radioactive wastes on freshwater and marine ecosystems especially pose very serious threats yet there is negligible monitoring of the influences of radiation on aquatic ecosystems.

Table 9.3 *Comparison of biochemical oxygen demand with two biotic indices*

Biochemical oxygen demand (mg/l)	General characteristics	Designation	Trent index	Chandler index
2	Salmonids, plecopterans, ephemeropterans, trichopterans, amphipods	Very clean	9–10	900+
2–3	Coarse fish; all above groups represented	Clean	7–10	500–900
2–3	Coarse fish; above groups restricted except amphipods, trichopterans and baetids	Clean	6–8	300–500
3–5	Few fish; trichopterans restricted, baetids rare, *Asellus* spp. dominant, molluscs and leeches present	Fairly clean	5–6	110–400
5–10	As above, but no baetids; molluscs and leeches present	Doubtful	3–5	45–300
5–10	No fish; otherwise as above	Doubtful	2–4	15–80
10+	No fish; Oligochaeta and red chrionomids only	Bad	1–3	9–20
10+	None of above: only air-breathing *Eristalis tenax*	Bad	0–1	0–10

The biochemical oxygen demand (BOD) is one common standard applied to monitoring and surveillance of freshwater. Often considered to be an aspect of chemical monitoring, it is in reality based on a biological process and should therefore be regarded as a basis for biological monitoring. The BOD is the ability of a given volume of water to use up oxygen over a period of five days at a temperature of 18 °C. Organic matter in the sample of water decomposes and the amount of oxygen consumed is then calculated. The range of BOD values and meaning of those values is compared with qualitative indices or biotic scores such as the Trent biotic scores and Chandler biotic scores (Table 9.3). Biotic scores are based on the presence or absence of certain taxa and the score is weighted according to the known tolerance of those taxa to pollution. Another common, but quantitative index used to monitor water quality is the diversity index in its many forms.

The water quality index

Determination of water quality requires value judgements no matter how many variables are measured and often there is conflict about which variables should be used. With the aim of promoting effective communication regarding the variables used for measuring water quality yet retaining a simple measure, an index of water quality, the WQI, was developed in 1970 at the National Sanitation Foundation in the USA. The WQI (Brown et al., 1972) can be expressed by:

$$\mathrm{WQI} = \sum_{i=1}^{n} w_i q_i$$

where WQI is a number between 0 and 100; q_i the quality of the ith variable, a number between 0 and 100; w_i the unit weight of the ith variable, a number between 0 and 1; and n the number of variables.

After consultation with a panel having expertise in water-quality management, the following nine variables were selected for the index: dissolved oxygen, faecal coliforms, pH, BOD, nitrate, phosphate, temperature, turbidity and total solids. Most of these variables are common to other schemes for assessing water quality. Because different experts apply different weighting to each variable, water quality was presented as a series of graphs, based on the respondent's views of what constituted low or high quality. This procedure, known as 'normalization of environmental indicators' (that is, converting measured indicators into a number or index) was described above with reference to environmental indicators (p. 99). With respect to the WQI, there was a close agreement for some parameters and poor agreement for others when normalization was undertaken (Fig. 9.12). The WQI has been advocated as being suitable for general usage as a uniform method reflecting the quality of water.

Use of indicators in freshwater monitoring

For many years, biological monitoring of freshwater has included the use of variables such as species composition, species richness, diversity, similarity, productivity, biomass and biotic indices based on community structure. Key or critical species such as commercially valuable fish or game fish have been used to assess pollution in aquatic systems and as biological early warning systems (Cairns et al., 1979a,b; Gruber & Diamond 1988) in the same manner, the concept of indicator species has been widely applied. There have been some exciting developments combining the use of diatoms (as indicator species) and laser holography for monitoring water quality. There is no general rule as regards choice of organisms for monitoring or surveillance; monitoring has to be based on those organisms that are most likely to provide the most appropriate information for the questions being asked.

Toxicity testing

The use of aquatic organisms for assessment of the effects of pollutants and for continuous toxicity testing is not new and has been long practised in the

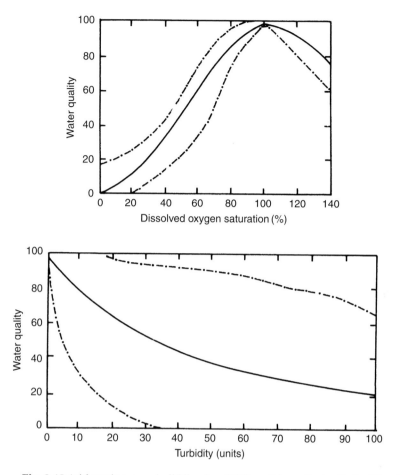

Fig. 9.12 Arithmetic means (solid lines) and 80% confidence limits (dashed lines) for two water quality parameters. (Modified from Brown *et al.*, 1972.)

USA in a very sophisticated manner. Toxicity tests have been used to provide data for several reasons including:

- to assess effects on organisms exposed to sublethal concentrations of a toxic substance for part or all of a life cycle
- to determine which pollutant is most toxic
- to determine which organism is most sensitive
- to assess changes in levels of toxicity
- to determine if the pollutant is within regulatory standards and to determine effects of periodic changes in concentrations of the pollutant.

Historically, toxicity tests have made much use of various species of adult fish, partly because more seems to be known about the ecology, physiology and behaviour of fish than other organisms. However, arguments against the use of fish in toxicity tests include the observation that many other taxa such as diatoms and macro-invertebrates are more sensitive to pollutants in water than are fish.

Many years ago in the USA, the EPA published criteria for determining which organisms should be used in toxicity testing. These and other suggested criteria include:

1.   The organism is representative of an ecologically important group (in terms of taxonomy, trophic level or niche).
2.   The organism occupies a position within a food chain leading to many other important species.
3.   The organism is widely available, is amenable to laboratory testing, easily maintained and genetically stable so uniform populations can be tested.
4.   There are adequate background data on the organism (physiology, genetics, taxonomy, role in the natural environment, behaviour).
5.   The response of the organism to the toxic substance should be consistent.
6.   The organisms should not be prone to disease, excessive levels of parasitism or physical damage.

Many of the environmental standards in the USA have been based on the use of single-species toxicity testing. However, although these tests provide many useful data and although the response of one very sensitive species may reflect other responses at higher levels of biological organization, the tests have been questioned, especially in the light of some advances in our knowledge of the complexities of ecosystems. Pontasch *et al.* (1989) argued, for example, that although single-species toxicity tests can efficiently examine relative toxicity, they may not be the most accurate method for predicting responses in receiving ecosystems because single-species tests do not take account of interactions between species and are sometimes conducted under physical and chemical conditions that are dissimilar to the natural environment.

Assessments of single versus multispecies toxicity testing were undertaken by Cairns and his colleagues in the 1980s (e.g. Cairns 1984, 1986); a consensus for the use of multispecies testing resulted in general support for the following.

1.   Multispecies tests can provide useful information for assessing the hazards of chemicals to aquatic and terrestrial organisms.
2.   The end points (response parameters) measured in multispecies tests are not as decisive as the end points measured in single species tests.

3. Multispecies tests can be powerful analytical tools. The capability of isolating ecological functions for study (while retaining a certain degree of complexity) is the strongest attribute of multispecies test.

4. Multispecies tests can be used as a 'conceptual bridge' between more-complex systems and less-complex systems.

5. Multispecies testing has the ability to isolate structural and functional components of ecosystems, which reduces variability.

6. Multispecies tests may be less sensitive to stress caused by a chemical when compared with the acute or chronic responses of uniform test organisms of a single species.

7. Multispecies tests are more complex than those conducted with single species and may exhibit more variability in structure and function.

8. The selection of the response parameter used to assess the effects of a chemical or effluent in a multispecies test is crucial because this selection can influence the interpretation of the results. Care must be taken to use state and rate variables correctly.

9. Because of the potential they offer in assessing the impact of chemicals on 'ecosystem level' functions, further research on multispecies toxicity tests is encouraged.

In a personal communication, J. Cairns once insisted that there was no reliable way, other than using an organism, to determine the integrated, collective impact of pollution on free-living aquatic organisms. That is almost certainly the case today. The rationale underlying the use of toxicity testing as outlined above has formed the basis of proposals for in-plant and in-stream monitoring of water quality. At its simplest, this means that testing of the water, by biological monitoring, would take place before the effluent discharge (in-plant) and at various locations in the stream below the effluent outfall. The costs of such sophisticated biological monitoring, especially automated biological monitoring using living organisms as sensors, often provide the reason for rejection of such an approach in favour of chemical monitoring or use of biotic indices.

A useful, thought-provoking set of questions for an assessment of the conceptual soundness of in-plant and in-stream biological monitoring was posed by Cairns and colleagues (1968). In an abbreviated form, these questions are as follows.

1. Will the system detect spills of lethal materials before they reach the outflow?

2. If only one organism is used for toxicity testing, will this species be more tolerant to some pollutants than to others?

3.   Is it possible to monitor abiotic variables and achieve the same results at lower cost and with greater efficiency?

4.   Since the biological monitoring cannot identify the pollutant, is it possible to combine successfully both biological and chemical monitoring?

5.   Will the behaviour of the organism be so reliable as not to cause 'false alarms' and possible expensive closing down of the plant?

6.   Should an organism indigenous to each system be used or should different species be used at different points in the in-plant and in-stream systems?

7.   Is it possible to use in-plant biological monitoring systems to detect the presence of pollutants that are either acutely lethal or not lethal in the short-term but have long-term effects by way of bioaccumulation?

8.   Are the in-plant biological monitoring systems only for a very large industry or is it possible to develop reliable, small in-plant biological monitoring that can be used by a person inexperienced in monitoring?

Classification of rivers and lakes for monitoring and surveillance

The concept of community structure has long been used as a basis for monitoring and surveillance of water quality. For instance, in Pennsylvania, Patrick (1949) used histograms showing comparative abundance of different species to assess the degree of pollution and she pioneered subsequent uses of the community structure concept in river classification and river-monitoring schemes.

Later, Savage (1982) showed that lakes in the UK could be classified on the basis of an analysis of affinities between communities of water boatmen (Corixidae). The method used by Savage involved the determination of similarity indices followed by cluster analysis. In brief, he showed a close relationship between species of Corixidae (*Sigara* spp. and a *Callicorixa* sp.), conductivity and organic load (eutrophic, rich in organic nutrients; oligotrophic, low in nutrients). Two species, *S. scotti* and *S. concinna*, were confined to waters of low and high conductivity, respectively. The species *S. falleni* and *C. praeusta* tended to occur in areas of water of intermediate size and high conductivity, the former being more numerous. The species *S. distincta* occurred in ponds and lakes of relatively low conductivity but is replaced by *S. scotti* at very low conductivity. The species *S. dorsalis* was found at all but the lowest conductivity levels but tended to be replaced by *S. falleni* in water bodies of high conductivity and intermediate size. On the basis of these results, Savage suggested that the species could usefully be arranged in three contrasting pairs, the first of each pair being more numerous in oligotrophic, low-conductivity conditions and the second in eutrophic, high-conductivity conditions. The pairs became *scotti–concinna*, *distincta–praeusta* and *dorsalis–falleni*.

Rather than basing a classification on either biotic or abiotic variables, there has been a tendency to use combinations of variables. In England and Wales, for example, what was known as the Department of the Environment and National Water Council BMWP was given the brief of developing a simple and efficient method for assessing biological quality of rivers. A simple scoring system with a broad classification based on macro-invertebrates was eventually recommended (see Ch. 5).

By way of contrast, a river classification system was developed in the early 1980s with mainly a simple chemical basis (Table 9.4). That classification system was based on a number of variables such as dissolved oxygen saturation levels, BOD and levels of ammonia. The classifications are normally based on a 95% basis (that is parameters which are expected to be achieved with 95% of the samples taken).

RIVPACS

Following the development of the BMWP score system, rapid appraisal of river quality was possible using identification at familylevels. Scientists at what was once the Institute of Freshwater Ecology and at the Freshwater Biological Association looked closely at the use of groups of invertebrates for monitoring freshwater communities and developed a river invertebrate prediction and classification system for the purpose of monitoring water quality and pollution.

That system or software package was aptly called RIVPACS (the River Invertebrate Prediction and Classification System) and was based on the assumption that the presence of certain taxonomic groups of invertebrates in rivers will depend on levels of certain physical and chemical variables. That is, the rationale was to assess a river on the basis of comparing the observed fauna with expected or 'target' fauna.

The programme to establish RIVPACS commenced in 1977 with two main objectives: to develop a classification system for unpolluted running-water sites in the UK based on macro-invertebrate fauna, and to develop procedures for prediction of fauna that would be expected on the basis of environmental variables (Wright *et al.*, 1988, 1989). The TWINSPAN program for analysis of multivariate data was used to establish classification of 30 site groups of fauna based on 370 sites from 61 river systems. A manual for RIVPACS (Furse *et al.*, 1986) provided details of the field and laboratory methods for taking the biological samples.

That technique provided equations for prediction of the probability of groups at new sites having sets of known environmental variables (Table 9.5). Any one of four sets of environmental variables may be used to predict the fauna at the

Table 9.4 *The UK Department of the Environment river classification*

River class[a]	Quality criteria; class-limiting criteria (95 percentile)[b]	Remarks	Current potential uses
1A	(i) Dissolved oxygen saturation greater than 80% (ii) BOD not greater than 3 mg/l (iii) Ammonia not greater than 0.4 mg/l (iv) Where the water is abstracted for drinking water, it complies with requirements for A2[c] water (v) Non-toxic to fish in EIFAC terms (or best estimates if EIFAC figures not available)	(i) Average BOD probably not greater than 80% (ii) Visible evidence of pollution should be absent	(i) Water of high quality suitable for potable supply abstractions and for all other abstractions (ii) Game or other high-class fisheries (iii) High amenity value
1B	(i) Dissolved oxygen greater than 60% saturation (ii) BOD not greater than 5 mg/l (iii) Ammonia not greater than 0.9 mg/l (iv) Where water is abstracted for drinking water, it complies with the requirements for A2[c] water (v) Non-toxic to fish in EIFAC terms (or best estimates if EIFAC figures not available)	(i) Average BOD probably not greater than 2 mg/l (ii) Average ammonia probably not greater than 0.5 mg/l (iii) Visible evidence of pollution should be absent (iv) Waters of high quality that cannot be placed in class 1A because of high proportion of high-quality effluent present or because of the effect of physical factors such as canalization, low gradient or eutrophication (v) Class 1A and class 1B together are essentially the class 1 of the RPS	Water of less high quality than class 1A but usable for substantially the same purposes
2	(i) Dissolved oxygen greater than 40% saturation (ii) BOD not greater than 9 mg/l	(i) Average BOD probably not greater than 5 mg/l (ii) Similar to class 2 of RPS	(i) Waters suitable for potable supply after advanced treatment (ii) Supporting reasonably good coarse fisheries

	(iii) Where water is abstracted for drinking water, it complies with the requirements for A3^c water		(iii) Moderate amenity value
	(iv) Non-toxic to fish in EIFAC terms (or best estimates if EIFAC figures not available)		
	(iii) Water not showing physical signs of pollution other than humic colouration and a little foaming below weirs		
3	(i) Dissolved oxygen greater than 10% saturation	Similar to class 3 of RPS	Waters that are polluted to an extent that fish are absent or only sporadically present; may be used for low-grade industrial abstraction purposes; considerable potential for further use if cleaned up
	(ii) Not likely to be anaerobic		
	(iii) BOD not greater than 17 mg/l		
4	Waters that are inferior to class 3 in terms of dissolved oxygen and likely to be anaerobic at times	Similar to class 4 of RPS	Waters that are grossly polluted and are likely to cause nuisance
X	Dissolved oxygen greater than 10% saturation		Insignificant watercourses and ditches not usable, where objective is simply to prevent nuisance developing

BOD, biochemical oxygen demand; EIFAC, European Inland Fisheries Advisory Commission; RPS, River Pollution Survey.

^{a} Under extreme weather conditions (e.g. flood, drought, freeze-up), or when dominated by plant growth or by aquatic plant decay, rivers usually in classes 1, 2 and 3 may have BODs, dissolved oxygen levels or ammonia content outside the stated levels for those classes. When this occurs the cause should be stated along with analytical results.

^{b} Ammonia figures expressed as NH_4; BOD has five-day carbonaceous values. In most instances this chemical classification will be suitable. However, it is restricted to a finite number of chemical determinants and there may be a few cases where the presence of a chemical substance other than those used in the classification markedly reduces the quality of the water. In such cases, the quality classification of the water should be downgraded on the basis of the biota actually present, and the reasons stated. EIFAC limits should be expressed as 95% percentile limits.

^{c} European Economic Community category A2 and A3 requirements are those specified in the Council Directive of 16 June 1975 concerning the Quality of Surface Water intended for Abstraction of Drinking Water in the Member States.

From UK Department of the Environment, 1985. Reproduced with the permission of the Controller of Her Majesty's Stationery Office.

313

Table 9.5 *RIVPACS species-level prediction for a site of high biological quality (Kings Farm, Moors River, Dorset)*
(a) River characteristics

Characteristic	Value
Mean width (m)	3.9
Mean depth (cm)	48.8
Substratum (%)	
Boulders + cobbles	0
Pebbles + gravel	9
Sand	16
Silt and clay	74
Mean substratum (phi)	6.01
Alkalinity (mg/l CaCO$_3$)	161.7
Altitude (m)	19
Distance from source (km)	17.0
Slope (m/km)	1.4
Discharge category	3
Mean air temperature (°C)	10.57
Annual air temperature range (°C)	12.39
Latitude (°N)	50.51
Longitude (°W)	1.51

Classification groups predicted using the above data: Gp 33, 72.1%; Gp 32, 17.8%; Gp 25, 4.3%; Gp 30, 3.8%.

(b) Predicted taxa in decreasing order of probability of capture

Taxa	Probability of capture (%)[a]
Glossiphonia complanata (L.)[b]	97.5
Micropsectra group	97.4
Erpobdella octoculata (L.)[b]	97.1
Gammarus pulex (L.)[b]	95.8
Elmis aenea (Muller)[b]	94.9
Hydracarina[b]	94.5
Cricotopus group[b]	94.3
Pisidium nitidum Jenyns[b]	92.6
Pisidium subtruncatum Malm[b]	90.4
Psammoryctides barbatus (Grube)[b]	89.8
Asellus aquaticus (L.)[b]	88.6
Potamopyrgus jenkinsi (Smith)[b]	87.4
Sphaerium corneum (L.)	87.2
Aulodrilus pluriseta (Piguet)[b]	86.1
Paratanytarsus group	86.0

Table 9.5 (*cont.*)

Taxa	Probability of capture (%)[a]
Thienemannimyia group[b]	85.4
Limnodrilus hoffmeisteri Claparede[b]	85.0
Helobdella stagnalis (L.)[b]	83.1
Sialis lutaria (L.)	81.4
Ceratopogonidae[b]	81.1
Polypedilum sp.[b]	78.5
Baetis vernus (Curtis)[b]	78.1
Lymnaea peregra (Muller)[b]	78.1
Physa fontinalis (L.)[b]	77.7
Sigara (*Sigara*) sp.	77.6
Stylaria lacustris (L.)[b]	75.5
Potamonectes depressus (Fabricius)[b]	75.3
Procladius sp.	75.3
Lumbriculus group[b]	74.6
Polycelis nigra group[b]	72.0
Microtendipes sp.[b]	72.0
Gyraulus albus (Muller)	70.6
Centroptilum luteolum (Muller)	70.2
Limnephilus lunatus (Curtis)[b]	69.5
Caenis luctuosa group[b]	68.9
Anisus vortex (L.)	66.1
Ephemerella ignita (Poda)[b]	65.2
Oulimnius tuberculatus (Muller)[b]	64.5
Hydroptila sp.[b]	64.4
Bithynia tentaculata (L.)	63.3
Eukiefferiella group[b]	62.4
Valvata piscinalis (Muller)	61.7
Ancylus fluviatilis Muller[b]	59.9
Crangonyx pseudogracilis (Bousfield)[b]	59.8
Piscicola geometra (L.)[b]	57.8
Potamothrix hammoniensis (Michaelsen)	56.1
Polycentropus flavomaculatus (Pictet)[b]	54.7
Halesus sp.[b]	52.0
Sigara falleni (Fieber)	51.9
Rhyacodrilus coccineus (Vejdovsky)[b]	51.4
Baetis rhodani (Pictet)[b]	50.7
Prodiamesa olivacea (Meigen)[b]	50.5

[a] At 50% probability, the observed number of taxa/expected number of taxa is 39.0/39.0, or 1.
[b] Observed taxa.
From data kindly provided by John Wright.

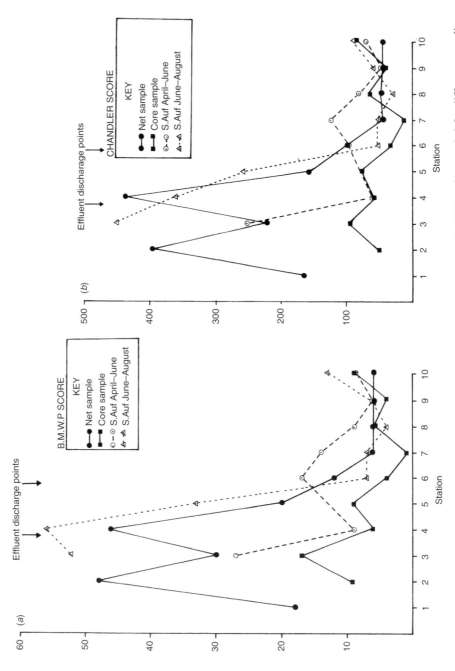

Fig. 9.13 The Biological Monitoring Working Party values (*a*) and the Chandler biotic indices (*b*) recorded for different sampling methods in the British Petroleum monitoring programme of Crymlyn Brook during 1982. S.Auf, standard Aufwuchs unit. (From Girton 1983.)

316

following taxonomic levels: species, all families (including logarithm categories of abundance) and restricted listings of BMWP families. An example for a river site in southern England gives the probability of capture of each group (in this instance to the 50% level of probability). The observed families are indicated in Fig. 9.13. A comparison of observed and predicted numbers of BMWP families is then used as a basis for assessment of environmental stress affecting river communities (Wright, 1995).

The establishment of the Countryside Survey 2000 has had important implications for monitoring the biological condition of streams and rivers in the UK. The data seem to suggest that there have been improvements between 1990 and 1998. The Countryside Survey 2000 has recorded an increase in the number of rare taxa in the streams surveyed between 1990 and 2000. Details are available on the Countryside Survey 2000 website.

9.7   Monitoring effects of oil refinery aqueous effluents

British Petroleum undertook freshwater biological monitoring of Crymlyn Bog in South Wales for many years with the aim of looking at the effects on the fauna of refinery aqueous effluents and other sources of pollution. One objective was to assess different biological monitoring techniques for sampling and analysis of data. For example, hand-net sampling of the fauna has been tried at several sampling sites on an annual basis (usually during mid-August). Three sampling sites were upstream of the effluent outfalls and seven were downstream. Sampling station No. 1 was subdivided into four sites, one above a quarry and three below a quarry outfall.

Sampling included both qualitative and quantitative methods. At each sampling station, three hand-net sweeps of 30 seconds duration were made of the benthos and surrounding macro-vegetation. Sediment and substratum material were washed through 710 μm mesh netting and the contents of the net preserved. Relative abundance was scored as follows: rare (1–2), occasional (3–10), common (11–50), abundant (51–100), very abundant (greater than 100).

Quantitative sampling was based on colonization samplers with a standard unit of substratum known as the standard Aufwuchs unit and also cylinder and core sampling of the benthos. Water chemistry was also recorded. In order to aid the interpretation of the data in terms of water quality, the number of taxa, the Trent biotic index and the Chandler score were calculated (Fig. 9.14). Standardized sampling methods proved difficult because of changes in substrata and, consequently, Sorensen's coefficient of similarity was used to compare material obtained with different sampling methods. A diversity index based on a simplification of the Shannon–Wiener index was employed.

Effluent discharge points

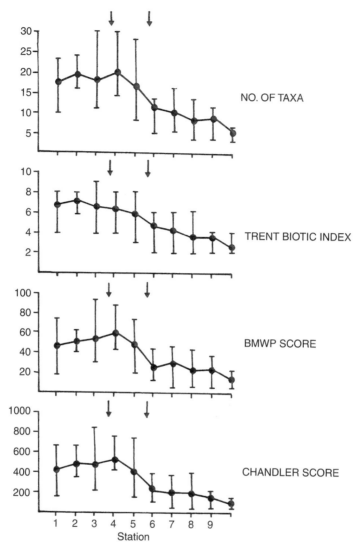

Fig. 9.14 Comparison of the number of taxa and biotic indices resulting from the British Petroleum monitoring programme of Crymlyn Brook, 1978–1982. The effluent discharge points are shown by the arrows. BMWP, Biological Monitoring Working Party. (From Girton, 1983.)

The different sampling methods, not surprisingly, collected different numbers of taxonomic groups. The analysis using Sorensen's similarity coefficient showed that similarity between colonization data and net data was highest and that there was poor similarity between net data and core data at stations 2, 4 and 7.

The results from the colonization sampling proved particularly successful in this kind of monitoring.

Overall, the decreasing number of taxa (Fig. 9.14) and the diminishing scores for the three biotic indices at the 10 sites was consistent. The first site was affected by ferruginous waste from a quarry, which eliminated species unable to withstand low oxygen levels and/or colloidal particles. The increased values of the indices and increase in taxa at station 2 was a result of decreasing effect of the quarry run-off but also reflected organic enrichment from agricultural run-off. Downstream of station 5, the refinery effluent had eliminated most pollution-sensitive species and the stream community was dominated largely by exploiter-type indicator groups such as Tubificidae, Chironomidae and *Lymnaea peregra*.

The BMWP biotic scores were similar in magnitude and resolution to the Chandler scores but the former seemed to be more sensitive (there was larger variation) when applied to the core samples (Fig. 9.14). In general, the conclusion was that the BMWP score had proved extremely successful, especially as it relied on a relatively crude level of identification. All biotic indices were, however, useful because they facilitated rapid interpretation of data.

A word of caution seems relevant here because the value of biotic scores and the relationship between biotic scores and other measures of water quality will vary from locality to locality. For example, Pinder & Farr (1987a,b) in their research on the surveillance of chalk streams found some interesting results when comparing biotic indices with BOD. They calculated both Chandler and BMWP scores, but also an average score per taxon was calculated for each biotic score by dividing the total score by the number of taxa contributing to it. They found that the Chandler score did not show a significant correlation with any measure of water quality, although the average Chandler score per taxon was significantly and negatively correlated with BOD and levels of dissolved organic carbon. Values of the BMWP score were significantly negatively correlated only with dissolved organic carbon, whereas the average BMWP score per taxon showed a highly significant negative correlation with BOD.

The British Petroleum monitoring programme (Girton, 1983) provided a generalized account but has also proved to be a useful basis for monitoring the responses of benthic populations to changes in levels of pollution. In the study, standardized sampling at all stations was difficult and the amount of material produced by various sampling methods at 13 stations and substations was large; this, in turn, demanded large amounts of laboratory time for sorting, identifying and analysing the collected data. Of relevance to planning a monitoring programme was the conclusion that there needed to be more selective sampling. This issue of amounts of data generated by even simple field methods was discussed in more detail in Ch. 7.

10

Ecological monitoring and environmental impact assessments

10.1   Introduction

There was a time when only the economics and health and safety of a new industrial or building development would be considered. Then came a requirement for social effects to be considered. Now it is commonplace for the economic, social and environmental effects to be considered. It is not possible for any project to avoid having some kind of effect on living organisms and the processes and properties of living systems. Ecological monitoring is, therefore, an integral part of any assessment of the effects of a project on the environment.

The term monitoring if used in association with environmental assessments for compliance monitoring and/or auditing is monitoring to determine whether or not agreed procedures have taken place or that the development complies with legal requirements. Compliance monitoring is, therefore, not the same as ecological monitoring. However, to comply there may be a requirement for ecological monitoring to take place over a certain period of time.

Auditing is a term that is sometimes used to refer to compliance review. Auditing can also refer to an important post-development component and that is testing the scientific accuracy of the environmental impact predictions. Some would argue that this is also part of the follow-up monitoring process.

Human impacts on the environment are not just one-off impacts at one time in one place. The effects of the impacts, projects or events can last for a long time and may have long-term effects on the biological and physical environment. The magnitude and extent of the impacts can be assessed using ecological monitoring programmes. Indeed, no assessment of impacts on nature can be undertaken without including some ecological monitoring.

What are the implications if there were no ecological monitoring? The disadvantage of not including ecological monitoring is that the extent and magnitude of impacts on the biological and human environment would not be assessed over time. One of the main benefits of ecological monitoring is that it provides data on which to base the detection and management of cumulative effects. One of the advantages of ecological monitoring is that it has the potential to add information, particularly in regions or countries where there is little known about the ecology.

The purpose of this chapter is threefold; first, to provide an introduction to the concept of EIAs; second, to give an account of ecological monitoring in the context of EIAS; and, third, to outline monitoring in the context of EIAs.

10.2   Environmental impact assessments and strategic environmental assessments

An EIA is a formalized procedure used to assess the impacts (whether good or bad) of a development or project on aspects of human welfare and the environment. It provides decision makers with the information they need to decide whether environmental impacts will be acceptable.

Much has been written about EIAs; there has been much research on EIAs and anyone interested in EIAs or practising in this area can join an EIA society. Perhaps one of the most comprehensive books on EIAs is that by Petts (1999). She has reviewed the history, practice in many countries and the content of EIAs. There is also a worldwide EIA organization, the International Association for Impact Assessment (IAIA). This organization advances innovation, development and communication of best practice in impact assessment. The website is http://www.iaia.org.

An EIA provides information that can be used to determine whether or not a development or project conforms with statutory national and international requirements or is simply perceived as being environmentally acceptable. An EIA should also act as a preventative management tool. If an aspect is not acceptable, then alternatives or modifications to the development may have to be considered. An EIA creates a demand for ecological monitoring.

The strategic environmental assessment (SEA) is becoming standard practice and has emerged because many EIAs have been reactive rather than the result of planning. There has been a need to have EIAs as part of the planning and policy-making process (Partidario, 2000). A SEA should provide a basis for avoiding impacts, for identifying cumulative effects and for assessing indirect effects of projects. In practice, a SEA is based on a process whereby the policies are agreed, planning follows and then project-based EIAs.

The EIA process formally emerged in the 1960s. There are now many countries that have legislation requiring an EIA to be undertaken prior to the approval or consent of certain developments. Indeed, in an international context, it is usual to have a requirement for an EIA before funding is agreed for a project. This requirement together with compliance with the Convention on Biological Diversity (Appendix) has shaped the nature of EIAs and has made certain that there is ecological monitoring.

The EIA process provides opportunities for looking at alternative sites, assessing the effects of different scales of developments and avoiding or modifying actions that may affect human health and safety. The main rationale underlying an EIA is, therefore, to ensure that the development will not infringe on human health and safety and that the development will not cause damage (over the life of the project and during decommissioning) to the natural and physical environment. The EIA provides decision makers with the information they need to decide whether the environmental impacts will be acceptable.

For many reasons, monitoring is an integral component of an EIA. In his book, Richard Morgan (1998) listed 12 reasons for including monitoring, including:

- it may provide an early warning of unpredicted impacts
- to check that mitigation has been successful
- to audit impact prediction
- to document actual impacts for use in similar situations elsewhere.

Despite apparent widespread agreement amongst environmental biologists that ecological monitoring should be part of the post-EIA process, there continue to be few examples of post-EIA ecological monitoring. Monitoring effluents at source is usually a requirement of an industrial development but the effects of toxic effluents on plants and animals (especially non-domestic or non-economic organisms) are rarely monitored.

This lack of ecological monitoring may have come about partly as a result of a lack of demands for monitoring in the legislation. Other contributing factors may include the costs of ecological monitoring and difficulties in identifying who is responsible for the monitoring.

However, even in the absence of legal requirements, some industrial organizations have established post-EIA biological monitoring programmes, some of which have proved to be of value in relation to environmental restoration and habitat renewal.

10.3   History of environmental impact and strategic environmental assessments

The term EIA is now more than 40 years old, but basic screening processes for environmental impacts arising from developments have been undertaken over a much longer time. In the USA, the National Environment Policy Act passed by the US Congress in 1969 required federal agencies to 'include in every recommendation or report on proposals for legislation and other major Federal actions significantly affecting the quality of the human environment a detailed statement … on the effects of that action'. It was this Act that first brought in a requirement for EIAs for federally funded projects in the USA. Many thousands of EIAs have now been prepared in the USA and there has been much debate about their effectiveness. Several scientific journals contain articles that assess both the requirements of EIAs and their effectiveness.

Since the early 1970s, many countries, including Australia, Canada, France, Ireland, the EEC and New Zealand have introduced EIA legislation despite much criticism to the effect that existing legislation was adequate for EIAs. A requirements for an EIA is also included in the International Convention on Biological Diversity (Article 14).

In the UK, the foundations for environmental assessment came with a planning system dating back to the 1947 Town and Country Planning Act. The government initially resisted any introduction of EIA legislation but in 1981 an environment subcommittee of the EEC recommended that EIAs had a role to play in the UK and accordingly a draft EEC Directive was prepared. This Directive was adopted by the Community in 1985 and became law in 1988. The Directive was amended in 1997 (97/11/EC).

Although a directive is a community law that is binding as to the results to be achieved, it is left to the member states as to the choice and methods of implementation. The EIA has now been expanded to include plans and programmes within the Strategic Environmental Assessment Communication of the EEC.

The methods of implementation in the UK have been widely discussed, as have the two lists of projects; list one included schedule 1 projects for which an EIA is required in every case; list two included schedule 2 projects for which an EIA is required only if the particular project in question is judged likely to give rise to significant environmental effects.

This EEC Council Directive (on the assessment of the effects of certain public and private projects on the environment) has 14 descriptive articles and three annexes: one lists projects for which EIAs will be mandatory; one lists projects for which EIAs may be carried out at the discretion of a member state; the third annex sets out the information required in an EIA. Although the word monitoring does not appear, monitoring would be presumed because in the introduction it says: 'Whereas the effects of a project on the environment must be assessed in order to take account of concerns to protect human health, to contribute by means of a better environment to the quality of life, to ensure maintenance of the diversity of species and to maintain the reproductive capacity of the ecosystem as a basic resource for life'. In Annex Three it says: '5. A description of the measures envisaged to prevent, reduce and where possible offset any significant adverse effects on the environment'. Both of these statements would suggest that biological monitoring would be required although words such as significant and maintenance suggest that a precise or objective approach to monitoring would not necessarily be adopted.

The word significant is not uncommon in environmental legislation. For example, significant appears in the New Zealand 1991 Resource Management Act, which is primarily a strategic planning statute but also operates at the resource consent level. It refers to 'avoiding, remedying, or mitigating any adverse affects of activities on the environment' (Section 5.2(c)). Section 6(c) refers to 'The protection of areas of significant indigenous vegetation and significant habitats of indigenous fauna'.

The Fourth Schedule of that Act deals with assessments of effects on the environment. Paragraph 1 (i) of the Fourth Schedule has implications for monitoring (Box 10.1): 'Where the scale or significance of the activity's effect are such that monitoring is required, a description of how, once the proposal is approved, effects will be monitored and by whom'. Further reference to monitoring appears in Section 35, where it says:

1.   Every local authority shall gather such information, and undertake or commission such research, as is necessary to carry out effectively its functions under this Act.
2.   Every local authority shall monitor:
   (a) The state of the whole or any part of the environment of its region or district to the extent that is appropriate to enable the local authority to effectively carry out its functions under the Act.

10.4   The process of environmental impact assessment

A prerequisite for a good EIA is the preparation of a conceptual plan or framework. That is, there needs to be a decision on what issues to focus on. Conceptual frameworks have been developed as part of many EIAs, ranging from those EIAs designed for the Antarctic (Benninghoff & Bonner, 1985) to those designed for impacts of electricity-generating stations on the natural environment in temperate, coastal regions (Bamber, 1989).

A preliminary outline EIA or scoping can be an important cost-effective stage where technical details of the development are prepared along with preliminary environmental information. The aim is to identify the scope and depth to which impact assessments should be undertaken and to focus attention on information gaps. Another important function of the preliminary EIA is to identify the expertise and personnel eventually required to achieve the environmental assessments and the ecological monitoring.

Box 10.1   The New Zealand Resource Management Act 1991: the Fourth Schedule *Assessment of Effects on the Environment*

1.   **Matters that should be included in an assessment of effects on the environment** – Subject to the provisions of any policy statement or plan, an assessment of effects on the environment for the purposes of section 88 (6) (b) should included –

(a)   A description of the proposal:

(b)   Where it is likely that an activity will result in any significant adverse effect on the environment, a description of any possible alternative locations or methods for undertaking the activity:

(c)   Where an application is made for a discharge permit, a demonstration of how the proposed option is the best practicable option:

(d)   An assessment of the actual or potential effect on the environment of the proposed activity:

(e)   Where the activity includes the use of hazardous substances and installations, an assessment of any risks to the environment which are likely to arise from such use:

(f)   Where the activity includes the discharge of any contaminant, a description of –

   (i)   The nature of the discharge and the sensitivity of the proposed receiving environment to adverse effects; and

   (ii)   Any possible alternative methods of discharge including discharge into any other receiving environment:

(g)   A description of the mitigation measures (safeguards and contingency plans where relevant) to be undertaken to help prevent or reduce the actual or potential effect:

(h)   An identification of those persons interested in or affected by the proposal, the consultation undertaken, and any response to the views of those consulted:

(i)   Where the scale or significance of the activity's effect are such
that monitoring is required, a description of how, once
the proposal is approved, effect will be monitored and by whom.

2.   **Matters that should be considered when preparing an
assessment of effects on the environment** – Subject to the
provisions of any policy statement or plan, any person preparing an
assessment of the effects on the environment should consider the
following matters:

(a)   Any effect on those in the neighbourhood and, where relevant,
the wider community including any socio-economic and
cultural effects:

(b)   Any physical effect on the locality, including any landscape
and visual effects:

(c)   Any effect on ecosystems, including effects on plants or animals
and any physical disturbance of habitats in the vicinity:

(d)   Any effect on natural and physical resources having aesthetic,
recreational, scientific, historical, spiritual, or cultural, or
other special value for present or future generations:

(e)   Any discharge of contaminants into the environment, including
any unreasonable emission of noise and options for the treatment and disposal of contaminants:

(f)   Any risk to the neighbourhood, the wider community, or the
environment through natural hazards or the use of hazardous
substances or hazardous installations.

The precise steps for an EIA (Fig. 10.1) are dependent on good practice and on
legislation, but in general there is a sequence of steps commencing with a
project proposal and screening; a decision is then taken as to whether or not
an EIA should be undertaken. The consultation phase should continue throughout the whole process.

The consultation phase can be of considerable benefit to the developer,
particularly with regard to the planning of the subsequent environmental
assessments and in preparing a smooth passage for the whole environmental
process. The consultation should continue throughout the whole EIA process.

A project outline would usually incorporate information about the location
and physical limits to the development. Whereas defining the physical limits of
the site can be done objectively, defining the area affected by other impacts such
as airborne effluents cannot be undertaken in a definitive manner without
detailed and prolonged studies and so may not be undertaken at all.

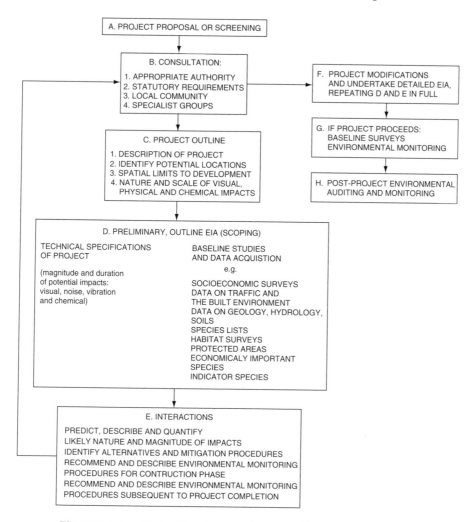

A. PROJECT PROPOSAL OR SCREENING

B. CONSULTATION:

1. APPROPRIATE AUTHORITY
2. STATUTORY REQUIREMENTS
3. LOCAL COMMUNITY
4. SPECIALIST GROUPS

F. PROJECT MODIFICATIONS
AND UNDERTAKE DETAILED EIA,
REPEATING D AND E IN FULL

G. IF PROJECT PROCEEDS:
BASELINE SURVEYS
ENVIRONMENTAL MONITORING

C. PROJECT OUTLINE

1. DESCRIPTION OF PROJECT
2. IDENTIFY POTENTIAL LOCATIONS
3. SPATIAL LIMITS TO DEVELOPMENT
4. NATURE AND SCALE OF VISUAL,
   PHYSICAL AND CHEMICAL IMPACTS

H. POST-PROJECT ENVIRONMENTAL
AUDITING AND MONITORING

D. PRELIMINARY, OUTLINE EIA (SCOPING)

TECHNICAL SPECIFICATIONS
OF PROJECT

(magnitude and duration
of potential impacts:
visual, noise, vibration
and chemical)

BASELINE STUDIES
AND DATA ACQUISTION
e.g.

SOCIOECONOMIC SURVEYS
DATA ON TRAFFIC AND
THE BUILT ENVIRONMENT
DATA ON GEOLOGY, HYDROLOGY,
SOILS
SPECIES LISTS
HABITAT SURVEYS
PROTECTED AREAS
ECONOMICALY IMPORTANT
SPECIES
INDICATOR SPECIES

E. INTERACTIONS

PREDICT, DESCRIBE AND QUANTIFY
LIKELY NATURE AND MAGNITUDE OF IMPACTS
IDENTIFY ALTERNATIVES AND MITIGATION PROCEDURES
RECOMMEND AND DESCRIBE ENVIRONMENTAL MONITORING
PROCEDURES FOR CONTRUCTION PHASE
RECOMMEND AND DESCRIBE ENVIRONMENTAL MONITORING
PROCEDURES SUBSEQUENT TO PROJECT COMPLETION

Fig. 10.1 A simplified outline for an environmental impact assessment (EIA) showing where monitoring could be included.

Scoping includes gathering of baseline data. A description of the site could include information on physical, demographic and related infrastructure features. Areas of conservation or landscape interest would be identified but biological information is usually confined to protected species, sensitive species, taxa of commercial interest, taxa of scientific importance and taxa perceived to be important by the local community.

Potential impacts may be easy to identify but then comes the more difficult task of assessing the interactions between impacts and the various components of the environment. This part of the process has traditionally

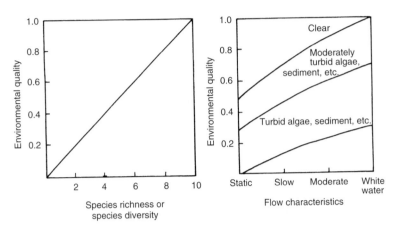

Fig. 10.2 Value function graphs employed in the Battelle Environmental Evaluation System. (After Canter, 1977.)

been directed largely towards the human environment or those aspects of the environment that are of immediate concern to human use of the environment.

The role of GIS in EIAs and in particular the role in determining 'significance' has been well researched (e.g. Antunes *et al.*, 2001). A spatial impact assessment methodology developed by Antunes and colleagues was based on the assumption that the importance of environmental impacts is dependent, in part, on the spatial distribution of the effects and the affected physical and biological environment. For each environmental component, Antunes and colleagues proposed that impact indices should be calculated and that these should be based on the spatial distribution of the impacts.

The emphasis on spatial aspects has been partly responsible for the types of tool used in identifying potential impacts at the scoping stage: flow diagrams or networks, matrices, and checklists. Of these, checklists are seemingly the most simple, but various methods have been refined to quantify the impacts. For example, in the 1970s, the Battelle Environmental Evaluation Scheme (Canter, 1977), which is particularly comprehensive, was designed to identify impacts arising from water resources development. The impacts were grouped into four main categories (ecology, environmental pollution, aesthetics, human interest) and these, in turn, were divided into components and parameters. There was an interesting next step in which each component of the checklist was transformed (see previous example of normalization in Fig. 3.2, p. 99) into a level of environmental quality ranging from 0 (poor) to 1.0 (very good). Examples of these transformations for species diversity, dissolved oxygen (indicated by the presence of algae etc.) and physical appearance of water are shown in Fig. 10.2.

The principle seemed good in that there was an attempt to quantify parameters of the environment and changes in these parameters could be quantified in terms of changing environmental quality. The relationship does, of course, require a subjective judgement as to what equates with different levels of environmental quality. There are, for example, some plant communities, such as heathlands, that characteristically have a low plant-species diversity and any increase in diversity would indicate a diminished quality and not an increase in quality of the heathland community.

Other well-known methods for identification of impacts at the scoping stage have been based on matrices, some of which can be used to give a subjective assessment of the magnitude and importance of the impact. The Leopold matrix (Fig. 10.3) is perhaps, historically, the most well-known matrix to be used in EIAs (Parker & Howard, 1977). In that matrix, environmental characteristics and project actions were recorded in an open matrix with up to 8800 interactions. First, the matrix was used to record those impacts (interactions) that were likely to occur; if the impact was 'significant' then a diagonal line was entered in a box. The interaction was then scored from 1 (low) to 10 (high) for both magnitude of the interaction (Fig. 10.3, upper left) and the relative importance (Fig. 10.3, lower right). Despite the criticisms that have been levelled at the Leopold matrix (for example, it does not cater for secondary or tertiary interactions) this kind of approach has a cost-effective role to play in providing a basis for monitoring changes in ecological systems.

The word interactions is used here to emphasize the integral role of ecology and, therefore, ecological monitoring. This is the stage where ecological skills are put into practice and where data are gathered to determine what the effects are, their magnitude, duration and location.

There have been some follow-up studies and reviews of the ecological content and of the ecological monitoring component of EIAs. It is worrying to find that the biological and ecological components of EIAs continues to leave a lot to be desired. This is especially so because in 1994 the Canadian Government and the International Association for Impact Assessment convened a meeting at which the effectiveness and efficiency of EIAs was examined because early studies advocated the need for good ecological impact assessments (EcIA). For example Beanlands & Dunker in 1984 commented on ecological frameworks for EIAs and even at that time were critical of the sporadic applications of ecological principles in impact assessments. Their work has often been cited but it is worth repeating that on the basis of their review of 30 EIA statements prepared in Canada, they made a number of recommendations for new initiatives and also drew attention to monitoring as a key EIA requirement. They suggested that an EIA should be required to have the following features.

Fig. 10.3 An example of a condensed version of the Leopold matrix for urban development.

1. Identify, early on, an initial set of valued ecosystem components to provide a focus for subsequent activities
2. Define a context within which the significance of changes in the valued ecosystem components can be determined
3. Show clear temporal and spatial contexts for the study and analysis of expected changes in the valued ecosystem components
4. Develop an explicit strategy for investigating the interactions between the project and each valued ecosystem component, and demonstrate how the strategy is to be used to coordinate the individual studies undertaken
5. State impact predictions explicitly and accompany them with the basis upon which they were made
6. Demonstrate and detail a commitment to a well-defined programme for monitoring project effects.

Treweek *et al.* (1993) have reviewed ecological assessments of proposed road projects in the UK and concluded that many statements failed to provide the data necessary to predict potential ecological impacts, and what was even worse, very few attempted to quantify the impacts. Not surprisingly Jo Treweek and colleagues recommended that EcIAs should be carried out earlier in the planning and design of new roads. On the basis of the shortcomings of EcIAs, Treweek (1999) has also drawn attention for the need for expert reviews.

10.5   Ecological impact assessments

The EIA address a wide range of impacts. One of these is the ecological impact or the impact on the ecology and biology (Box 10.2). An EcIA is a formalized process for identifying and quantifying impacts of a proposed project on ecosystems, their components, properties and processes. It should be undertaken as part of an EIA but this is often not the case.

Box 10.2   Stages in an ecological impact assessment

1.   Legal, policy, planning and consultation issues

The legislative context. What is legally required? Land tenure and ownership.
Constraints. Are there constraints imposed by local regulators or councils?
Public consultation. Who to consult and how and when?
Defining the context. What are the spatial and temporal parameters and limits to the ecological impact assessment (EcIA)?

Focusing. Decide what can be done within the time and resources available.

Policies. Are there policies for mitigation and compensation?

2. Scoping

A particularly important stage is that of scoping or early screening. If done well, the scoping stage can provide a sound basis and save time and resources.

Within the scoping exercise the following would be considered:

- agreement about the spatial boundaries or at least reaching agreement about the spatial boundaries for the project
- language and communication
- the proposed timetable for the development
- presence of protected habitats and protected species
- identifying constraints; these may be administrative, financial and technical
- what can be achieved in the time available for the EcIA (what is desirable, what is practical)?

There needs to be agreement about the time available and when the EcIA is to take place. The proposed project may have a direct impact on some species and habitats but over what area and during what times (year) will the EcIA be undertaken. There needs to be a balance between that time available and the depth of the EcIA in terms of scientific integrity.

It is not practical to undertake a comprehensive EcIA where every species is taken into consideration and where every ecosystem component, property and process is taken into consideration. Therefore, it follows that there has to be a selection and there needs to be agreed criteria for that selection. Priority species would include those that were protected, species in decline and those categorized as rare or endangered.

Ecological monitoring should follow on from environmental impact assessments (EIAs) or at least be included in EIAs. There seems to be very few reports that deal with this aspect, especially ecological monitoring.

3. Ecological impact assessments: the essential elements

(i) Statutory requirements

Identify the statutory requirements for an EcIA. Are there any international agreements? Is there a requirement to meet the needs of the

Convention on Biological Diversity? Is the project likely to impact on sites protected by international agreements?

Does the national legislation include protected areas and or protected species? If so are there protected areas and/or species in the area?

What is considered 'significant' in terms of likely impacts on the environment?

(ii) Consultation

Are there any organizations that must legally be consulted (such as planning authorities, conservation authorities)?

Identify the interested parties (that is the individuals or groups who can affect or who are affected by the impacts on nature).

(iii) Definitions of technical terms

Multidisciplinary teams and groups of people are involved in EcIAs within a general framework of EIAs. Not surprisingly, language presents a difficult challenge. Simple terms such as 'natural', 'landscape' and 'environment' will mean different things to different people. Terms such as 'environment', 'ecosystem', 'biological community', 'biological diversity' and 'populations' should be defined clearly so as to avoid misunderstandings at a later date.

(iv) Information technology

Different EcIAs may be undertaken at different scales using many datasets. The use of information technology such as GIS will almost certainly be an important element of the spatial analysis and the analysis of the interactions between the road and the environment. Computer-aided design is another tool that may be used to analyse visual impacts on the environment.

(v) Spatial extent of the ecological impact

There needs to be agreement on the spatial extent of the EcIA. There also needs to be agreement on whether or not associated activities will be included in the impact assessment.

(vi) Recognition of the functions and values of nature

Nature has intrinsic values but nature also provides sources, sinks and services that are used by humans. Mangrove swamps, for example, provide a habitat for a range of aquatic species and support mangrove biological communities. The ecosystem helps to prevent flooding of inland areas and provides a sources of food and fuel for humans. The swamps are used to absorb wastes (the swamp is both a source and an ecological sink). Wetlands in general contribute to water quality and quantity management. They support wildlife and recreational activities.

(vii) Sources of ecological information

How will the baseline ecological information be obtained? It may have to be commissioned and collected via surveys and the establishment of new databases (primary information) or it could be obtained from existing databases (secondary information). The sources of the secondary information need to be identified. These could be in the form of national or regional biological databases. Alternatively, the information could be held by councils, government departments or local natural history organizations.

(viii) Gathering primary ecological information

There are many methods used by ecologists for collecting primary ecological information. All these methods have their limitations and also biases towards certain groups of organisms. The limitations and biases should be considered when the ecological information is analysed and presented.

Ecological surveys can be in the form of direct observation, that is simply walking the route in a random fashion and recording what is seen. There are two main limitations. This method will detect only those organisms, habitats and communities present and visible to the observer at that time. Typically, this would include many kinds of plant (grasses, herbs, shrubs and trees), mammal, bird and butterfly. At the same time, other information can be collected on the topography, geology, soil types, aquatic features and climate. At any moment in time there will be organisms not visible to the observer and, therefore, not recorded. Secondly, some organisms that are present and visible may be overlooked because of the random survey method used.

There is a range of methods and equipment that can help to make surveys more comprehensive, whether that be for organisms in the soil, aquatic organisms or those high in the tree canopies above ground. Similarly, there are survey methods that enable information to be collected in a more systematic and compressive manner.

(ix) Surveys and sampling effort

Not surprisingly, the amount of 'sampling effort' will affect the results of the survey. For example, the number of bird species recorded via a one hour random walk along the proposed route is likely to be different to the results from a series of random walks undertaken for two hours each on one day in each week for three weeks.

(x) Temporal and spatial variation

One random walk for recording birds undertaken during one afternoon in mid summer will be likely to give different results to a series of walks undertaken during different times of day and night and throughout the year. Animals come and go and some plant species such as spring-flowering bulbs may be present at only certain times of year.

The distribution and movements of plants and animals over time and in different parts of the landscape will have implications for the comprehensiveness and accuracy of the information being collected.

(xi) Species

A common unit in biological and ecological studies is the species. For this reason, most ecological surveys would be on species (rather than on populations, subspecies, families or higher units of taxonomic organization). Baseline information will, therefore, almost certainly include the species. However, it is impractical to have a definitive account of all species. It is because of this difficulty that ecological surveys very often do not attempt to obtain information on all groups of organisms. Commonly the groups chosen are those most easy to survey. The issue is not so much which groups are selected but the criteria for selecting or not selecting a group. The criteria could be based on any of the following:

- legally protected species
- species in Red Data books (lists of species compiled on the basis of an aggregation of ecological and other criteria; these are the species in greatest need of conservation)
- species considered or shown to be 'rare'
- keystone species (species that are key to an ecosystem: their extinction has implications for the survival of many other species)
- species with specialist requirements
- species considered to be culturally important
- species considered to be economically important
- easily surveyed species
- indicator species (those species whose present, absence or condition tells us something about the state of the habitat or the environment)
- umbrella species (those species that, if conserved, will benefit the conservation of other species using the same habitat).

A rationale and criteria should be provided for the selection of species to be surveyed.

(xii) Habitats and communities

Loss of habitats and fragmentation of habitats is the most serious impact arising from road projects. It may be possible to obtain reasonably comprehensive information on habitats and or biotic communities within the proposed route for the road. However, the terms 'habitat' and 'biological community' are often used in a general and subjective manner. A habitat, ecologically speaking, is that place where a population of a certain species lives. Within that habitat, there will be other species and there may be more that one biological community.

A biological community is made up of a collection of plant and animal species. For example, we generally talk of wetland communities, woodland communities or coastal marine communities. However, any one of these general communities could be made up of many subcommunities. In some parts of the world, particularly Europe, biological communities have been classified into different types, based partly on the particular mix or the assemblage of species present. For example, woodlands and scrubland in the UK have been classified into 25 different kinds. Of those, four kinds of oak woodland have been recognized (Rodwell *et al.*, 1991).

There may be more than just one type of wetland, woodland or grassland.

(xiii) Ecological processes

Examples of ecological processes are:

- litter (leaf fall etc.) accumulation
- decomposition
- respiration
- bioaccumulation
- productivity
- nutrient accumulation: eutrophication.

It would be impractical to undertake an assessment of the effects of all ecological processes. Therefore, a selection has to be made and before that some criteria for selection of processes to be assessed have to be agreed.

(xiv) Data reliability and confidence limits

All ecological survey methods have their limitations. The extent to which any ecological information is representative of an area or the extent to which it is complete must be made known. There should be an independent review of the methods used.

(xv) Methods for identifying the impacts

Impacts can be direct, indirect, associated, cumulative or synergistic. What methods can be used to ensure that the identification of impacts is as comprehensive as possible? The range of methods used to identify impacts in EIAs and EcIAs includes checklists, flow diagrams and matrices.

Ecological impacts should be quantified. For example, the area of loss of a particular plant community as a result of land-take could be estimated or measured. More difficult are measures of losses of plants and animals as a result of the loss of habitat.

(xvi) What is significant and how to decide priorities?

Are all impacts equal in terms of their importance or are some impacts relatively more important than some others? Is the loss of a heritage site of greater importance that the loss of a habitat of a rare species. These are issues that need to be discussed and there has to be an agreed rationale and process for deciding priorities.

(xvii) Ecological mitigation and compensation

Ecological impacts should be avoided where at all possible. If not, then proposals for remedying the impacts or mitigation and compensation measures must be devised, presented, assessed and adopted.

Ecological and conservation considerations come first but then they have to be considered in a wider context of what is practical, not just what is desirable. The monetary costs of such measures must be considered together with the environmental and conservation costs.

(xviii) Ecological monitoring

What needs to be monitored, how often and by who? Remedying and mitigating the effects may require some follow-up studies. Monitoring ecological change over time is a commitment that requires not only ecological expertise but also an infrastructure.

(xix) Communicating the results

Who is the intended audience and what is the most useful way or presenting the information?

An EcIA results in a lot of information, some technically complex. The interested parties in an EcIA may not all have the expertise to understand readily all the information. It is necessary, therefore, that consideration be given to the method used for communication the information, whether that be written or verbal.

Whereas ecological monitoring may or may not be required as part of an EIA, there is no doubt at all that it is an integral part of an EcIA (Box 10.2). However, many issues remain. The costs of ecological monitoring may be used as an excuse to try to avoid any monitoring at all. Should the developer pay for the ecological monitoring as part of the overall costs of the development?

There are several issues with regard to ecological monitoring and EIAs (Table 10.1). One of these is the fact that very little is known about some taxonomic groups. Within some taxa, few species may have been named. In addition, ecological datasets are often incomplete, partly because it has not been practical to collect the data in the time available. An overall issue has been the continuing emphasis on species and species inventories rather than an ecological systems approach. Once again, time constraints often force the EcIA to deal only with species lists of a few taxa.

Finally, although EcIAs can identify likely impacts and although ecologists may recommend specific actions, those actions may not be acted on. It appears that, all too often, the motions of an EcIA are carried out but the remedial requirements and follow-up are not undertaken.

10.6   Examples of ecological monitoring in environmental impact assessment

Ecological monitoring of coastal communities

Most examples of ecological monitoring have been in industrial contexts. For example, the CONCAWE group (the oil companies' European organization for environment, health and safety) was established in the Hague in 1963. The CONCAWE Secretariat is now in Brussels. The emphasis of its work is within the technical and economic studies relevant to oil refining, distribution and marketing in Europe.

However, work undertaken under the umbrella of CONCAWE has in the past identified some ecological monitoring procedures (including ideas for sampling

Table 10.1 *Environmental impact assessments: issues identified for the purposes of discussion although many have no simple answers*

Issue	Features
Taxonomic skills	The number of people with taxonomic skills is declining; this has implications for survey and inventory work
Lack of information	In some regions and some countries there is very little known about the ecology of the area where the proposed development is to take place
Species and ecosystems	Too much emphasis is placed on simple species inventories rather than the effects on ecological systems
Reliability of information	It is common for inventories to be assembled without any way of auditing the material
Standards of professional practice	How competent is the ecologist? Only more recently have there been opportunities for practitioners in ecology and environmental management to seek personal certification
Where is the ecology and ecological monitoring in the process?	All too often the ecology is an 'add on' or is undertaken in a very superficial manner
How many times and for how long?	Perhaps the most challenging issue is the frequency of recording information and the duration; sometimes the effects of a project on the ecology of an area may not be detected for many years
Who pays for the ecological monitoring?	The developer or the regulator?
What if the ecological monitoring demonstrates failure of the mitigation or an inaccurate assessment of the impacts?	What happens and who is responsible?

techniques and data analysis) for aqueous effluents from petroleum refineries (CONCAWE, 1982). Ideas for data analysis incorporated the use of biotic indices, diversity indices, similarity indices and simple diagrammatic presentations showing relative abundance. Obviously, there is as much choice of methods and analysis as there is variation in coastal communities and conditions. The variation in coastal communities means that different monitoring methods are required for the different communities. Generally speaking, data collection can be of two kinds: removal of material (destructive sampling), and non-destructive recording techniques. The latter includes satellite imagery, aerial photography, stereophotography and visual recording.

One other interesting consideration, and one which tends to be forgotten until faced with samples of material, is the identification of specimens. There are, for example, many coastal areas, such as on some of the coasts of Alaska, where the fauna and flora have not been studied in detail. Taxonomic keys may not be available for some groups, and some groups such as marine algae may include many species that are simply difficult to identify. Taxonomic difficulties highlight the dependence of biological and ecological monitoring on taxonomy, and it is regrettable that there seem to be fewer and fewer taxonomists graduating from institutions of higher education.

To illustrate some of the applications of methods used for monitoring ecological effects of oil spills, we can refer briefly to a selection of case studies based on different types of community such as rocky shores, salt marshes, mudflats and sub-tidal communities (rocky and soft sediments).

Ecological monitoring of rocky shores

The monitoring of the biological effects of the development of the oil terminal at Sullom Voe in the Shetlands is one example of a carefully designed part of a large EIA (Dicks, 1989). The biological monitoring commenced in 1978 and was a large integrated programme directed at a range of taxonomic groups and habitats including rocky shores. At each site selected for surveying and later monitoring, permanent markers identified the sampling area (today GPS would be used). In this instance, the permanent markers were paint spots and many photographs were also taken to ensure that the sampling site could readily be located. The data were collected (during spring tides) at 20 cm intervals along transects from low water up to the lower limit of the flowering plants. Abundance of lichens, algae and animals was recorded within a 3 m wide strip. Photographs were also used to supplement the visual inspections of the monitoring sites.

These abundance scales (Table 10.2) have also been used, since 1972, to monitor organisms of rocky shores in the vicinity of a refinery at Mongstad in Norway. In addition, size-class distributions of the mollusc *Patella vulgata* have been monitored at some sites to investigate levels of population recruitment.

It is perhaps important to note that the levels of monitoring described above can detect only gross changes in the ecosystems and communities, but nevertheless this monitoring, as an integral component of a much larger biological monitoring programme, has provided important data from which to assess the impacts during construction and even during extreme perturbations caused by oil spill clean-up operations.

Table 10.2 *Abundance values for ecological monitoring of rocky shores*

1. Flowering plants and lichens
   - Ex. More than 80% cover
   - S.   50–80% cover
   - A.   20–50% cover
   - C.   1–20% cover
   - F.   Large scattered patches
   - O.   Widely scattered patches, all small
   - R.   Only one or two patches

2. Seaweeds
   - Ex. More than 90% cover
   - S.   60–90% cover
   - A.   30–60% cover
   - C.   5–30% cover
   - F.   Less than 5% cover, zone still apparent
   - O.   Scattered plants, zone indistinct
   - R.   Only one or two plants

3. Barnacles (excluding *Balanus perforatus*), *Littorina neritoides*, small forms of *L. saxatilis*
   - Ex. More than $4/cm^2$
   - S. $3–4/cm^2$
   - A. $1–3/cm^2$
   - C. $10–100/10\ cm^2$
   - F. $1–10/10\ cm^2$, never more than 10 cm apart
   - O. $1–100/cm^2$, few within 10 cm of each other
   - R.   Less than $1/m^2$

4. *Balanus perforatus*
   - Ex. More than $3/cm^2$
   - S. $1–3/cm^2$
   - A. $10–100/10\ cm^2$
   - C. $1–100/m^2$
   - F. $10–100/m^2$
   - O. $1–10/m^2$
   - R.   Less than $1/m^2$

5. Limpets and winkles (excluding *Littorina neritoides* and small forms of *Littorina saxatilis*)
   - Ex. More than $200/m^2$
   - S. $100–200/m^2$
   - A. $50–100/m^2$
   - C. $10–50/m^2$

Table 10.2 (*cont.*)

F. $1–10/m^2$
O. $1–10/10 \ m^2$
R.   Less than $1/10 \ m^2$

6. Gastropods (excluding limpets and winkles)
   Ex.   More than $100/m^2$
   S. $50–100/m^2$
   A. $10–50/m^2$
   C. $1–10/m^2$, locally sometimes more
   F.   Less than $1/m^2$, locally sometimes more
   O.   Always less than $1/m^2$
   R.   Less than $1/10 \ m^2$

7. Mussels
   Ex.   More than 80% cover
   S.   50–60% cover
   A.   20–50% cover
   C.   Large patches, but less than 20% cover
   F.   Many scattered individuals and small patches
   O.   Scattered individuals, no patches
   R.   Less than $1/m^2$

8. *Pomatoceros triqueter*
   Ex   –
   S   –
   A.   More than 50 tubes/$10 \ cm^2$
   C. $1–50$ tubes/$10 \ cm^2$
   F. $10–100$ tubes/m^2
   O. $1–10$ tubes/m^2
   R.   Less than 1 tube/m^2

9. *Spirorbis* spp.
   Ex   –
   S   –
   A. $5/cm^2$ or more on 50% of suitable surfaces
   C. $5/cm^2$ or more on 5–50% of suitable surfaces
   F. $1–5/cm^2$ on 1–5% of suitable surfaces
   O.   Less than $1/cm^2$
   R.   Less than $1/m^2$

Ex, extensive; S, superabundant; A, abundant; C, common;
F, frequent; O, occasional; R, rare.

Ecological monitoring of salt marshes and mudflats

The monitoring that has been undertaken on the Spartina salt marshes adjacent to the Fawley Refinery on Southampton Water has already been described in detail (see p. 133) as an example of ecosystem monitoring.

Ecological monitoring of subtidal communities

As early as 1975, ecologists at Stirling University (Scotland) were monitoring the effects of aqueous effluents on intertidal soft sediment communities around the oil refinery at Grangemouth (east coast of Scotland). There, the monitoring was based on mud samples taken at points along 18 permanent transects (Cowell, 1978), and abundance and distribution of sentinel indicator species (see Ch. 5), such as the mollusc *Hydrobia ulvae* were recorded. Based on the species composition, distribution and abundance of various organisms, it was possible at that time to make subjective assessments of 'pollution zones' of the mudflats that then formed the basis of pollution zone maps. These maps contributed to an assessment of the quality and extent of the management operations and resulted in improved management operations.

At Naantali in Finland, a more detailed ecological monitoring programme was established as a basis for monitoring what were severe impacts from a refinery on the organisms of subtidal soft sediments (Table 10.3; see also Fig. 4.2). Poor management of refinery effluents over a period of about 15 years until 1972 had quite severe visible effects on adjacent coastal fauna. Following the commencement of an ecological monitoring programme in 1970, there was an introduction of a new treatment process in 1973 for the effluents. The sampling was undertaken by staff from the University of Turku and took place within permanent monitoring sites located by fixing on inshore landmarks. Sampling was undertaken at control sites as well as adjacent to the outflow from the refinery.

The usual abundance records of selected taxa have been recorded along with measurements of biomass. Species richness and species diversity of benthic littoral macroorganisms were also recorded both before and after the new effluent treatment had been introduced. Results from one of the inshore sites shows the effect of controlling and treating the effluent.

10.7   Research and development

Apart from any technical and logistic problems of monitoring, the spatial and temporal changes in communities pose one of the greatest problems for ecological monitoring. In other words, if the organisms were distributed evenly

Table 10.3 *The benthic macrofauna at a sampling site near Naantali harbour (1965–1976)*

	1965	1967	1970	1972	1973	1974	1975	1976
No. of species	4	4	3	4	5	9	9	9
Density (individuals/m²)	250	530	400	70	1650	1470	2634	1970
Biomass (g/m²)	6.5	7.9	17.9	0.4	27.6	29.6	157.4	282.5
Diversity	1.6	1.3	1.2	1.2	0.9	0.7	0.9	n.a.
Species richness	0.5	0.5	0.5	0.5	0.4	0.7	1.0	n.a.
Evenness	0.8	0.7	0.6	0.8	0.4	0.3	0.3	n.a.

n.a., not available.

on the shores and did not change over time, then monitoring would be made much easier. Monitoring change over time requires a decision about frequency of sampling: too little sampling may not reveal seasonal or shorter patterns and too much sampling may damage (via physical disturbance) the habitat. Spatial patterns are no easier to detect. As well as being accurate, the sampling area has to be adequate and representative. On the basis of rocky shore monitoring around the Isle of Man (UK), Hartnoll & Hawkins (1980) suggested that, where resources are limited, the concentration of effort by sampling larger areas could be offset by sampling at a reduced number of stations. Clearly, long-term ecological studies could provide data of much value to the design of effective coastal monitoring programmes.

Much attention has been given to methods of scoring abundance, but there has been less attention to research that may be required for establishing natural spatial and temporal changes, especially long-term changes in coastal communities. This has been noted in relation to long-term ecological studies still continuing today by ecologists of British Petroleum at the Rafinor oil refinery at Mongstad on Fernsfjord in Norway. It has become very clear after many years of monitoring that change in biological communities is the rule rather than the exception. The main problem there has not been so much detecting change but in assessing its significance in attributing a cause.

Methods for ecological monitoring of the effects of pollution have been available for many years. For example, the research by Christie (1980) on rocky subtidal communities in Norway was designed to combine the use of stereophotography, field experiments and manipulation as a method for monitoring effects of chronic pollution. Two experiments and a control were established in a site of about 2.0 m². The aim of one experiment was to record the recolonization of a small cleared area and the second experiment used exclusion cages in order to examine the effects of predation. Photographs of the percentage cover of different organisms was the main source of data.

After 14 months, the results of these experiments showed that predation by *Coryphella*, *Asterias* and *Psammechinus* spp. was the most important factor affecting the structure of the community. Christie concluded that these predators are 'key' species and, therefore, an effective monitoring programme would need to record the abundance and effects of these predators on the subtidal community.

One of the areas in greatest need of development is post-EIA monitoring and ecological monitoring of reclamation and restoration. For example, ecological monitoring is a crucial part of reclamation of mining activities, particularly as mine reclamation has become compulsory in some countries. In one study, regulatory performance standards and monitoring requirements were used (Smyth & Deardon, 1998) to evaluate reclamation success in western North America. They concluded that reclamation practices and regulatory control would be improved greatly by 'effective' long-term monitoring.

Appendix

The following text is the 1992 Convention on Biological Diversity (excluding the preamble and annexes).

This Convention was adopted on 5 June 1992 at the UN Conference on Environment and Development in Rio de Janeiro, Brazil. The objective of this Convention is to conserve biological diversity and to use biological diversity in a sustainable and equitable manner.

Throughout this Convention there is reference to ecological monitoring or at least there are implications for ecological monitoring.

Article 1

Objectives

The objectives of this Convention, to be pursued in accordance with its relevant provisions, are the conservation of biological diversity, the sustainable use of its components and the fair and equitable sharing of the benefits arising out of the utilisation of genetic resources, including by appropriate access to genetic resources and by appropriate transfer of relevant technologies, taking into account all rights over those resources and to technologies, and by appropriate funding.

Article 2

Use of Terms

For the purposes of this Convention:
'Biological diversity' means the variability among living organisms from all sources including, *inter alia*, terrestrial, marine and other aquatic ecosystems and the ecological complexes of which they are part; this includes diversity within species, between species and of ecosystems.

'Biological resources' includes genetic resources, organisms of parts thereof, populations, or any other biotic component of ecosystems with actual or potential use or value for humanity.

'Biotechnology' means any technological application that uses biological systems, living organisms, or derivatives thereof, to make or modify products or processes for specific use.

'Country of origin of genetic resources' means the country which possesses those genetic resources in in-situ conditions.

'Country providing genetic resources' means the country supplying genetic resources collected from in-situ sources, including populations of both wild and domesticated species, or taken from ex-situ sources, which may or may not have originated in that country.

'Domesticated or cultivated species' means species in which the evolutionary process has been influenced by humans to meet their needs.

'Ecosystem' means a dynamic complex of plant, animal and micro-organism communities and their non-living environment interacting as a functional unit.

'Ex-situ conservation' means the conservation of components of biological diversity outside their natural habitats.

'Genetic material' means any material of plant, animal, microbial or other origin containing functional units of heredity.

'Genetic resources' means genetic material of actual or potential value.

'Habitat' means the place or type of site where an organism or population naturally occurs.

'In-situ conditions' means conditions where genetic resources exist within ecosystems and natural habitats, and, in the case of domesticated or cultivated species, in the surroundings where they have developed their distinctive properties.

'In-situ conservation' means the conservation of ecosystems and natural habitats and the maintenance and recovery of viable populations of species in their natural surroundings and, in the case of domesticated or cultivated species, in the surroundings where they have developed their distinctive properties.

'Protected area' means a geographically defined area which is designated or regulated and managed to achieve specific conservation objectives.

'Regional economic integration organisation' means an organisation constituted by sovereign States of a given region, to which its Member States have transferred competence in respect of matters governed by this Convention and which has been duly authorised, in accordance with its internal procedures, to sign, ratify, accept, approve or accede to it.

'Sustainable use' means the use of components of biological diversity in a way and at a rate that does not lead to the long-term decline of biological diversity, thereby maintaining its potential to meet the needs and aspirations of present and future generations.

'Technology' includes biotechnology.

Article 3

Principle

States have, in accordance with the Charter of the United Nations and the principles of international law, the sovereign right to exploit their own resources pursuant to their own environmental policies, and the responsibility to ensure that activities within their jurisdiction or control do not cause damage to the environment of other States or of areas beyond the limits of national jurisdiction.

Article 4

Jurisdictional Scope

Subject to the rights of other States, and except as otherwise expressly provided in this Convention, the provisions of this Convention apply, in relation to each Contracting Party:

a.   In the case of components of biological diversity, in areas within the limits of its national jurisdiction; and

b.   In the case of processes and activities, regardless of where their effects occur, carried out under its jurisdiction or control, within the area of its national jurisdiction or beyond the limits of national jurisdiction.

Article 5

Co-operation

Each Contracting Party shall, as far as possible and as appropriate, co-operate with other Contracting Parties, directly or, where appropriate, through competent international organisations, in respect of areas beyond national jurisdiction and on other matters of mutual interest, for the conservation and sustainable use of biological diversity.

Article 6

General Measures for Conservation and Sustainable Use

Each Contracting Party shall, in accordance with its particular conditions and capabilities:

a.   Develop national strategies, plans or programmes for the conservation and sustainable use of biological diversity or adapt for this purpose existing strategies, plans or programmes which shall reflect, *inter alia*, the measures set out in this Convention relevant to the Contracting Party concerned; and

b.   Integrate, as far as possible and as appropriate, the conservation and sustainable use of biological diversity into relevant sectoral or cross-sectoral plans, programmes and policies.

Article 7

Identification and Monitoring

Each Contracting Party shall, as far as possible and as appropriate, in particular for the purposes of Articles 8 to 10:

a.   Identify components of biological diversity important for its conservation and sustainable use having regard to the indicative list of categories set down in Annex I;

b.   Monitor, through sampling and other techniques, the components of biological diversity identified pursuant to subparagraph (a) above, paying particular attention to those requiring urgent conservation measures and those which offer the greatest potential for sustainable use;

c.   Identify processes and categories of activities which have or are likely to have significant adverse impacts on the conservation and sustainable use of biological diversity, and monitor their effects through sampling and other techniques; and

d.   Maintain and organise, by any mechanism data, derived from identification and monitoring activities pursuant to subparagraphs (a), (b) and (c) above.

Article 8

In-situ Conservation

Each Contracting Party shall, as far as possible and as appropriate:

a.   Establish a system of protected areas or areas where special measures need to be taken to conserve biological diversity;

b.   Develop, where necessary, guidelines for the selection, establishment and management of protected areas or areas where special measures need to be taken to conserve biological diversity;

c.   Regulate or manage biological resources important for the conservation of biological diversity whether within or outside protected areas, with a view to ensuring their conservation and sustainable use;

d.   Promote the protection of ecosystems, natural habitats and the maintenance of viable populations of species in natural surroundings;

e.   Promote environmentally sound and sustainable development in areas adjacent to protected areas with a view to furthering protection of these areas;

f.   Rehabilitate and restore degraded ecosystems and promote the recovery of threatened species, *inter alia*, through the development and implementation of plans or other management strategies;

g.   Establish or maintain means to regulate, manage or control the risks associated with the use and release of living modified organisms resulting from biotechnology which are likely to have adverse environmental impacts that could affect the conservation and sustainable use of biological diversity, taking also into account the risks to human health;

h.   Prevent the introduction of, control or eradicate those alien species which threaten ecosystems, habitats or species;

i.   Endeavour to provide the conditions needed for compatibility between present uses and the conservation of biological diversity and the sustainable use of its components;

j.   Subject to its national legislation, respect, preserve and maintain knowledge, innovations and practices of indigenous and local communities embodying traditional lifestyles relevant for the conservation and sustainable use of biological diversity and promote their wider application with the approval and involvement of the holders of such knowledge, innovations and practices and encourage the equitable sharing of the benefits arising from the utilisation of such knowledge, innovations and practices;

k. Develop or maintain necessary legislation and/or other regulatory provisions for the protection of threatened species and populations;

l. Where a significant adverse effect on biological diversity has been determined pursuant to Article 7, regulate or manage the relevant processes and categories of activities; and

m. Co-operate in providing financial and other support for in-situ conservation outlined in subparagraphs (a) to (l) above, particularly to developing countries.

Article 9

Ex-situ Conservation

Each Contracting Party shall, as far as possible and as appropriate, and predominantly for the purpose of complementing in-situ measures:

a. Adopt measures for the ex-situ conservation of components of biological diversity, preferably in the country of origin of such components;

b. Establish and maintain facilities for ex-situ conservation of and research on plants, animals and micro-organisms, preferably in the country of origin of genetic resources;

c. Adopt measures for the recovery and rehabilitation of threatened species and for their reintroduction into their natural habitats under appropriate conditions;

d. Regulate and manage collection of biological resources from natural habitats for ex-situ conservation purposes so as not to threaten ecosystems and in-situ populations of species, except where special temporary ex-situ measures are required under subparagraph (c) above; and

e. Co-operate in providing financial and other support for ex-situ conservation outlined in subparagraphs (a) to (d) above and in the establishment and maintenance of ex-situ conservation facilities in developing countries.

Article 10

Sustainable Use of Components of Biological Diversity

Each Contracting Party shall, as far as possible and as appropriate:

a. Integrate consideration of the conservation and sustainable use of biological resources into national decision making;

b. Adopt measures relating to the use of biological resources to avoid or minimise adverse impacts on biological diversity;

c. Protect and encourage customary use of biological resources in accordance with traditional cultural practices that are compatible with conservation or sustainable use requirements;

d.   Support local populations to develop and implement remedial action in degraded areas where biological diversity has been reduced; and

e.   Encourage co-operation between its governmental authorities and its private sector in developing methods for sustainable use of biological resources.

Article 11

Incentive Measures

Each Contracting Party shall, as far as possible and as appropriate, adopt economically and socially sound measures that act as incentives for the conservation and sustainable use of components of biological diversity.

Article 12

Research and Training

The Contracting Parties, taking into account the special needs of developing countries, shall:

a.   Establish and maintain programmes for scientific and technical education and training in measures for the identification, conservation and sustainable use of biological diversity and its components and provide support for such education and training for the specific needs of developing countries;

b.   Promote and encourage research which contributes to the conservation and sustainable use of biological diversity, particularly in developing countries, *inter alia*, in accordance with decisions of the Conference of the Parties taken in consequence of recommendations of the Subsidiary Body on Scientific, Technical and Technological Advice; and

c.   In keeping with the provisions of Articles 16, 18 and 20, promote and co-operate in the use of scientific advances in biological diversity research in developing methods for conservation and sustainable use of biological resources.

Article 13

Public Education and Awareness

The Contracting Parties shall:

a.   Promote and encourage understanding of the importance of, and the measures required for, the conservation of biological diversity, as well as its propagation through media, and the inclusion of these topics in educational programmes; and

b.   Co-operate, as appropriate, with other States and international organisations in developing educational and public awareness programmes, with respect to conservation and sustainable use of biological diversity.

Article 14

Impact Assessment and Minimising Adverse Impacts

1) Each Contracting Party, as far as possible and as appropriate, shall:

   a. Introduce appropriate procedures requiring environmental impact assessment of its proposed projects that are likely to have significant adverse effects on biological diversity with a view to avoiding or minimising such effects and, where appropriate, allow for public participation in such procedures;

   b. Introduce appropriate arrangements to ensure that the environmental consequences of its programmes and policies that are likely to have significant adverse impacts on biological diversity are duly taken into account;

   c. Promote, on the basis of reciprocity, notification, exchange of information and consultation on activities under their jurisdiction or control which are likely to significantly affect adversely the biological diversity of other States or areas beyond the limits of national jurisdiction, by encouraging the conclusion of bilateral, regional or multilateral arrangements, as appropriate;

   d. In the case of imminent or grave danger or damage, originating under its jurisdiction or control, to biological diversity within the area under jurisdiction of other States or in areas beyond the limits of national jurisdiction, notify immediately the potentially affected States of such danger or damage, as well as initiate action to prevent or minimise such danger or damage; and

   e. Promote national arrangements for emergency responses to activities or events, whether caused naturally or otherwise, which present a grave and imminent danger to biological diversity and encourage international co-operation to supplement such national efforts and, where appropriate and agreed by the States or regional economic integration organisations concerned, to establish joint contingency plans.

2) The Conference of the Parties shall examine, on the basis of studies to be carried out, the issue of liability and redress, including restoration and compensation, for damage to biological diversity, except where such liability is a purely internal matter.

Article 15

Access to Genetic Resources

1) Recognising the sovereign rights of States over their natural resources, the authority to determine access to genetic resources rests with the national governments and is subject to national legislation.

2) Each Contracting Party shall endeavour to create conditions to facilitate access to genetic resources for environmentally sound uses by other Contracting Parties and not to impose restrictions that run counter to the objectives of this Convention.

3) For the purpose of this Convention, the genetic resources being provided by a Contracting Party, as referred to in this Article and Articles 16 and 19, are only

those that are provided by Contracting Parties that are countries of origin of such resources or by the Parties that have acquired the genetic resources in accordance with this Convention.

4) Access, where granted, shall be on mutually agreed terms and subject to the provisions of this Article.

5) Access to genetic resources shall be subject to prior informed consent of the Contracting Party providing such resources, unless otherwise determined by that Party.

6) Each Contracting Party shall endeavour to develop and carry out scientific research based on genetic resources provided by other Contracting Parties with the full participation of, and where possible in, such Contracting Parties.

7) Each Contracting Party shall take legislative, administrative or policy measures, as appropriate, and in accordance with Articles 16 and 19 and, where necessary, through the financial mechanism established by Articles 20 and 21 with the aim of sharing in a fair and equitable way the results of research and development and the benefits arising from the commercial and other utilisation of genetic resources with the Contracting Party providing such resources. Such sharing shall be upon mutually agreed terms.

Article 16

Access to and Transfer of Technology

1) Each Contracting Party, recognising that technology includes biotechnology, and that both access to and transfer of technology among Contracting Parties are essential elements for the attainment of the objectives of this Convention, undertakes subject to the provisions of this Article to provide and/or facilitate access for and transfer to other Contracting Parties of technologies that are relevant to the conservation and sustainable use of biological diversity or make use of genetic resources and do not cause significant damage to the environment.

2) Access to and transfer of technology referred to in paragraph 1 above to developing countries shall be provided and/or facilitated under fair and most favourable terms, including on concessional and preferential terms where mutually agreed, and, where necessary, in accordance with the financial mechanism established by Articles 20 and 21. In the case of technology subject to patents and other intellectual property rights, such access and transfer shall be provided on terms which recognise and are consistent with the adequate and effective protection of intellectual property rights. The application of this paragraph shall be consistent with paragraphs 3, 4 and 5 below.

3) Each Contracting Party shall take legislative, administrative or policy measures, as appropriate, with the aim that Contracting Parties, in particular those that are developing countries, which provide genetic resources are provided access to and transfer of technology which makes use of those resources, on mutually agreed terms, including technology protected by patents and other intellectual property

rights, where necessary, through the provisions of Articles 20 and 21 and in accordance with international law and consistent with paragraphs 4 and 5 below.

4) Each Contracting Party shall take legislative, administrative or policy measures, as appropriate, with the aim that the private sector facilitates access to, joint development and transfer of technology referred to in paragraph 1 above for the benefit of both governmental institutions and the private sector of developing countries and in this regard shall abide by the obligations included in paragraphs 1, 2 and 3 above.

5) The Contracting Parties, recognising that patents and other intellectual property rights may have an influence on the implementation of this Convention, shall co-operate in this regard subject to national legislation and international law in order to ensure that such rights are supportive of and do not run counter to its objectives.

Article 17

Exchange of Information

1) The Contracting Parties shall facilitate the exchange of information, from all publicly available sources, relevant to the conservation and sustainable use of biological diversity, taking into account the special needs of developing countries.

2) Such exchange of information shall include exchange of results of technical, scientific and socio-economic research, as well as information on training and surveying programmes, specialised knowledge, indigenous and traditional knowledge as such and in combination with the technologies referred to in Article 16, paragraph 1. It shall also, where feasible, include repatriation of information.

Article 18

Technical and Scientific Co-operation

1) The Contracting Parties shall promote international technical and scientific co-operation in the field of conservation and sustainable use of biological diversity, where necessary, through the appropriate international and national institutions.

2) Each Contracting Party shall promote technical and scientific co-operation with other Contracting Parties, in particular developing countries, in implementing this Convention, *inter alia*, through the development and implementation of national policies. In promoting such co-operation, special attention should be given to the development and strengthening of national capabilities, by means of human resources development and institution building.

3) The Conference of the Parties, at its first meeting, shall determine how to establish a clearing-house mechanism to promote and facilitate technical and scientific co-operation.

4) The Contracting Parties shall, in accordance with national legislation and policies, encourage and develop methods of co-operation for the development and use of technologies, including indigenous and traditional technologies, in pursuance of the

objectives of this Convention. For this purpose, the Contracting Parties shall also promote co-operation in the training of personnel and exchange of experts.

5)   The Contracting Parties shall, subject to mutual agreement, promote the establishment of joint research programmes and joint ventures for the development of technologies relevant to the objectives of this Convention.

Article 19

Handling of Biotechnology and Distribution of its Benefits

1)   Each Contracting Party shall take legislative, administrative or policy measures, as appropriate, to provide for the effective participation in biotechnological research activities by those Contracting Parties, especially developing countries, which provide the genetic resources for such research, and where feasible in such Contracting Parties.

2)   Each Contracting Party shall take all practicable measures to promote and advance priority access on a fair and equitable basis by Contracting Parties, especially developing countries, to the results and benefits arising from biotechnologies based upon genetic resources provided by those Contracting Parties. Such access shall be on mutually agreed terms.

3)   The Parties shall consider the need for and modalities of a protocol setting out appropriate procedures, including, in particular, advance informed agreement, in the field of the safe transfer, handling and use of any living modified organism resulting from biotechnology that may have adverse effect on the conservation and sustainable use of biological diversity.

4)   Each Contracting Party shall, directly or by requiring any natural or legal person under its jurisdiction providing the organisms referred to in paragraph 3 above, provide any available information about the use and safety regulations required by that Contracting Party in handling such organisms, as well as any available information on the potential adverse impact of the specific organisms concerned to the Contracting Party into which those organisms are to be introduced.

Article 20

Financial Resources

1)   Each Contracting Party undertakes to provide, in accordance with its capabilities, financial support and incentives in respect of those national activities which are intended to achieve the objectives of this Convention, in accordance with its national plans, priorities and programmes.

2)   The developed country Parties shall provide new and additional financial resources to enable developing country Parties to meet the agreed full incremental costs to them of implementing measures which fulfil the obligations of this Convention and to

benefit from its provisions and which costs are agreed between a developing country Party and the institutional structure referred to in Article 21, in accordance with policy, strategy, programme priorities and eligibility criteria and an indicative list of incremental costs established by the Conference of the Parties. Other Parties, including countries undergoing the process of transition to a market economy, may voluntarily assume the obligations of the developed country Parties. For the purpose of this Article, the Conference of the Parties, shall at its first meeting establish a list of developed country Parties and other Parties which voluntarily assume the obligations of the developed country Parties. The Conference of the Parties shall periodically review and if necessary amend the list. Contributions from other countries and sources on a voluntary basis would also be encouraged. The implementation of these commitments shall take into account the need for adequacy, predictability and timely flow of funds and the importance of burden-sharing among the contributing Parties included in the list.

3)   The developed country Parties may also provide, and developing country Parties avail themselves of, financial resources related to the implementation of this Convention through bilateral, regional and other multilateral channels.

4)   The extent to which developing country Parties will effectively implement their commitments under this Convention will depend on the effective implementation by developed country Parties of their commitments under this Convention related to financial resources and transfer of technology and will take fully into account the fact that economic and social development and eradication of poverty are the first and overriding priorities of the developing country Parties.

5)   The Parties shall take full account of the specific needs and special situation of least developed countries in their actions with regard to funding and transfer of technology.

6)   The Contracting Parties shall also take into consideration the special conditions resulting from the dependence on, distribution and location of, biological diversity within developing country Parties, in particular small island States.

7)   Consideration shall also be given to the special situation of developing countries, including those that are most environmentally vulnerable, such as those with arid and semi-arid zones, coastal and mountainous areas.

Article 21

Financial Mechanism

1)   There shall be a mechanism for the provision of financial resources to developing country Parties for purposes of this Convention on a grant or concessional basis the essential elements of which are described in this Article. The mechanism shall function under the authority and guidance of, and accountable to, the Conference of the Parties for purposes of this Convention. The operations of the mechanism shall be carried out by such institutional structure as may be decided upon by the Conference of the parties at its first meeting. For purposes of this Convention, the Conference of

the Parties shall determine the policy, strategy, programme priorities and eligibility criteria relating to the access to and utilisation of such resources. The contributions shall be such as to take into account the need for predictability, adequacy and timely flow of funds referred to in Article 20 in accordance with the amount of resources needed to be decided periodically by the Conference of the Parties and the importance of burden-sharing among the contributing Parties included in the list referred to in Article 20, paragraph 2. Voluntary contributions may also be made by the developed country Parties and by other countries and sources. The mechanism shall operate within a democratic and transparent system of governance.

2) Pursuant to the objectives of this Convention, the Conference of the Parties shall at its first meeting determine the policy, strategy and programme priorities, as well as detailed criteria and guidelines for eligibility for access to and utilisation of the financial resources including monitoring and evaluation on a regular basis of such utilisation. The Conference of the Parties shall decide on the arrangements to give effect to paragraph 1 above after consultation with the institutional structure entrusted with the operation of the financial mechanism.

3) The Conference of the Parties shall review the effectiveness of the mechanism established under this Article, including the criteria and guidelines referred to in paragraph 2 above, not less than two years after the entry into force of this Convention and thereafter on a regular basis. Based on such review, it shall take appropriate action to improve the effectiveness of the mechanism if necessary.

4) The Contracting Parties shall consider strengthening existing financial institutions to provide financial resources for the conservation and sustainable use of biological diversity.

Article 22

Relationship with Other International Conventions

1) The provisions of this Convention shall not affect the rights and obligations of any Contracting Party deriving from any existing international agreement, except where the exercise of those rights and obligations would cause a serious damage or threat to biological diversity.

2) Contracting Parties shall implement this Convention with respect to the marine environment consistently with the rights and obligations of States under the law of the sea.

Article 23

Conference of the Parties

1) A Conference of the Parties is hereby established. The first meeting of the Conference of the Parties shall be convened by the Executive Director of the United

Nations Environment Programme not later than one year after the entry into force of this Convention. Thereafter, ordinary meetings of the Conference of the Parties shall be held at regular intervals to be determined by the Conference at its first meeting.

2) Extraordinary meetings of the Conference of the Parties shall be held at such other times as may be deemed necessary by the Conference, or at the written request of any Party, provided that, within six months of the request being communicated to them by the Secretariat, it is supported by at least one third of the Parties.

3) The Conference of the Parties shall by consensus agree upon and adopt rules of procedure for itself and for any subsidiary body it may establish, as well as financial rules governing the funding of the Secretariat. At each ordinary meeting, it shall adopt a budget for the financial period until the next ordinary meeting.

4) The Conference of the Parties shall keep under review the implementation of this Convention, and, for this purpose, shall:

a. Establish the form and the intervals for transmitting the information to be submitted in accordance with Article 26 and consider such information as well as reports submitted by any subsidiary body;

b. Review scientific, technical and technological advice on biological diversity provided in accordance with Article 25;

c. Consider and adopt, as required, protocols in accordance with Article 28;

d. Consider and adopt, as required, in accordance with Articles 29 and 30, amendments to this Convention and its Annexes;

e. Consider amendments to any protocol, as well as to any Annexes thereto, and, if so decided, recommend their adoption to the parties to the protocol concerned;

f. Consider and adopt, as required, in accordance with Article 30, additional annexes to this Convention;

g. Establish such subsidiary bodies, particularly to provide scientific and technical advice, as are deemed necessary for the implementation of this Convention;

h. Contact, through the Secretariat, the executive bodies of conventions dealing with matters covered by this Convention with a view to establish appropriate forms of co-operation with them; and

i. Consider and undertake any additional action that may be required for the achievement of the purposes of this Convention in the light of experience gained in its operation.

5) The United Nations, its specialized agencies and the International Atomic Energy Agency, as well as any State not Party to this Convention, may be represented as observers at meetings of the Conference of the Parties. And any other body or agency, whether governmental or non-governmental, qualified in fields relating to conservation and sustainable use of biological diversity, which has informed the Secretariat of its wish to be represented as an observer at a meeting of the Conference of the Parties, may be admitted unless at least one third of the Parties present object. The admission and participation of observers shall be subject to the rules of procedure adopted by the Conference of the Parties.

Article 24

Secretariat

1) A secretariat is hereby established. Its functions shall be:
    a. To arrange for and service meetings of the Conference of the Parties provided for in Article 23;
    b. To perform the functions assigned to it by any protocol;
    c. To prepare reports on the execution of its functions under this Convention and present them to the Conference of the Parties;
    d. To co-ordinate with other relevant international bodies and, in particular to enter into such administrative and contractual arrangements as may be required for the effective discharge of its functions; and
    e. To perform such other functions as may be determined by the Conference of the Parties.
2) At its first ordinary meeting, the Conference of the Parties shall designate the secretariat from amongst those existing competent international organisations which have signified their willingness to carry out the secretariat functions under this Convention.

Article 25

Subsidiary Body on Scientific, Technical and Technological Advice

1) A subsidiary body for the provision of scientific, technical and technological advice is hereby established to provide the Conference of the Parties and, as appropriate, its other subsidiary bodies with timely advice relating to the implementation of this Convention. This body shall be open to participation by all Parties and shall be multidisciplinary. It shall comprise government representatives competent in the relevant field of expertise. It shall report regularly to the Conference of the Parties on all aspects of its work.
2) Under the authority of and in accordance with guidelines laid down by the Conference of the Parties, and upon its request, this body shall:
    a. Provide scientific and technical assessments of the status of biological diversity;
    b. Prepare scientific and technical assessments of the effects of types of measures taken in accordance with the provisions of this Convention;
    c. Identify innovative, efficient and state-of-the-art technologies and know-how relating to the conservation and sustainable use of biological diversity and advise on the ways and means of promoting development and/or transferring such technologies;
    d. Provide advice on scientific programmes and international co-operation in research and development related to conservation and sustainable use of biological diversity; and

e.   Respond to scientific, technical, technological and methodological questions that the Conference of the Parties and its subsidiary bodies may put to the body.

3)   The functions, terms of reference, organisation and operation of this body may be further elaborated by the Conference of the Parties.

Article 26

Reports

Each Contracting Party shall, at intervals to be determined by the Conference of the Parties, present to the Conference of the Parties, reports on measures which it has taken for the implementation of the provisions of this Convention and their effectiveness in meeting the objectives of this Convention.

Article 27

Settlement of Disputes

1)   In the event of a dispute between Contracting Parties concerning the interpretation or application of this Convention, the Parties concerned shall seek solution by negotiation.

2)   If the Parties concerned cannot reach agreement by negotiation, they may jointly seek the good offices of, or request mediation by, a third party.

3)   When ratifying, accepting, approving or acceding to this Convention, or at any time thereafter, a State or regional economic integration organisation may declare in writing to the Depositary that for a dispute not resolved in accordance with paragraph 1 or paragraph 2 above, it accepts one or both of the following means of dispute settlement as compulsory:

a.   Arbitration in accordance with the procedure laid down in Part 1 of Annex II;

b.   Submission of the dispute to the International Court of Justice.

4)   If the Parties to the dispute have not, in accordance with paragraph 3 above, accepted the same or any procedure, the dispute shall be submitted to conciliation in accordance with Part 2 of Annex II unless the Parties otherwise agree.

5)   The provisions of this Article shall apply with respect to any protocol except as otherwise provided in the protocol concerned.

Article 28

Adoption of Protocols

1)   The Contracting Parties shall co-operate in the formulation and adoption of protocols to this Convention.

2)   Protocols shall be adopted at a meeting of the Conference of the Parties.

3)   The text of any proposed protocol shall be communicated to the Contracting Parties by the Secretariat at least six months before such a meeting.

Article 29

Amendment of the Convention or Protocols

1)   Amendments to this Convention may be proposed by any Contracting Party. Amendments to any protocol may be proposed by any Party to that protocol.

2)   Amendments to this Convention shall be adopted at a meeting of the Conference of the Parties. Amendments to any protocol shall be adopted at a meeting of the Parties to the Protocol in question. The text of any proposed amendment to this Convention or to any protocol, except as may otherwise be provided in such protocol, shall be communicated to the Parties to the instrument in question by the secretariat at least six months before the meeting at which it is proposed for adoption. The secretariat shall also communicate proposed amendments to the signatories to this Convention for information.

3)   The Parties shall make every effort to reach agreement on any proposed amendment to this Convention or to any protocol by consensus. If all efforts at consensus have been exhausted, and no agreement reached, the amendments shall as a last resort be adopted by a two-thirds majority vote of the Parties to the instrument in question present and voting at the meeting, and shall be submitted by the Depositary to all Parties for ratification, acceptance or approval.

4)   Ratification, acceptance or approval of amendments shall be notified to the Depositary in writing. Amendments adopted in accordance with paragraph 3 above shall enter into force among Parties having accepted them on the ninetieth day after the deposit of instruments of ratification, acceptance or approval by at least two thirds of the Contracting Parties to this Convention or of the Parties to the protocol concerned, except as may otherwise be provided in such protocol. Thereafter the amendments shall enter into force for any other Party on the ninetieth day after that Party deposits its instrument of ratification, acceptance or approval of the amendments.

5)   For the purposes of this Article, 'Parties present and voting' means Parties present and casting an affirmative or negative vote.

Article 30

Adoption and Amendment of Annexes

1)   The Annexes to this Convention or to any protocol shall form an integral part of the Convention or of such protocol, as the case may be, and, unless expressly provided otherwise, a reference to this Convention or its protocols constitutes at the same time a reference to any Annexes thereto. Such Annexes shall be restricted to procedural, scientific, technical and administrative matters.

2)   Except as may be otherwise provided in any protocol with respect to its Annexes, the following procedure shall apply to the proposal, adoption and entry into force of additional Annexes to this Convention or of Annexes to any protocol:

   a.   Annexes to this Convention or to any protocol shall be proposed and adopted according to the procedure laid down in Article 29;

   b.   Any Party that is unable to approve an additional Annex to this Convention or an Annex to any protocol to which it is Party shall so notify the Depositary, in writing, within one year from the date of the communication of the adoption by the Depositary. The Depositary shall without delay notify all Parties of any such notification received. A Party may at any time withdraw a previous declaration of objection and the Annexes shall thereupon enter into force for that Party subject to subparagraph (c) below;

   c.   On the expiry of one year from the date of the communication of the adoption by the Depositary, the Annex shall enter into force for all Parties to this Convention or to any protocol concerned which have nosubmitted a notification in accordance with the provisions of subparagraph (b) above.

3)   The proposal, adoption and entry into force of amendments to Annexes to this Convention or to any protocol shall be subject to the same procedure as for the proposal, adoption and entry into force of Annexes to the Convention or Annexes to any protocol.

4)   If an additional Annex or an amendment to an Annex is related to an amendment to this Convention or to any protocol, the additional Annex or amendment shall not enter into force until such time as the amendment to the Convention or to the protocol concerned enters into force.

Article 31

Right to Vote

1)   Except as provided for in paragraph 2 below, each Contracting Party to this Convention or to any protocol shall have one vote.

2)   Regional economic integration organisations, in matters within their competence, shall exercise their right to vote with a number of votes equal to the number of their Member States which are Contracting Parties to this Convention or the relevant protocol. Such organisations shall not exercise their right to vote if their Member States exercise theirs, and vice versa.

Article 32

Relationship between this Convention and Its Protocols

1)   A State or a regional economic integration organisation may not become a Party to a protocol unless it is, or becomes at the same time, a Contracting Party to this Convention.

2)   Decisions under any protocol shall be taken only by the Parties to the protocol concerned. Any Contracting Party that has not ratified, accepted or approved a protocol may participate as an observer in any meeting of the Parties to that protocol.

Article 33

Signature

This Convention shall be open for signature at Rio de Janeiro by all States and any regional economic integration organisation from 5 June 1992 until 14 June 1992, and at the United Nations Headquarters in New York from 15 June 1992 to 4 June 1993.

Article 34

Ratification, Acceptance or Approval

1)   This Convention and any protocol shall be subject to ratification, acceptance or approval by States and by regional economic integration organisations. Instruments of ratification, acceptance or approval shall be deposited with the Depositary.

2)   Any organisation referred to in paragraph 1 above which becomes a Contracting Party to this Convention or any protocol without any of its Member States being a Contracting Party shall be bound by all the obligations under the Convention or the protocol, as the case may be. In the case of such organisations, one or more of whose Member States is a Contracting Party to this Convention or relevant protocol, the organisation and its Member States shall decide on their respective responsibilities for the performance of their obligations under the Convention or protocol, as the case may be. In such cases, the organisation and the Member States shall not be entitled to exercise rights under the Convention or relevant protocol concurrently.

3)   In their instruments of ratification, acceptance or approval, the organisations referred to in paragraph 1 above shall declare the extent of their competence with respect to the matters governed by the Convention or the relevant protocol. These organisations shall also inform the Depositary of any relevant modification in the extent of their competence.

Article 35

Accession

1)   This Convention and any protocol shall be open for accession by States and by regional economic integration organisations from the date on which the Convention or the protocol concerned is closed for signature. The instruments of accession shall be deposited with the Depositary.

2)   In their instruments of accession, the organisations referred to in paragraph 1 above shall declare the extent of their competence with respect to the matters governed by

the Convention or the relevant protocol. These organisations shall also inform the Depositary of any relevant modification in the extent of their competence.

3) The provisions of Article 34, paragraph 2, shall apply to regional economic integration organisations which accede to this Convention or any protocol.

Article 36

Entry into Force

1) This Convention shall enter into force on the ninetieth day after the date of deposit of the thirtieth instrument of ratification, acceptance, approval or accession.

2) Any protocol shall enter into force on the ninetieth day after the date of deposit of the number of instruments of ratification, acceptance, approval or accession, specified in that protocol, has been deposited.

3) For each Contracting Party which ratifies, accepts or approves the Convention or accedes thereto after the deposit of the thirtieth instrument of ratification, acceptance, approval or accession, it shall enter into force on the ninetieth day after the date of deposit by such Contracting Party of its instrument of ratification, acceptance, approval or accession.

4) Any protocol, except as otherwise provided in such protocol, shall enter into force for a Contracting Party that ratifies, accepts or approves that protocol or accedes thereto after its entry into force pursuant to paragraph 2 above, on the ninetieth day after the date on which that Contracting Party deposits its instrument of ratification, acceptance, approval or accession, or on the date on which this Convention enters into force for that Contracting Party, whichever shall be the later.

5) For the purposes of paragraphs 1 and 2 above, any instrument deposited by a regional economic integration organisation shall not be counted as additional to those deposited by Member States of such organisation.

Article 37

Reservations

No reservations may be made to this Convention.

Article 38

Withdrawals

1) At any time after two years from the date on which this Convention has entered into force for a Contracting Party, that Contracting Party may withdraw from the Convention by giving written notification to the Depositary.

2) Any such withdrawal shall take place upon expiry of one year after the date of its receipt by the Depositary, or on such later date as may be specified in the notification of the withdrawal.

3)   Any Contracting Party which withdraws from this Convention shall be considered as also having withdrawn from any protocol to which it is Party

Article 39

Financial Interim Arrangements

Provided that it has been fully restructured in accordance with the requirements of Article 21, the Global Environment Facility of the United Nations Development Programme, the United Nations Environment Programme and the International Bank for Reconstruction and Development shall be the institutional structure referred to in Article 21 on an interim basis, for the period between the entry into force of this Convention and the first meeting of the Conference of the Parties or until the Conference of the Parties decides which institutional structure will be designated in accordance with Article 21.

Article 40

Secretariat Interim Arrangements

The secretariat to be provided by the Executive Director of the United Nations Environment Programme shall be the secretariat referred to in Article 24, paragraph 2, on an interim basis for the period between the entry into force of this Convention and the first meeting of the Conference of the Parties.

Article 41

Depositary

The Secretary-General of the United Nations shall assume the functions of Depositary of this Convention and any protocols.

Article 42

Authentic Texts

The original of this Convention, of which the Arabic, Chinese, English, French, Russian and Spanish texts are equally authentic, shall be deposited with the Secretary-General of the United Nations.

References

Allen, R. B., Bellingham, P. J. & Wiser, S. K. (2003a). Forest biodiversity assessment for reporting conservation performance. *Science for Conservation 216*. Wellington, New Zealand: Department of Conservation.

Allen, R. B., Bellingham, P. J. & Wiser, S. K. (2003b). Developing a forest biodiversity monitoring approach for New Zealand. *New Zealand Journal of Ecology*, **27**, 207–220.

Allred, D. M. (1975). *Great Basin Naturalist*, **35**, 405–406.

Andersen, A. N. (1990). The use of ant communities to evaluate change in Australian terrestrial ecosystems: a review and a recipe. *Proceedings of the Ecological Society of Australia*, **16**, 347–357.

Andersen, A. N., Hoffmann, B. D., Muller, W. J. & Griffiths, A. D. (2002). Using ants as bioindicators in land management: simplifying assessment of ant community responses. *Journal of Applied Ecology*, **39**, 8–17.

Andre, H. M., Bolly, C. & Lebrun, P. H. (1982). Monitoring and mapping air pollution through an animal indicator: a new and quick method. *Journal of Applied Ecology*, **19**, 107–111

Angermeier, P. L. & Karr, J. R. (1986). Applying an index of biotic integrity based on stream-fish communities: considerations in sampling and interpretation. *North American Journal of Fisheries Management*, **6**, 418–429.

Antunes, P., Santos, R. & Jordao, L. (2001). The application of Geographical Information Systems to determine environmental impact significance. *Environmental Impact Assessment Review*, **21**, 511–535.

Armstrong, D. W. & Cook, R. M. (1986). Proposal for a revised use of IYFS indices for calibrating VPA. *ICES*, CM 1986/G, 3.

Au, J., Bagchi, P., Chen, B. *et al.* (2000). Methodology for public monitoring of total coliforms, *Escherichia coli* and toxicity in waterways by Canadian high school students. *Journal of Environmental Management*, **58**, 213–230.

Austin, J. E., Buhl, T. K., Guntenspergen, G. R., Norling, W. & Sklebar, H. T. (2001). Duck populations as indicators of landscape condition in the prairie pothole region. *Environmental Monitoring and Assessment*, **69**, 29–47.

Bailey, R. G. (2002). *Ecoregion-based design for sustainability*. New York: Springer.

Baillie, S. R. (1990). Integrated population monitoring of breeding birds in Britain and Ireland. *Ibis*, **132**, 151–166.

Bamber, R. N. (1989). *Environmental Impact Assessment: the Example of CEGB Power Stations and Marine Biology. (Report RD/L/3524/R89.)* Leatherhead, UK: Central Electricity Board.

Barbour, M. T., Gerritsen, J., Snyder, B. D. & Stribling, J. B. (1999). *Rapid Bioassessment Protocols for use in Streams and Wadeable Rivers: Periphyton, Benthic Macroinvertebrates and Fish. 2nd edn. (EPA 841–B–99–002.)* Washington, DC: US Environmental Protection Agency, Office of Water.

Batty, L. (1989). *Biologist*, **36**, 151–154.

Bauerle B. (1975) *Copiea* **11** 366–368.

Beanlands, G. E. & Dunker, P. N. (1984). An ecological framework for environmental impact assessment. *Journal of Environmental Management*, **18**, 267–277.

Beebee, T. J. C. (2002). Amphibian phenology and climate change. *Conservation Biology*, **16**, 1454–1455.

Belbin, L., Gibson, J. A. E., Davis, C., Wats, D. & McIvor, E. (2003). State of the environment reporting: an Antarctic case study. *Polar Record*, **39**, 193–201.

Belsky, A. J. (1985). Long-term vegetation monitoring in the Serengeti National Park, Tanzania. *Journal of Applied Ecology*, **22**, 449–460.

Ben-Eliahu, M. N. & Safriel, U. N. (1982). A comparison between species diversities of polychaetes from tropical and temperate structurally similar intertidal habitats. *Journal of Biogeography*, **9**, 371–390.

Benninghoff, W. S. & Bonner, W. N. (1985). *Man's Impact on the Antarctic Environment: A Procedure for Evaluating Impacts from Scientific and Logistic Activities.* Cambridge, UK: SCAR and the Scott-Polar Institute.

Bernstein, B. B. & Weisberg, S. B. (2003). Southern California's marine monitoring system ten years after the National Research Council Evaluation. *Environmental Monitoring and Assessment*, **81**, 3–14.

Bibby, C. J., Burgess, N. D., Hill, D. A. & Mustor, S. H. (2000). *Bird Census Techniques*, 2nd edn. London: Academic Press.

Bicknell, K. B., Ball, R. J., Cullen, R. and Bigsby, H. R. (1998). A new methodology for ecological footprint with an application to the New Zealand economy. *Ecological Economics*, **27**, 149–190.

Biggs, B. J. F. & Kilroy, C. (2000). *Stream Periphyton Monitoring Manual*. Christchurch: NIWA.

Biggs, B. F. J., Kilroy, C., Mulcock, C. & Scarsbrook, M. (2002). *New Zealand Stream Monitoring and Assessment Kit. (Stream Monitoring Manual Version Two. National Institute of Water and Atmospheric Research Technical Publication 44.)* Christchurch: NIWA.

Boddicker, M., Rodriguez, J. J. & Amanzo, J. (2002). Indices for assessment and monitoring of large mammals within an adaptive management framework. *Environmental Monitoring and Assessment*, **76**, 105–123.

Bonn, A., Rodriques, A. S. L. and Gaston, K. J. (2002). Threatened and endemic species: are they good indicators of patterns of biodiversity on a national scale. *Ecology Letters*, **5**, 733–741.

Bonner, W. N. (1984). Conservation and the Antarctic. In Laws, R. M. *Antarctic Ecology*, vol. 2. London: Academic Press, pp. 821–850.

Boothroyd, I. & Stark, J. D. (2000). Use of invertebrates in monitoring. In: Collier, K. J. & Winterbourn, M. J. (eds.) *New Zealand Stream Invertebrates: Ecology and Implications for Management*. Christchurch: New Zealand Limnological Society, pp. 344–373.

Borodin, A. M. (1978). *Red Data Book of USSR*. Moscow: Lesnaya Promyshlenost.

Bowles, M. L., Hess, W. J., DeMauro, M. M. & Hiebert, R. D. (1986). Endangered plant inventory and monitoring strategies at Indiana Dunes National Lakeshore. *Natural Areas Journal*, **6**, 18–26.

Brakefield, P. M. (1990). A decline of melanism in the peppered moth *Biston betularia* in the Netherlands. *Biological Journal of the Linnean Society*, **39**, 327–334.

Brillouin, L. (1960). *Science and Information Theory*, 2nd edn. New York: Academic Press.

Brock, D. A. (1977). Comparison of community similarity indices. *Journal of the Water Pollution Control Federation*, **49**, 2488-2494.

Brown, J. H. & Heske, E. J. (1990). Temporal changes in a Chihuahuan desert rodent community. *Oikos*, **59**, 290–302.

Brown, L. R., Larsen, J. & Fischlowitz-Roberts, B. (2000). *State of the World 2000*. London: Earthscan.

Brown, R. M., McClelland, N. I., Deininger, R. A. & O'Connor, M. F. (1972). A water quality index crashing the psychological barrier. In Thomas, W. A. (ed.) *Indicators of Environmental Quality*. New York: Plenum Press, pp. 173-182.

Brydges, T. (2001). Ecological change and the challenges for monitoring. *Environmental Monitoring and Assessment*, **67**, 89–96.

Brydges, T. & Lumb, A. (1998). Canada's ecological monitoring and assessment network: where we are at and where we are going. *Environmental Monitoring and Assessment*, **51**, 595–603.

Bunce, R. G. H. (1982). *A Field Key for Classifying British Woodland Vegetation*, Part 1. Cambridge, UK: Institute of Terrestrial Ecology.

Cairns, J. (1979). Biological monitoring concept and scope. In Cairns, J., Patil, G. P. & Waters W. E. (eds.) *Environmental Biomonitoring, Assessment, Prediction, and Management: Certain Case Studies and Related Quantitative Issues*, Fairland, MD: International Co-operative Publishing House, pp. 3–20.

Cairns, J. (1984). Are single species toxicity tests alone adequate for estimating environmental hazard? *Environmental Monitoring and Assessment*, **4**, 259–273.

Cairns, J. (1986). Multispecies toxicity testing: a new information base for hazard evaluation. *Current Practices in Environmental Science and Engineering*, **2**, 37–49.

Cairns, J., Albaugh, D. W., Busey, F. & Chanay, M. D. (1968). The sequential comparison index: a simplified method for non-biologists to estimate relative differences in biological diversity in stream pollution studies. *Journal of Water Pollution Control Federation*, **40**, 1607–1613.

Cairns, J., Kuhn, D. L. & Plafkin, J. L. (1979a). Protozoan colonization of artificial substrates. *American Society for Testing and Materials, Special Technical Publication*, **690**, 34–57.

Cairns, J., Patil, G. P. & Waters, W. E. (1979b). *Environmental Biomonitoring, Assessment, Prediction, and Management: Certain Case Studies and Related Quantitative Issues.* Fairland, MD: International Co-operative Publishing House.

Canter, L. W. (1977). *Environmental Impact Assessment.* London: McGraw-Hill.

Carignan, V. & Villard, M.-A. (2002). Selecting indicator species to monitor ecological integrity: a review. *Environmental Monitoring and Assessment*, **78**, 45–61.

Carson, R. (1962). *Silent Spring.* Boston, MA: Houghton Mifflin. [Republished 2002]

Caughley, G., Dublin, H. & Parker, L. (1990). Projected decline of the African elephant. *Biological Conservation*, **54**, 157–164.

Chandler, J. R. (1970). A biological approach to water quality management. *Journal of Water Pollution Control*, **69**, 415–422.

Chessman, B. C. (1995). Rapid assessment of rivers using macroinvertebrates: a procedure based on habitat-specific sampling, family-level identification and a biotic index. *Australian Journal of Ecology*, **20**, 122–129.

Chessman, B. C. (2003). New sensitivity grades for Australian river macroinvertebrates. *Marine and Freshwater Research*, **54**, 95–103.

Christie, H. (1980). Methods for ecological monitoring: biological interactions in a rocky sub-tidal community. *Helgolander Meeresuntersuchungen*, **33**, 473–483.

Collins, J. P. & Storfer, A. (2003). Global amphibian declines: sorting the hypotheses. *Diversity and Distributions*, **9**, 89–98.

COMNAP (1998). Summary of environmental monitoring activities in Antarctica. Hobart, Australia: Council of Managers of National Antarctic Programmes.

CONCAWE (1982). *Ecological Monitoring of Aqueous Effluents from Petroleum Refineries.* (*Report no. 8/82.*) The Hague: CONCAWE.

Cook, L. M., Rigby, K. D. & Seaward, M. R. D. (1990). Melanic moths and changes in epiphytic vegetation in north-west England and north Wales. *Biological Journal of the Linnean Society*, **39**, 343–354.

Coomes, D. A., Allen, R. B., Scott, N. A., Goulding, C. & Beets, P. (2002). Designing systems to monitor carbon stocks in forests and shrublands. *Forest Ecology and Management*, **164**, 89–108.

Cordonnery, L. (1999). Implementing the Protocol on Environmental Protection to the Antarctic Treaty: future applications of geographic information systems within the Committee for Environmental Protection. *Journal of Environmental Management*, **56**, 285–298.

Council of Europe (1977). *List of Rare, Threatened and Endemic Plants in Europe.* (*Nature and Environment Series*, No. 14.) Strasbourg: Council of Europe.

Cowell, E. B. (1978). Ecological monitoring as a management tool in industry. *Ocean Management*, **4**, 273–285.

Cruickshank, M. M., Tomlinson, R. W. and Trew, S. (2000). Application of CORINE land-cover mapping to estimate carbon stored in the vegetation of Ireland. *Journal of Environmental Management*, **58**, 269–287.

Curtis, T. G. F. & McGough, H. N. (1988). *The Irish Red Data Book.* Dublin: The Stationary Office.

Dacre, J. C. & Scott, D. (1973). Effects of dieldrin on brown trout in field and laboratory studies. *New Zealand Journal of Marine and Freshwater Research*, **7**, 235–246.

Danielsen, F., Balete, D. S., Poulsen, M. K. *et al.* (2000). A simple system for monitoring biodiversity in protected areas of a developing country. *Biodiversity and Conservation*, **9**, 1671–1705.

Davis, M. B. (1989). Retrospective studies. In Likens, G. E. (ed.) *Long-term Studies in Ecology: Approaches and Alternatives*. New York: Springer Verlag, pp. 71–89.

Debinski, D. M., Jakubauskas, M. E. & Kindscher, K. (2000). Montane meadows as indicators of environmental change. *Environmental Monitoring and Assessment*, **64**, 213–225.

Dennis, B., Patil, G. P. & Rossi, O. (1979). The sensitivity of ecological diversity indices to the presence of pollutants in aquatic communities. In Cairns, J., Patil, G. P. & Waters, W. E. (eds.) *Environmental Biomonitoring, Assessment, Prediction and Management: Certain Case Studies and Related Quantitative Issues*. Baltimore, MD: International Cooperative Publishing House, pp. 379–413.

Department of the Environment (1985). *Methods of Biological Sampling. A Colonization Sampler for Collecting Macro-invertebrate Indicators of Water Quality in Lowland Rivers*. London: HMSO.

Department of the Environment (1986). *DoE Digest of Environmental Protection and Water Statistics*, No. 9. London: HMSO.

Diamond, J., Collins, M. & Gruber, D. (1988). An overview of automated biomonitoring: past developments and future needs. In Gruber, D. & Diamond, J. (eds.) *Automated Biomonitoring: Living Sensors as Environmental Monitors*. Chichester, UK: Ellis Horwood, pp. 23–39.

Dicks, B. (1976). The effects of refinery effluents: the case history of a saltmarsh. In Baker, J. M. (ed.) *Marine Ecology and Oil Pollution*. Barking, UK: Applied Science, pp. 227–245.

(1989). *Ecological Impacts of the Oil Industry. Proceedings of an International Meeting Organized by the Institute of Petroleum*, London, UK: November 1987. Chichester Wiley, on behalf of the Institute of Petroleum.

Dicks, B. & Hartley, J. P. (1982). The effects of repeated small oil spillages and chronic discharges. *Philosophical Transactions of the Royal Society of London*, **B297**, 285–307.

Dicks, B. & Iball, K. (1981). Ten years of saltmarsh monitoring: the case history of a Southampton water saltmarsh and a changing refinery effluent discharge. In *Proceedings of the 1981 Oil Spill Conference (Presentation, Behaviour, Control, Cleanup)*. Washington, DC: American Petroleum Institute, pp. 361–374.

Diggle, P. J. (1990). *Time Series. A Biostatistical Introduction*. Oxford, UK: Oxford Scientific.

Dills, G. & Rogers, D. T. (1974). Macroinvertebrate community structure as an indicator of acid mine pollution. *Environmental Pollution*, **6**, 239–262.

Downes, B. J., Barmuta, L. A., Fairweather, P. G. *et al.* (2002). *Monitoring Ecological Impacts*. Cambridge, UK: Cambridge University Press.

Dufrene, M. & Legendre, P. (1997). Species assemblages and indicator species: the need for a flexible asymmetrical approach. *Ecological Monographs*, **67**, 345–366.

Dunnett, N. P. & Willis, A. J. (2000). Dynamics of *Chamerion angustifolium* in grassland vegetation over a thirty-nine year period. *Plant Ecology*, **148**, 43–50.

Dunwiddie, P. W. & Kuntz, R. C. (2001). Long-term trends of Bald Eagles in winter on the Skagit River, Washington. *Journal of Wildlife Management*, **65**, 290–299.

Dzwonko, Z. (2001). Assessment of light and soil conditions in ancient and recent woodlands by Ellenberg indicator values. *Journal of Applied Ecology*, **38**, 942–951.

Ehrlich, P. R., Ehrlich, A. E. & Holdren, J. P. (1977). *Ecoscience, Populations, Resources, Environment*. San Francisco, CA: Freeman.

Engstrom, D. R., Swain, E. B. & Kingston, J. C. (1985). A palaeolimnological record of human disturbance from Harvey's Lake, Vermont: geochemistry, pigments and diatoms. *Freshwater Ecology*, **15**, 261–288.

Fairbrothers, D. E. & Hough, M. Y. (1975). Rare or endangered vascular plants of New Jersey. *New Jersey State Museum Scientific Notes*, **14**, 1–53.

Fisher, R. A., Corbet, A. S. & Williams, C. B. (1943). The relation between the number of species and the number of individuals in a random sample of an animal population. *Journal of Animal Ecology*, **12**, 42–58.

Flower, R. J. & Battarbee, R. W. (1983). Diatom evidence for recent acidification of two Scottish lochs. *Nature*, **305**, 130–133.

Franklin, J. F., Bledsoe, C. S. & Callahan, J. T. (1990). An expanded network of scientists, sites, and programs can provide crucial comparative analyses. *BioScience*, **40**, 509–523.

Fuhlendorf, S. D. & Smeins, F. E. (1996). Spatial scale influence in longterm temporal patterns of a semi-arid grassland. *Landscape Ecology*, **11**, 107–113.

Fujita, T., Itaya, A., Miura, M., Manabe, T. & Yamamoto, S. (2003). Long-term canopy dynamics analyzed by aerial photographs in a temperate old-growth evergreen broad-leaved forest. *Journal of Ecology*, **91**, 686–693.

Fuller, R. J., Marchant, J. J. & Morgan, R. A. (1985). How representative of agriculture practice in Britain are Common Bird Census farmland plots? *Bird Study*, **32**, 56–70.

Furse, M. T., Moss, D., Wright, J. F., Armitage, P. D. & Gunn, R. J. M. (1986). *A Practical Manual for the Classification and Prediction of Macroinvertebrate Communities in Running Water in Great Britain. (Preliminary Version.)* East Stoke, UK: FBA River Lab.

Gaines, K. F., Romanek, C. S., Boring, C. S. *et al.* (2002). Using Raccoons as an indicator species for metal accumulation across trophic levels: a stable isotope approach. *Journal of Wildlife Management*, **66**, 811–821.

GEMS (1988). *The Global Environment Monitoring System. (Sahel Series Main Report. AG: EP/ SEN/OOl Technical Report.)* Rome: UNEP/FAO.

Gibbons, D. W., Reid, J. B. & Chapman, R. A. (1993). *The New Atlas of Breeding Birds in Britain and Ireland: 1988-1991*. London: T. & A. D. Poyser.

Girton, C. (1983). *Freshwater Biological Monitoring of the Crymlyn Bog at BP Oil Llandarcy Refinery Limited, Neath, West Glamorgan, 1978-1982*. London: BP Environmental Control Centre.

Given, D. R. (1983). *Conservation of Plant Species and Habitats. (A Symposium held at 15th Pacific Science Congress*, Dunedin, New Zealand, February 1983.) Wellington: Nature Conservation Council.

Given, D. R. (1989). Monitoring of threatened plants. In Craig, B. (ed.) *Proceedings of a Symposium on Environmental Monitoring in New Zealand with Emphasis on Protected Natural Areas*. Wellington: Department of Conservation, pp. 192-198.

Gold, A., Green, L. & Herron, E. (1994). *FY92: Volunteer Monitoring Program for Lake Water Quality Assessment. University of Rhode Island Cooperative Extension's Watershed Watch Program*. Kingston, RI: Department of Natural Resources Science, University of Rhode Island.

Goldman, C. R. (1974). *Eutrophication of Lake Tahoe Emphasizing Water Quality. (Report No. EPA-660/3-74-034, Ecological Research Series.)* Washington, DC: US Environmental Protection Agency.

Gooneratne, R., Shaw, W. & Dalton, P. (1999). *Proceedings of The Pollution Effects Biomarkers in Environmental Toxicology Conference*, University of Canterbury, New Zealand.

Green, R. H. (1979). *Sampling Design and Statistical Methods for Environmental Biology*. New York: John Wiley.

Groombridge, B. (1992). *Global Biodiversity, Status of the Earth's Living Resources*. (A Report of the World Conservation Monitoring Centre.) London: Chapman & Hall.

Gruber, D. S. & Diamond, J. M. (1988). *Automated Biomonitoring: Living Sensors as Environmental Monitors*. Chichester, UK: Ellis Horwood.

Haefner, P. A. (1970). The effect of low dissolved oxygen concentrations on the temperature–salinity tolerance of the sand shrimp, *Crangon septemspinosa*. *Physiological Zoology*, **43**, 30–37.

Haines-Young, R. H., Barr, C. J., Black, H. I. J. *et al.* (2000) *Accounting for Nature: Assessing Habitats in the UK Countryside*. London: UK Department of the Environment, Transport and the Regions.

Hall, A. V., de Winter, B. & Oosterhout, S. A. M. (1980). *Threatened Plants of Southern Africa, Pretoria. (South African National Scientific Programmes* Report No. 45.) Tygerborg: Council for Scientific and Industrial Research.

Hanna, R. G. & Muir, G. L. (1990). *Environmental Monitoring and Assessment*, **14**, 211–222.

Harcombe, P. A., Bill, C. J., Fulton, M. *et al.* (2002). Stand dynamics over 18 years in a southern mixed hardwood forest, Texas, USA. *Journal of Ecology*, **90**, 947–957.

Hardy, A. C. (1956). *The Open Sea. Its Natural History: the World of Plankton*. London: Collins.

Harrington, R., Tatchell, G. M. & Bale, J. S. (1990). Weather, life cycle strategy and spring populations of aphids. *Acta Phytopathologica et Entomologica Hungarica*, **25**, 423–432.

Harrington, R., Bale, J. S. & Tatchell, G. M. (1995). Aphids in a changing climate. In Harrington, R. and Stork, N. E. (eds.) *Insects in a Changing Environment*. London: Academic Press, pp. 125–155.

Harrington, R., Verrier, P., Denholm, C. *et al.* (2002). 'EXAMINE' (Exploitation of Aphid Monitoring in Europe): an EU Thematic Network for the study of global change impacts on aphids. In *Aphids in a New Millenium. Proceedings, of the 6th International Aphid Symposium*, Rennes, September 2001, pp. 45–49.

Hartnoll, R. G. & Hawkins, S. J. (1980). Monitoring rocky-shore communities: a critical look at spatial and temporal variation. *Helgolander Meeresuntersuchungen*, **33**, 484–494.

Hawkes, H. A. (1979). Invertebrates as indicators of river water quality. In James, A. & Evison, L. (eds.) *Biological Indicators of Water Quality*. Chichester, UK: Wiley, pp. 2.1–2.45.

Hawksworth, D. L. & Rose, F. (1976). *Lichens as Pollution Monitors*. London: Edward Arnold.

Hellawell, J. M. (1977). Change in natural and managed ecosystems: detection, measurement and assessment. *Proceedings of the Royal Society of London*, **B197**, 1-57.

Hellawell, J. M. (1978). *Biological Surveillance of Rivers*. Stevenage, UK: NERC.

Herzig, R., Liebendorfer, L., Urech, M. & Ammann, K. (1989). Passive biomonitoring with lichens as a part of an integrated biological measuring system for monitoring air pollution in Switzerland. *International Journal of Environmental and Analytical Chemistry*, **35**, 4357.

Heywood, V. H. (ed.) (1995). *Global Biodiversity Assessment*. Cambridge, UK: Cambridge University Press.

Hill, M. O. (1973). Diversity and evenness: a unifying notation and its consequences. *Ecology*, **54**, 427-432.

Hill, M. O. & Radford, G. L. (1986). *Register of Permanent Vegetation Plots*. Abbots Ripton, UK: ITE.

Hill, M. O., Roy, D. B., Mountford, J. O. & Bunce, R. G. H. (2000). Extending Ellenberg's indicator values to a new area: an algorithmic approach. *Journal of Applied Ecology*, **37**, 3-15.

Hilenshoff, W. L. (1977). Use of arthopods to evaluate water quality of streams. *Technical Bulletin of the Wisconsin Department of Natural Resources*. No. 100. Madison, WI: Wisconsin Department of Natural Resources.

Hilsenhoff, W. L. (1987). An improved biotic index of organic stream pollution. *Great Lakes Entomologist*, **20**, 31-39.

Hobbie, J. E. (2003). Scientific accomplishments of the long term ecological research program: an introduction. *BioScience*, **53**, 17-20.

Hobbie, J. E., Carpenter, S. R., Grimm, N. B., Gosz, J. R. & Seastedt, T. R. (2003). The US long term ecological research program. *BioScience*, **53**, 21-32.

Hoenicke, R., Davis, J. A., Gunther, A. *et al.* (2003). Effective application of monitoring information: the case of San Francisco Bay. *Environmental Monitoring and Assessment*, **81**, 15-25.

Holdgate, M. W., Kassas, M., & White, G. F. (eds.) (1982). *The World Environment, 1972-1982: A Report by the United Nations Environment Programme*. Dublin: Tycooly.

Holzner, W. (1982). In Holzner, W. & Numata, N. (eds.) *Biology and Ecology of Weeds*. The Hague: J. W. Junk, pp. 187-190.

Hooper, M. D. (1970). Dating hedges. *Area*, **4**, 63-65.

Hope, C. & Parker, J. (1995). Environmental indices for France, Italy and the UK. *European Environment*, **5**, 13-19.

Hopkins, B. (1962). Vegetation of the Olokemeji Forest Reserve, Nigeria. *Journal of Ecology*, **50**, 559-598.

Hopkins, B. (1999). The effect of shade and weather on daffodils *Narcissus pseudonarcissus* in West Dean Woods, West Sussex. *English Nature Research Reports*, No. 340.

Hruby, T. (1987). Using similarity measures in benthic impact assessments. *Environmental Monitoring and Assessment*, **8**, 163-180.

Hughes, R. M., Whittier, T. R., Thiele, S. A. *et al.* (1992). Lake and stream indicators for the United States Environmental Protection Agency's Environmental Monitoring and Assessment Program. In McKenzie, D. H. (ed.) *Ecological Indicators*. Essex, UK: Elsevier pp. 305–335.

Huhta, V. (1979). Evaluation of different similarity indices as measures of succession in arthropod communities of the forest floor after clear-cutting. *Oecologia*, **41**, 11–23.

Hulbert, I. A. R. & French, J. (2001). The accuracy of GPS for wildlife telemetry and habitat mapping. *Journal of Applied Ecology*, **38**, 869–878.

Hunsaker, C. T., Carpenter, D. & Messer, J. (1990). Ecological indicators for regional monitoring. *Bulletin of the Ecological Society of America*, **71**, 165–172.

Hunsaker, C. T., Goodchild, M. F., Friedl, M. A. & Case, T. J. (eds.) (2001). *Spatila Uncertainty in Ecology. Implications for Remote Sensing and GIS*. New York: Springer Verlag.

Hynes, H. B. N. (1960). *The Biology of Polluted Waters*. Liverpool: Liverpool University Press.

Inhaber, H. (1976). *Environmental Indices*. New York: Wiley.

IUCN (1980). *World Conservation Strategy: Living Resource Conservation for Sustainable Development*. Gland, Switzerland: IUCN UNEP-WWF.

IUCN (1989a). *From Strategy to Action. The IUCN Response to the Report of the World Commission on Environment and Development*. Gland, Switzerland: IUCN.

IUCN (1989b). *The IUCN Sahel Studies*. Nairobi: IUCN Regional Office for Eastern Africa.

IUCN (1991). *World Conservation Strategy: Caring for the Earth*. Gland, Switzerland: IUCN.

IUCN (1997). *The Pocket Guide to IUCN 1996–1997*. Gland, Switzerland: IUCN.

Jaccard, P. (1902). Lois de distribution florale dans la zone alpine. *Bulletin de la Societe Vaudoise de Sciences Naturelles*, **38**, 69.

Jensen, J. R., Ramsey, E. W., Holmes, J. M. *et al.* (1990). Environmental sensitivity index (ESI) mapping for oil spills using remote sensing and geographical information system technology. *International Journal of Geographical Information Systems*, **4**, 181–201.

Johnson, D. L., Ambrose, S. H., Bassett, T. J. *et al.* (1997). Meanings of environmental terms. *Journal of Environmental Quality*, **26**, 581–589.

Jones, K. C. & Samiullah, Y. (1985). Deer antlers as pollution monitors. *Deer*, **6**, 253–255.

Jones, L., Fredricksen, L. and Wates, T. (2002). *Environmental Indicators Critical Issues Bulletin*. Vancouver: The Fraser Institute.

Kalkhoven, J. T. R., Stumpel, A. H. P. & Stumpel-Rienks, S. E. (1976). Landelijke Milieukartering. (Environmental survey of the Netherlands. A landscape ecological survey of the natural environment in the Netherlands for physical planning on a national level.) *Rijksinstituut voor Natuurbeheer (Research Institute for Nature Management)* No. 9. 's-Gravenhage: Staatsuitgeverij.

Karr, J. R. (1981). Assessment of biotic integrity using fish communities. *Fisheries*, **6**, 21–27.

Karr, J. R. (1987). Biological monitoring and environmental assessment: a conceptual framework. *Environmental Management*, **11**, 249–256.

Kates, R. (2001). Sustainability science. *Science*, **292**, 641–642.

Kerry, K. R. & Hempel, G. (1990). *Antarctic Ecosystems: Ecological Change and Conservation*. Berlin: Springer-Verlag.

Kershaw, L. J. & Morton, J. K. (1976). Rare and potentially endangered species in the Canadian flora a preliminary list of vascular plants. *Canadian Botanical Association Bulletin*, **9**, 26–30.

Kery, M. & Gregg, K. B. (2003). Effects of life-state on detectability in a demographic study of the terrestrial orchid *Cleistes bifaria*. *Journal of Ecology*, **91**, 265–273.

Khudyakov, I. I. (1965). The vegetation cover as an indicator of the chemical composition and depth of groundwaters. In Chikishev, A. G. (ed.) *Plant Indicators of Soils, Rocks, and Subsurface Waters*. (Authorized translation from the Russian of the *Proceedings of the Conference on Indicational Geobotany*, 1961.) New York: Consultant Bureau Enterprises.

Kidron, M. & Segal, R. (1995). *The State of the World Atlas*. London: Penguin.

Kirby, K. J. & Thomas, R. C. (2000). Changes in the ground flora in Wytham Woods, southern England from 1974 to 1991: implications for nature conservation. *Journal of Vegetation Science*, **11**, 871–880.

Klimetzek, D. & Yue, C. (1997). Climate and forest insect outbreaks. *Biologia Bratislava*, **52**, 153–157.

Kreisel, W. E. (1984). Representation of the environmental quality profile of a metropolitan area. *Environmental Monitoring and Assessment*, **4**, 15–33.

Laake, J. L., Buckland, S. T., Anderson, D. R. & Burnham, K. P. (1994). *DISTANCE User's Guide V2.1*. Fort Collins, CO: Colorado Cooperative Fish and Wildlife Research Unit, Colorado State University.

Lack, D. (1966). *Population Studies of Birds*. Oxford: Claredon Press.

Lack, D. (ed.) (1986). *The Atlas of Wintering Birds in Britain and Ireland*. London: T. & A. D. Poyser.

Lambie, J. (1997). Community-based environmental monitoring; conditions essential for long term provision of quality data. M.Sc. Thesis, Lincoln University, New Zealand.

Lambshead, P. J. D., Platt, H. M. & Shaw, K. M. (1983). The detection of differences among assemblages of marine benthic species based on assessment of dominance and diversity. *Journal of Natural History*, **17**, 859–874.

Leatherland, T. M. & Burton, J. D. (1974). The occurrence of some trace metals in coastal organisms with particular reference to the Solent area. *Journal of the Marine Biology Association of the UK*, **54**, 457–468.

Lees, D. R., Creed, E. R. & Duckett, J. G. (1973). Atmospheric pollution and industrial melanism. *Heredity*, **30**, 227–232.

Leigh, J. & Boden, R, (1979). Australian flora in the endangered species convention CITES. *Australian National Parks and Wildlife Service Special Publication*, No. 3. Canberra: National Parks and Wildlife Service.

Leppakoski, E. (1979). The use of zoobenthos in evaluating effects of pollution in brackish-water environments. In Hytteborn, H. (ed.) *The Use of Ecological Variables in Environmental Monitoring (Report PM 1151.)* Solna, Sweden: The National Swedish Environment Protection Board.

Liddle, M. J. (1975). A selective review of the ecological effects of human trampling on natural ecosystems. *Biological Conservation*, **7**, 17–36.

Likens, G. E. (1989). *Long-term Studies in Ecology: Approaches and Alternatives*. New York: Springer Verlag.

Linstone, H. A & Turoff, M (1975). *The Delphi Method: Techniques and Applications*. Boston, MA: Addison-Wesley.

Llanso, R. J., Dauer, D. M., Volstad, J. H. & Scott, L. C. (2003). Application of the benthic index of biotic integrity to environmental monitoring in Chesapeake Bay. *Environmental Monitoring and Assessment*, **81**, 163–174.

Loh, J. (ed.) (2002). *Living Planet Report*. Gland, Switzerland: World Wildlife Fund.

Lomborg, B. (2001). *The Sceptical Environmentalist: Measuring the Real State of the World*. Cambridge UK: Cambridge University Press.

Ludwig, J. A., Bastin, G. N., Eager, R. W. *et al.* (2000). Monitoring Australian rangeland sites using landscape function indicators and ground- and remote-based techniques. *Environmental Monitoring and Assessment*, **64**, 167–178.

Luff, M. L., Eyre, M. B. & Rushton, S. P. (1989). Classification and ordination of habitats of ground beetles (Coleoptera, Carabidae) in north-east England. *Journal of Biogeography*, **16**, 121–130.

MacArthur, R. & Wilson, E. O. (1967). *The Theory of Island Biogeography*. Princeton, NJ: Princeton University Press.

Macaulay, E. D. M., Tatchell, G. M. & Taylor, L. R. (1988) The Rothamsted Insect Survey '12-metre' suction trap. *Bulletin of Entomological Research*, **78**, 121–129.

Mallett, T. J. (1999). A critical appraisal of environmental indices. M.Sc. Thesis, Lincoln University, New Zealand.

Malyshev, V. L. & Soboleuska, K. A. (1980). *Rare and Endangered Plant Species of Siberia*. [In Russian.] Moscow: Hayka.

Manners, J. G. & Edwards, P. J. (1986). Death of old beech trees in the New Forest. *Hampshire Field Club and Archaeological Society Proceedings*, **42**, 155–156.

Manning, W. J. R., & Feder, W. A. (1980). *Biomonitoring Air Pollutants with Plants*. London: Applied Science.

Marchant, J. H., Freeman, S. N., Crick, H. Q. P. & Beaven, L. P. (2004). The BTO Heronries Census of England and Wales 1828–2000: new indices and a comparison of analytical methods. *Ibis*, **146**, 323–324.

Margalef, R. (1951). Diversidad de especies en las comunidades naturales. *Publicaciones del Instituto de Biologia Aplicada* (Barcelona), **6**, 59–72.

Martin, F. W., Pospahala, R. S. & Nichols, J. D. (1979). Assessment and population monitoring of North American Migratory Birds. In Cairns, J., Patil, G. P. & Waters, W. E. (eds.) *Environmental Monitoring, Assessment, Prediction and Management: Certain Case Studies and Related Quantitative Issues*. Fairland, MD: International Cooperative Publishing House pp. 187–239.

Massachusetts Institute of Technology (1970). *Man's Impact on the Global Environment: Assessment and Recommendations for Action. A Report of the Study of Critical Environmental Problems (SCEP)*. Cambridge, MA: MIT Press.

Matthews, R. (ed.) (1987). *Conservation Monitoring and Management. A Report on the Monitoring and Management of Wildlife Habitats on Demonstration Farms*. Cheltenham, UK: Countryside Commission.

McCormick, P. V., Cairns, J., Belanger, S. E., & Smith, E. P. (1991). *Aquatic Toxicology*, **20**, 41–70.

McGeoch, M. A., van Rensburg, B. J. & Botes, A. (2002). The verification and application of bioindicators: a case study of dung beetles in a savanna ecosystem. *Journal of Applied Ecology*, **39**, 661–672.

Meadows, D. H., Meadows, D. L., Randers, J. & Behrens, W. W. (1972). *Limits to Growth*. London: Earth Island and Potomac Associates.

Meire, P. M. & Dereu, J. (1990). Use of the abundance/biomass comparisons method for detecting environmental stress: some considerations based on intertidal macrozoobenthos and bird communities. *Journal of Applied Ecology*, **27**, 210–223.

Menhinick, E. F. (1964). A comparison of some species–individuals diversity indices applied to samples of field insects. *Ecology*, **45**, 859–861.

Morgan, R. K. (1998). *Environmental Impact Assessment*. Dordrecht: Kluwer Academic.

Morista, M. (1959). Measuring of interspecific association and similarity between communities. *Memoirs Faculty of Science Kyushu University, Series E, Biology*, **3**, 65–80.

Morris, H. L., Samiullah, Y. & Burton, M. S. A. (1988). Strategies for biological monitoring: the European experience. In Seeliger, U., Lacerda, L. D. & Patchineelam, S. R. (eds.) *Metals in Coastal Environments of Latin America*, Berlin: Springer Verlag, pp. 286–292.

Morris, M. G. (1969). Associations of aquatic Heteroptera at Woodwalton Fen, Huntingdonshire, and their use in characterising aquatic biotopes. *Journal of Applied Ecology*, **6**, 359–373.

Moss, D. (1985). Some statistical checks on the BTO Common Bird Census Index 20 years on. In Taylor, K., Fuller, R. J. & Lack, P. C. (eds.) *Bird Census and Atlas Studies*, Tring, UK: British Trust for Ornithology, pp. 175–179.

Mountford, E. P., Peterken, G. F., Edwards, P. J. & Manners, J. G. (1999). Long-term change in growth, mortality and regeneration of trees in Denny Wood, an old-growth wood-pasture in the New Forest (UK). *Perspectives in Plant Ecology, Evolution and Systematics*, **2**, 223–272.

Mountford, M. D. (1962). An index of similarity and it application to classifcatory problems. In P. W. Murphy ed. *Progress in Soil Zoology*, London: Butterworths, pp. 43–50.

Murphy, P. M. (1978). The temporal variability in biotic indices. *Environmental Pollution*, **17**, 227–236.

Nagendra, H. & Gadgil, M. (1999). Satellite imagery as a tool for monitoring species diversity: an assessment. *Journal of Applied Ecology*, **36**, 388–397.

National Swedish Environmental Protection Board (1985). *Monitor 1985. The National Swedish Environmental Monitoring Programme PMK*. Stockholm: National Swedish Environmental Protection Board.

National Water Council (1981). *River Quality: the 1980 Survey and Future Outlook*. London: National Water Council.

Niklaus, P. A., Leadley, P. W., Schmid, B. & Korner, C. (2001). A long-term field study on biodiversity X elevated CO_2 interactions in grassland. *Ecological Monographs*, **71**, 341–356.

NOAA (1990). *Ocean Pollution Monitoring, Research, and Assessment. A Bibliography of NOAA-sponsored Reports and Publication*. Silver-Spring, MD: Ocean Assessments Division, Office of Oceanography and Marine Assessment National Ocean service.

Norman, F. I. (2000). Adélie Penguin colonies in eastern Prydz Bay: 'biological indicators' of exploration history and political change. *Polar Record*, **36**, 215–232.

Northing, P., Walters, K., Barker, I. *et al.* (2002). Use of the internet for provision of user specific support for decisions on the control of aphid-borne viruses. In *Aphids in a New Millenium, Proceedings of the 6th International Aphid Symposium*, Rennes, France, September 2001.

Noss, R. F. (1990). Indicators for monitoring biodiversity: a hierarchial approach. *Conservation Biology*, **4**, 355–364.

Noss, R. F. (1999). Assessing and monitoring forest biodiversity: a suggested framework and indicators. *Forest Ecology and Management*, **115**, 135–146.

NSF (1977). *Long-term Ecological Measurements*. (Report of a Conference, Woods Hole, Massachusetts, March, 1977.) Arlington, VA: National Science Foundation Directorate for Biological, Behavioural and Social Sciences. Division of Environmental Biology.

NSF (1978). *A Pilot Program for Long-term Observation and Study of Ecosystems in the United States*. (Report of a Second Conference on Long-term Ecological Measurements, Woods Hole, MA, February, 1978.) Arlington, VA: National Science Foundation, Directorate for Biological, Behavioural and Social Sciences. Division of Environmental Biology.

O'Connell, T. J., Jackson, L. E. & Brooks, R. P. (1998). A bird community index of biotic integrity for the mid-Atlantic highlands. *Environmental Monitoring and Assessment*, **51**, 145–156.

O'Connor, R. J. (1981). The influence of observer and analyst efficiency in mapping method censuses. *Studies in Avian Biology*, **6**, 372–376.

O'Connor, R. J. & Hicks, R. K. (1980). The influence of weather conditions on the detection of birds during common birds census fieldwork. *Bird Study*, **27**, 137–151.

Oden, J. (1971) Nederbordens forsuning-ett generellt hot mot ekosystemen. In Mysterud, (ed.) *Forurensning og Biologisk Milijovern*, Oslo: Oslo University Press, pp. 63–98.

Odum, E. P. (1971). *Fundamentals of Ecology*. 3rd edn. Philadelphia, PA: Saunders.

OECD (1979). *The State of the Environment in Member Countries*. Paris: OECD.

OECD (1994). *Environmental Indicators. OECD Core Set*. Paris: OECD.

OECD (2001a). *OECD Environmental Indicators 2001. Towards Sustainable Development*. Paris: OECD.

OECD (2001b). *OECD Environmental Outlook*. Paris: OECD.

Oglesby, R. T. (1977). Fish yield as a monitoring parameter and its prediction for lakes. In Alabaster, J. S. (ed.) *Biological Monitoring of Inland Fisheries*. London: Applied Science, pp. 195–205.

Oliver, J. & Beattie, A. J. (1996). Designing a cost-effective invertebrate survey: a test of methods for rapid assessment of biodiversity. *Ecological Applications*, **6**, 594–607.

Olsgard, F., Brattegard, T. & Holthe, T. (2003). Polychaetes as surrogates for marine biodiversity: lower taxonomic resolution and indicator groups. *Biodiversity and Conservation*, **12**, 1033–1049.

O'Neil, R. V., Hunsaker, C. T., Timmins, S. P. *et al.* (1996). Scale problems in reporting landscape pattern at the regional scale. *Landscape Ecology*, **11**, 169–180.

O'Neil, R. V., Hunsaker, C. T., Jones, K. B. *et al.* (1997). Monitoring environmental quality at the landscape scale. *BioScience*, **47**, 513–519.

Opschoor, H. (2000). The ecological footprint: measuring rod or metaphor? *Ecological Economics*, **32**, 363–365.

Orio, A. A. (1989). Modern chemical technologies for assessment and solution of environmental problems. In Botkin, D. B., Caswell, M. F., Estes, J. E. & Orio, A. A. (eds.) *Changing the Global Environment. Perspective on Human Involvement*. London: Academic Press, pp. 169–184.

Orio, A. A. & Botkin, D. B. (1986). Man's role in changing the global environment. *Science of the Total Environment*, **55**, 1400; **56**, 1416.

Palmer, M. E. (1986). A survey of rare plant monitoring: programs, regions, and species priority. *Natural Areas Journal*, **6**, 27–42.

Palmer, M. E. (1987). A critical look at rare plant monitoring in the United States. *Biological Conservation*, **39**, 113–127.

Parker, B. C. & Howard, R. V. (1977). The first environmental impact monitoring and assessment in Antarctica. The Dry Valley drilling project. *Biological Conservation*, **12**, 163–177.

Parker, J. (1991). Environmental reporting and environmental indices. Ph. D. Thesis, Cambridge University, Cambridge.

Parks, N. (2002). Measuring climate change. *BioScience*, **52**, 3.

Parr, T. W., Ferretti, M., Simpson, I. C., Forsius, M. & Kovacs-Lang, E. (2002). Towards a long-term integrated monitoring programme in Europe: network design in theory and practice. *Environmental Monitoring and Assessment*, **78**, 253–290.

Partidario, M. R. (2000). Elements of an SEA framework: improving the added-value of SEA. *Environmental Impact Assessment Review*, **20**, 647–663.

Patrick, R. (1949). A proposed biological measure of stream condition based on a survey of the Conestaga Basin, Lancaster County, Pennsylvania. *Proceedings of the Academy of Natural Sciences USA*, **101**, 377–381.

Patrick, R. (1972). Aquatic communities as indices of pollution. In Thomas, W. A. (ed.) *Indicators of Environmental Quality*. New York: Plenum Press, pp. 93–100.

Pattengill-Semmens, C. V. & Semmens, B. X. (2003). Conservation and management applications of the REEF volunteer fish monitoring program. *Environmental Monitoring and Assessment*, **81**, 43–50.

Peach, W. J., Baillie, S. R. & Underhill, L. (1991). Survival of British Sedge Warblers *Acrocephalus schoenobaenus* in relation to West African rainfall. *Ibis*, **133**, 300–305.

Pearce, F. (2003). Can ocean friendly labels save dwindling stocks? *New Scientist*, **178**, 5.

Pearson, D. L. (1995). Selecting indicator taxa for the quantitative assessment of biodiversity. In Hawksworth, D. L. (ed.) *Biodiversity Measurement and Estimation*. London: Chapman & Hall, pp. 75–79.

Pearson, D. L. & Carroll, S. S. (1998). Global patterns of species richness: spatial models for conservation planning using bioindicators and precipitation data. *Conservation Biology*, **12**, 809–821.

Perring, F. H. (1992). BSBI distribution maps scheme: the first forty years. In Harding, P. T. (ed.) *Biological Recording of Changes in British Wildlife*. London: HMSO, pp. 1–4. NERC/BRC 25th Anniversary Vol.

Perring, F. H. & Farrell, L. (1983). *British Red Data Books, 1, Vascular Plants*, 2nd edn., Nettleham, UK: RSNC.

Perring, F. H. & Walters, S. M. (eds.) (1962). *Atlas of the British Flora*. London: BSBL/Nelson.

Peterken, G. F. (1974). A method for assessing woodland flora for conservation using indicator species. *Biological Conservation*, **6**, 239–245.

Peterken, G. F. & Backmeroff, C. (1988). *Long-term Monitoring in Unmanaged Woodland Nature Reserves*. (*Research and Survey in Nature Conservation*, No. 9.) Peterborough: Nature Conservancy Council.

Peterken, G. F. & Mountford, E. P. (1998). Long-term changes in an unmanaged population of wych elm subjected to Dutch elm disease. *Journal of Ecology*, **86**, 205–218.

Petts, J. (1999). *Handbook of Environmental Impact Assessment*. Cambridge: Cambridge University Press.

Phillips, D. J. H. (1980). *Quantitative Aquatic Biological Indicators. Their Use to Monitor Trace Metal and Organochlorine Pollution*. London: Applied Science.

Pielou, E. C. (1966). The measurement of diversity in different types of biological collections. *Journal of Theoretical Ecology*, **13**, 131–144.

Pilegaard, K. (1978). Airborne metals and SO_2 monitored by epiphytic lichens in an industrial area. *Environmental Pollution*, **17**, 81–92.

Pinder, L. C. V. & Farr, I. S. (1987a). Biological surveillance of water quality 2. Temporal and spatial variation in the macroinvertebrate fauna of the River Frome, a Dorset chalk stream. *Archives of Hydrobiology*, **109**, 321–331.

Pinder, L. C. V. & Farr, I. S. (1987b). Biological surveillance of water quality 3. The influence of organic enrichment on the macroinvertebrate fauna of a small chalk stream. *Archives of Hydrobiology*, **109**, 619–637.

Pinkham, C. F. A. & Pearson, J. G. (1976). Applications of a new coefficient of similarity to pollution surveys. *Journal of Water Pollution Control Federation*, **48**, 717–723.

Pinnegar, J. K., Jennings, S., O'Brien, C. M. & Polunin, N. V. C. (2002). Long-term changes in the trophic level of the Celtic sea fish community and fish market price distribution. *Journal of Applied Ecology*, **39**, 377–390.

Plumptre, A. J. (2000). Monitoring mammal populations with line transect techniques in African forests. *Journal of Applied Ecology*, **37**, 356–368.

Pollard, E. (1982). Monitoring butterfly abundance in relation to the management of a nature reserve. *Biological Conservation*, **24**, 317–328.

Pollard, E., Hooper, M. D. & Moore, N. W. (1974). *Hedges*. London: Collins.

Pollard, E., Hall, M. L. & Bibby, T. J. (1986). *Monitoring the Abundance of Butterflies*. (*Research and Survey in Nature Conservation*, No. 2.) Peterborough: Nature Conservancy Council.

Pollard, E., Moss, D. & Yates, T. J. (1995). Population trends of common British butterflies at monitored sites. *Journal of Applied Ecology*, **32**, 9–16.

Pontasch, K., Niederlehner, B. R. & Cairns, J. (1989). Comparisons of single species, microcosm and field responses to a complex effluent. *Environmental Toxicology and Chemistry*, **8**, 521–532.

Rapport, D. J. (2000). Ecological footprints and ecosystem health: complementary approaches to a sustainable future. *Ecological Economics*, **32**, 367–370.

Ratcliffe, D. (1980). *The Peregrine Falcon*. Carlton, UK: T. & A. D. Poyser.

Rees, W. E. (2000). Eco-footprint analysis: merits and brickbats. *Ecological Economics*, **32**, 371–374.

Reynolds, C. M. (1979). The heronries census: 1972–1977 population changes and a review. *Bird Study*, **26**, 7–12.

Reynolds, R. T., Scott, J. M. & Nussbaum, R. A. (1980). A variable circular-plot method for estimating bird numbers. *Condor*, **82**, 309–313.

Rhind, D. W., Wyatt, B. K., Briggs, D. J. & Wiggins, J. (1986). The creation of an environmental information system for the European Community. *Nachrichten aus Karten und Vermessungswesen*, Series 11, **44**, 147.

Ribe, R. G. (1986). A test of uniqueness and diversity visual assessment factors using judgement-independent measures. *Landscape Research*, **11**, 13–18.

Richey, J. S., Mar, B. W. & Horner, R. R. (1985). The Delphi technique in environmental assessment I. Implementation and effectiveness. *Journal of Environmental Management*, **21**, 135–146.

Ringius, G. (2002). *International Conference on Ecosystem Health: a Blueprint for Design and Implementation of a Global Climate Change Monitoring Node in the Quetico–Superior Region of Northwestern Ontario*. Atikokan, Ontario: Quetico Centre.

Roberts, R. D. & Roberts, T. M. (1984). *Planning and Ecology*. London: Chapman & Hall.

Robertson, G. P., Coleman, D. C., Bledsoe, C. S. & Sollins, P. (1999). *Standard Soil Methods for Long-term Ecological Research*. Oxford: Oxford University Press.

Rodwell, J. S. (ed.) (1991). *British Plant Communities*, vol. 1: *Woodlands and Scrub*. Cambridge: Cambridge University Press.

Root, T. (1988). *Atlas of Wintering North American Birds. An Analysis of Christmas Bird Count Data*. Chicago, IL: The University of Chicago Press.

Rosenberg, D. M., Resh, V. H., Balling, S. S. *et al.* (1981). Recent trends in environmental impact assessment. *Canadian Journal of Fisheries and Aquatic Sciences*, **38**, 591–624.

Rose, C. I. & Hawksworth, D. L. (1981). Lichen recolonization in London's cleaner air. *Nature*, **289**, 289–292.

Rothery, P., Wanless, S. & Harris, M. P. (1988). Analysis of counts from monitoring Guillemots in Britain and Ireland. *Journal of Animal Ecology*, **57**, 1–19.

Rowell, A. (1996). *Green Backlash*. London: Routledge.

Sackman, H. (1975). *Delphi Critique*. Lanham, MD: Lexington Books.

Samiullah, Y. (1985). Biological effects of marine oil pollution. *Oil and Petrochemical Pollution*, **2**, 235–264.

Samiullah, Y. (1986). *Bee Craft*, **68**, 5–11.

Sandhu, S., Jackson, L., Austin, K. *et al.* (1998). *Monitoring Ecological Condition at Regional Scales*. Dordrecht: Kluwer Academic.

Savage, A. A. (1982). Use of water boatmen (Corixidae) in the classification of lakes. *Biological Conservation*, **23**, 55–70.

Savage, A. A. & Pratt, M. M. (1976). Corixidae (water boatmen) of the northwest Midland meres. *Field Studies*, **4**, 465–476.

Savitsky, B. & Davis, B. A. (1990). Environmental sensitivity index (ESI) mapping for oil spills using remote sensing and geographic information system technology. *International Journal of Geographical Information Systems*, **A**, 181–201.

Sawicka-Kapusta, K. (1979). Roe deer antlers as bioindicators of environmental pollution in southern Poland. *Environmental Pollution*, **19**, 283–293.

SCAR (1996). *Monitoring of Environmental Impacts from Science and Operations in Antarctica.* Oslo: SCAR and COMNAP.

SCEP (1970). *Man's Impact on the Global Environment. Report of the Study of Critical Environmental Problems (SCEP).* Cambridge, MA: MIT Press.

Schiller, A., Hunsaker, C. T., Kane, M. A. *et al.* (2001). Communicating ecological indicators to decision makers and the public. *Conservation Ecology*, **5**, 19–28.

Schindler, D. W., Mills, K. H., Malley, D. F. *et al.* (1985). Long-term ecosystem stress: the effects of years of experimental acidification on a small lake. *Science*, **228**, 1395–1401.

Schmidt-Bleek, F. (1994). *Carnoules Declaration of the Factor Ten Club.* Berlin: Wuppertal Institute.

Schneider, G. (1989). The CORINE programme. *Naturopa*, **61**, 15.

Scott, D. (1989). Biological monitoring of rivers. In Craig, B. (ed.) *Proceedings of a Symposium on Environmental Monitoring in New Zealand with Emphasis on Protected Natural Areas*, Wellington: Department of Conservation.

Scott, W. A., Adamson, J. K., Rollinson, J. & Parr, T. W. (2002). Monitoring of aquatic macrophytes for detection of long-term changes in river systems. *Environmental Monitoring and Assessment*, **73**, 131–153.

Scullion, J. & Edwards, R. W. (1980). The effects of coal industry pollutants on the macro-invertebrate fauna of a small river in the South Wales coalfield. *Freshwater Biology*, **10**, 141–162.

Seager, J. (1995). *The State of the Environment Atlas.* London: Penguin.

Shannon, C. E. & Weaver, W. (1962). *The Mathematical Theory of Communication.* Urbana, IL: University of Illinois Press.

Sharrock, J. T. R. (1974). The changing status of breeding birds in Britain and Ireland. In Hawksworth, D. L. (ed.) *The Changing Flora and Fauna of Britain.* London: Academic Press, pp. 204–220.

Sharrock, J. T. R. (1976) *The Atlas of Breeding Birds in Britain and Ireland.* Berkhampsted: Poyser.

Shears, J. R. (1988). The use of airborne thematic mapper imagery in monitoring saltmarsh vegetation and marsh recovery from oil refinery effluent: the case study of Fawley saltmarsh, Hampshire. In *Proceedings of the NERC 1987 Airborne Campaign Workshop*, Southampton, December 1988. Swindon, UK: NERC, pp. 99–119.

Shears, J. R. (1989). The application of airborne remote sensing to the monitoring of coastal saltmarshes. In Barrett, E. C. & Brown, K. A. (eds.) *Remote Sensing for Operational Applications. (Technical Contents of the 15th Annual Conference of the Remote Sensing Society.)* Nottingham, UK: Remote Sensing Society, pp. 371–379.

Simon, J. L. & Kahn, H. (1984). *The Resourceful Earth: A Response to Global 2000.* Oxford: Basil Blackwell.

Simon, T. P., (1999). Introduction: biological integrity and use of ecological health concepts for application to water resource characterization. In Simon, T. P. (ed.) *Assessing the Sustainability and Biological Integrity of Water Resources Using Fish Communities*. London: CRC Press, pp. 3–14.

Skye, E. (1979). Lichens as biological indicators of air pollution. *Annual Reviews of Phytopathology*, **17**, 325–341.

Sladecek, V. (1973). The reality of three British biotic indices. *Water Research*, **7**, 95–1002.

Sladecek, V. (1979). Continental systems for the assessment of river water quality. In James, A. & Evison, L. (eds.) *Biological Indicators of Water Quality*. Chichester, UK: Wiley, pp. 3.1–3.32.

Slocombe, D. S. (1992). Environmental monitoring for protected areas: a review and prospect. *Environmental Monitoring and Assessment*, **21**, 49–78.

Smart, S. M., Bunce, R. G. H., Black, H. J. *et al.* (2001). *Measuring Long Term Ecological Change in British Woodlands* (1997–2000). (*English Nature Research Report* No. 461.) Peterborough: English Nature.

Smith, B. (1986). Evaluation of different similarity indices applied to data from the Rothamsted Insect Survey. M.Sc. Thesis, University of York, UK.

Smith, D. G. (1987). *Water Quality Indexes for Use in New Zealand Rivers and Streams*. (*Water Quality Centre Publication*, 12.) Wellington: Water Quality Centre, Ministry of Works and Development.

Smithsonian Institution (1970). *National and International Environmental Monitoring Activities: a Directory*. Cambridge, MA: Smithsonian Institution Center for Short-lived Phenomena.

Smyth, C. R. & Dearden, P. (1998). Performance standards and monitoring requirements of surface coal mine reclamation success in mountainous jurisdictions of western North America: a review. *Journal of Environmental Management*, **53**, 209–229.

Sorensen, T. A. (1948). A method of establishing groups of equal amplitude in plant sociology based on similarity of species content, and its application to analyses of the vegetation on Danish commons. *Biologiske Skrifter*, **5**, 1–34.

Spellerberg, I. F. (1981). *Ecological Evaluation for Conservation*. London: Arnold.

Spellerberg, I. F. (1991). A biogeographical basis of conservation. In Spellerberg, I. F., Goldsmith, F. B. & Morris, M. J. (eds.) *The Scientific Management of Temperate Communities for Conservation*. Oxford: Blackwell Scientific, pp. 293–322.

Spellerberg, I. F. (1992). *Evaluation and Assessment for Conservation*. London: Chapman & Hall.

Spellerberg, I. F. (2002). *Ecological Effects of Roads*. Enfield, CT: Science Publishers.

Spellerberg, I. F. & Fedor, P, J. (2003). A tribute to Claude Shannon (1916–2001) and a plea for more rigorous use of species richness, species diversity and the 'Shannon-Wiener' index. *Global Ecology and Biogeography*, **12**, 177–179.

Stark, J. D. (1985). *A Macroinverebrate Community Index of Water Quality for Stony Streams*. (*Water and Soil Miscellaneous Publications* 87.) Wellington: New Zealand National Water and Soil Conservation Authority.

Stark, J. D. (1993). Performance of the macroinvertebrate community index: effects of sampling method, sample replication, water depth, current velocity, and substratum on index values. *New Zealand Journal of Marine and Freshwater Research*, **27**, 463–478.

Stark, J. D. (1998). SQMCI: a biotic index for freshwater macroinvertebrate coded-abundance data. *New Zealand Journal of Marine and Freshwater Research*, **32**, 55–66.

Stein, B. A., Kutner, L. S., & Adams, J. S. (2000). *Precious Heritage. The Status of Biological Diversity in the United States*. Oxford: Oxford University Press.

Stevenson, R. J., Pan, Y. & Vaithiyanathan, P. (2002). Ecological assessment and indicator development in wetlands: the case of algae in the Everglades, USA. *Verhandlungen Internationalen Vereinigung Limnologie*, **28**, 1248–1252.

Stonehouse, B. (ed.) (2002). *Encyclopedia of Antarctica and the Southern Oceans*. New York: Wiley.

Stout, B. B. (1993). The good, the bad and the ugly of monitoring programs: defining questions and establishing objectives. *Environmental and Assessment*, **26**, 91–98.

Strayer, D., Glitzenstein, J. S., Jones, C. G. et al. (1986). *Long-term Ecological Studies: an Illustrated Account of their Design, Operation, and Importance to Ecology*. (Occasional Publication of the Institute of Ecosystem Studies, No. 2) Millbrook, NY: The New York Botanical Garden.

Suter, G. W. (1993). A critique of ecosystem health concepts and indexes. *Environmental Toxicology and Chemistry*, **12**, 1533–1539.

Sykes, J. M. & Lane, A. M. J. (eds.) (1996). *The UK Environmental Change Network: Protocols for Standard Measurements at Terrestrial Sites*. London: The Stationary Office.

Symstad, A. J., Chapin, F. S., Wall, D. H. et al. (2003). Long-term and large-scale perspectives on the relationship between biodiversity and ecosystem functioning. *BioScience*, **53**, 89–98.

Syratt, W. J. & Richardson, M. G. (1981). Anti-oil pollution in Sullom Voe environmental considerations. *Proceedings of the Royal Society of Edinburgh*, **80B**, 35–51.

Tapper, S. (1992). *Game Heritage: an Ecological Review from Shooting and Gamekeeping Records*. Fordingbridge, UK: Game Conservancy.

Taylor, L. R. (1986). Synoptic dynamic, migration and the Rothamsted Insect Survey. *Journal of Animal Ecology*, **55**, 1–38.

Thomas, W. A. (1972). *Indicators of Environmental Quality*. New York: Plenum Press.

Thomas, W. A., Goldstein, G. & Wilcox, W. H. (1973). *Biological Indicators of Environmental Quality: A Bibliography of Abstracts*. Ann Arbor, MI: Ann Arbor Science.

Thomas, W. L. (ed.) (1956). *Man's Role in Changing the Face of the Earth*. Chicago: University of Chicago Press.

Tilman, D. (1989). Ecological experimentation: strengths and conceptual problems. In Likens, G. E. (ed.) *Long-term Studies in Ecology*. New York: Springer, pp. 136–157.

Tips, W. E. J. & Savasdisara, T. (1986). The influence of the environmental background of subjects on their landscape preference evaluation. *Landscape and Urban Planning*, **13**, 125–133.

Treweek, J. (1999). *Ecological Impact Assessment*. Oxford: Blackwell Science.

Treweek, J. R., Thompson, S., Veitch, N. & Japp, C. (1993). Ecological assessment of proposed road developments: a review of environmental statements. *Journal of Environmental Planning and Management*, **36**, 295–307.

Tscharntke, T., Gathmann, A. & Steffan-Dewenter, I. (1998). Bioindication using trap-nesting bees and wasps and their natural enemies: community structure and interactions. *Journal of Applied Ecology*, **35**, 708–719.

Tubbs, C. (1986). *The New Forest*. London: Collins.

UNDP, UNEP, World Bank and World Resources Institute (2000). *World Resources 2000–2001; People and Ecosystems: The Fraying Web of Life*. Washington, DC: World Resources Institute.

UNEP (1987). *United Nations Environment Programme, Environmental Data Report*. Oxford: Basil Blackwell.

UNEP (1993). *Guidelines for Country Studies on Biological Diversity*. Nairobi: United Nations Environment Programme.

UNEP (1999). *Global Environment Outlook 2000*. London: Earthscan.

UNEP (2000). *GEO 2000*. London: Earthscan.

UNEP/GEMS (1989). *The African Elephant. (UNEP/GEMS Environment Library No. 3.)* Nairobi: UNEP.

UNEP–WCMC (2002). *World Atlas of Biodiversity: Earth's Living Resources for the 21st Century*. www.unep-wcmc.org/information_services/publications/biodiversityatlas.

Urlich, S. & Brady, P. (2003). *Monitoring Terrestrial Habitats in Wellington Conservancy*. Wellington: Department of Conservation.

van Strien, A. J., van de Pavert, R., Moss, D. *et al.* (1997). The statistical power of two butterfly monitoring schemes to detect trends. *Journal of Applied Ecology*, **34**, 817–828.

Varley, G. C., Gradwell, G. R. & Hassell, M. P. (1973). *Insect Population Ecology*. Oxford: Blackwell Scientific.

Vaughan, H., Brydges, T., Fenech, A. & Lumb, A. (2001). Monitoring long-term ecological changes through the ecological monitoring and assessment network: science-based and policy relevant. *Environmental Monitoring and Assessment*, **67**, 3–28.

von Weizsaecker, E., Lovins, A. B. & Lovins, L. H. (1997). *Factor Four Doubling Wealth: Halving Resource Use*. London: Allen & Unwin.

Vos, J. B., Feenstra, J. F., de Boer, J., Braat, L. C. & van Baalen, J. (1985). *Indicators for the State of the Environment*. Amsterdam: Institute for Environmental Studies, Free University.

Vujakovic, P. (1987). Monitoring extensive 'buffer zones' in Africa: an application for satellite imagery. *Biological Conservation*, **39**, 195–207.

Wackernagel, M. & Rees, W. (1996). *Our Ecological Footprint. Reducing Human Impact on the Earth*. Gabriola Island: New Society.

Wackernagel, M. & Yount, J. D. (1998). The ecological footprint: an indicators of progress toward regional sustainability. *Environmental Monitoring and Assessment*, **51**, 511–529.

Walsh, P. D. & White, L. J. T. (1999). What it will take to monitor forest elephant populations. *Conservation Biology*, **13**, 1194–1202.

Ward, J. C. (1990). *Environmental Indicators for State of the Environment Reporting. (Information Paper No. 21.)* Lincoln: Centre for Resource Management, Lincoln University.

Warwick, R. M. (1986). A new method for detecting pollution effects on marine macrobenthic communities. *Marine Biology*, **92**, 557–562.

Waterhouse, E. (2001). *Ross Sea Region 2001: a State of the Environment Report for the Ross Sea Region of Antarctica*. Christchurch: Antarctic Institute.

WCED (1987). *Our Common Future*. Oxford: Oxford University Press.

Weatherhead, P. J. (1986). How unusual are unusual events? *American Naturalist*, **128**, 150–154.

Webb, W. M. (ed.) (1911). *A Nature Calendar by Gilbert White*. London: The Selbourne Society.

WEF (2001). *2001 Environmental Sustainability Index*. (An initiative of the Global Leaders of Tomorrow Environment Task Force, World Economic Forum in collaboration with the Yale Centre for Environmental Law and Policy (Yale University) and the Centre for International Earth Science Information Network (Columbia University). New York: World Economic Forum.

Wells, F., Metzeling, L. & Newall, P. (2002). Macroinvertebrate regionalisation for use in the management of aquatic ecosystems in Victoria, Australia. *Environmental Monitoring and Assessment*, **74**, 271–294.

Westman, W. E. (1985). *Ecology, Impact Assessment, and Environmental Planning*. New York: Wiley.

Westwood, S. S. C., Dunnet, G. M. & Hiscock, K. (1989). Monitoring the Sullom Voe Terminal. In Dicks, B. (ed.) *Ecological Impacts of the Oil Industry*. Chichester: Wiley, on behalf of the Institute of Petroleum, pp. 261–285.

Whitford, W. G., Soyza, A., Zee, J. W. V., Herrrick, J. E. & Havstad, K. M. (1998). Vegetation, solid and animal indicators of rangeland health. *Environmental Monitoring and Assessment*, **51**, 179–200.

Whittaker, R. H. (1954). The ecology of serpentine soils. *Ecology*, **35**, 258–288.

Wiener, N. (1939). The Ergodic theorem. *Duke Mathematical Journal*, **5**, no page numbers.

Wiener, N. (1948). *Cybernetics*. New York: Wiley.

Wiener, N. (1949). *The Interpolation, Extrapolation and Smoothing of Stationary Time Series*. New York: Wiley.

Wiens, J. A. (1989). Spatial scaling in ecology. *Functional Ecology*, **3**, 385–397.

Wiersma, G. B. (ed.) (2000). Monitoring ecological condition in the United States. (Proceedings of the Fourth Symposium on the Environmental Monitoring and Assessment Programme (EMAP), 1999.) *Environmental Monitoring and Assessment*, **64**, 1–447.

Wilhm, J. L. & Dorris, T. C. (1968). Biological parameters for water quality criteria. *BioScience*, **18**, 477–481.

Williams, C. B. (1964). *Patterns in the Balance of Nature*. New York: Academic Press.

Williams, G. R. & Given, D. R. (1981). *The Red Data Book of New Zealand*. Wellington: Nature Conservation Council.

Williams, N. V. & Dussart, G. B. J. (1976). A field course survey of three English river systems. *Journal of Biological Education*, **10**, 4–14.

Williamson, K. (1965). *Fair Isle and its Birds*. Edinburgh: Oliver & Boyd.

Willmot, A. (1980). The woody species of hedges with special reference to age in Church Broughton Parish, Derbyshire. *Journal of Ecology*, **68**, 269–285.

Wilson, K. J., Taylor, R. H. & Barton, K. J. (1990). The impact of man on Adélie Penguins at Cape Hallett, Antarctica. In Kerry, K. R. & Hempel, G. (eds.) *Antarctic Ecosystems. Ecological Change and Conservation*. Berlin: Springer-Verlag, pp. 183–190.

Winstanley, D., Spencer, R. & Williamson, K. (1974). Where have all the whitethroats gone? *Bird Study*, **21**, 1–14.

Witt, G. B., Berghammer, L. J., Beeron, R. S. J. and Moll, E. J. (2000). Retrospective monitoring of rangeland vegetation change: ecohistory from deposits of sheep dung associated with shearing sheds. *Austral Ecology*, **25**, 260–267.

Wolfe, D. A. (1988). Urban wastes in coastal waters: assimilative capacity and management. In Wolfe, D. A. & O'Connor, T. P. (eds.) *Oceanic Processes in Marine Pollution*. Vol. 5. *Urban Wastes in Coastal Marine Environments*. Malabar, FL: Robert E. Kreiger, pp. 1–20.

Wolfe, D. A. & O'Connor, J. S. (1986). Some limitations of indicators and their place in monitoring schemes. In *Oceans '86: Proceedings of the Third Monitoring Strategies Symposium*. Washington, DC: Marine Technology Society, pp. 878–884.

Wolfe, D. A., Champ, M. A., Flemer, D. A. & Mearns, A. J. (1987). Long-term biological data sets: their role in research, monitoring, and management of estuarine and coastal marine systems. *Estuaries*, **10**, 181–193.

Woodiwiss, F. S. (1964). The biological system of stream classification used by the Trent River Board. *Chemistry and Industry*, **11**, 443–447.

Woodley, S. (1993). Monitoring and measuring ecosystem integrity in Canadian national parks. In Woodley, S., Kay, J. & Francis, G. (eds.) *Ecological Integrity and the Management of Ecosystems*. Gainsville, FL: St. Lucia Press, pp. 155–175.

World Bank (2002). *Sleeping on our Own Mats: an Introductory Guide to Community-based Monitoring and Evaluation*. Washington, DC: World Bank.

Worldwatch Institute (2000). *State of the World*. (17th report). Washington, DC: Worldwatch Institute.

Wright, J. F. (1995). Development and the use of a system for predicting the macroinvertebrate fauna in flowing waters. *Australian Journal of Ecology*, **20**, 181–197.

Wright, J. F., Armitage, M. T., Furse, M. T. & Moss, D. (1988). A new approach to the biological surveillance of river quality using macroinvertebrates. *Verhandlungen Internationalen Vereinigung Limnologie*, **23**, 1548–1552.

Wright, J. F., Armitage, M. T., Furse, M. T. & Moss, D. (1989). Prediction of invertebrate communities using stream measurements. *Regulated Rivers: Research and Management*, **4**, 147–155.

Wright, J. F., Sutcliffe, D. W. & Furse, M. T. (eds.) (2000). *Assessing the Biological Quality of Fresh Waters. RIVPACS and Other Techniques*. Ambleside, UK: Freshwater Biological Association.

Yoccoz, N. G, Nichols, J. D. & Boulinier, T. (2001). Monitoring of biological diversity in space and time. *TRENDS in Ecology and Evolution*, **16**, 446–453.

Young, E. C. (1994). *Skua and Penguin Predator and Prey*. Cambridge, UK: Cambridge University Press.

Index